Towards a New European Bauhaus—Challenges in Design Education

Manuel Blanco Lage ·
Oya Atalay Franck ·
Nicolas Marine ·
Manuel Rodrigo de la O Cabrera
Editors

# Towards a New European Bauhaus—Challenges in Design Education

## EAAE Annual Conference—Madrid 2022

Springer

*Editors*

Manuel Blanco Lage
ETSAM School of Architecture
Universidad Politécnica de Madrid
Madrid, Spain

Nicolas Marine
ETSAM School of Architecture
Universidad Politécnica de Madrid
Madrid, Spain

Oya Atalay Franck
School Architecture, Design and Civil Engineering
Zurich University of Applied Sciences
Winterthur, Switzerland

Manuel Rodrigo de la O Cabrera
ETSAM School of Architecture
Universidad Politécnica de Madrid
Madrid, Spain

ISBN 978-3-031-40187-9 ISBN 978-3-031-40188-6 (eBook)
https://doi.org/10.1007/978-3-031-40188-6



This Springer imprint is published by the registered company Springer Nature Switzerland AG
The registered company address is: Gewerbestrasse 11, 6330 Cham, Switzerland

Paper in this product is recyclable.

# Preface

## Organization

The Annual Conference of the European Association for Architectural Education (EAAE/AEEA) was held in Madrid from 31 August to 2 September 2022, at the ETSAM School of Architecture of the Universidad Politécnica de Madrid.

The EAAE is a globally connected European network for architectural education. For more than 45 years, it has been promoting the exchange and dissemination of knowledge and experience in architectural culture, education and research. It represents more than 150 members from all over Europe and beyond.

The EAAE Annual Conference is held every year since its founding in 1975 and attracts professionals from a wide range of backgrounds who share a common interest in architectural education. The conference provides a platform for the exchange of ideas, experiences and research results, as well as an opportunity to build networks and maintain existing partnerships. Under the title "Towards a New European Bauhaus: Challenges in Architectural Education and Research," the 2022 Madrid ETSAM conference aimed to be at the forefront of the design debate. The New European Bauhaus is an open initiative launched in 2021 by the President of the European Commission, Ursula von der Leyen. It calls for new ideas about the living space we all share, taking into account local culture, but at the same time strengthening the links that unite the different European countries.

For this reason, the ETSAM School of Architecture was the perfect host for the event. This institution is the official link with the Spanish government, specifically with the General Directorate for Urban Agenda and Architecture, in order to analyse, promote and position our country's contribution to the New European Bauhaus on an international level. Furthermore, ETSAM is the first school of architecture created in Spain, dating back to the 1840s, although its origins can be traced back to the 1500s.

EAAE 2022 had a hybrid format to allow participants who could not travel to attend the conference. Although the conference was largely in-person, this was also important because it was possible to resume face-to-face meetings after the pandemic. The conference was organized through a single call for abstracts. We received a total of 110 submissions, and the acceptance rate was 40%. Submissions were divided into three tracks, each with three sessions. Many of these papers were accepted for inclusion in this volume. The final decision on acceptance or rejection of full papers was made after further peer-review following the conference.

The conference also included several cultural visits in and around the city of Madrid. In terms of keynote speakers, we were particularly honoured to welcome Alberto Campo Baeza (ETSAM School of Architecture, UPM), Carmen Pinós (Estudio Carme Pinós), Iñaki Ábalos and Renata Sentkiewicz (ETSAM School of Architecture, UPM), and Carme Pigem and Ramon Vilalta (RCR Architects). We were also privileged to have as special guests Guillermo Cisneros (Rector of the Universidad Politécnica de Madrid), Lauriane Bertrand (European Commission), Inaqui Carnicero (Director General of the Government's Urban Agenda and Architecture), Simeona Manova (European

Commission), Marcos Ros Sempere (Member of the European Parliament) and Marta Vall-Llosera (President of the Higher Council of Spanish Architects' Associations).

We would like to express our sincere gratitude to all the people whose work ensured the quality of the EAHN 2022 programme: the five Local Organizing Committee Members, eight Scientific Committee Members, 11 Session and Roundtable Chairs, 31 Reviewers, six Keynote and Invited Talks, two Research Assistants and 8 Student Volunteers.

We thank all the authors who chose EAAE 2022 as a forum to present and publish their research and who generously shared their results with the EAAE community. Last but not least, we are grateful for the support of the Spanish Ministry of Transport, Mobility and Urban Agenda (General Directorate of Urban Agenda and Architecture) and the sponsorship of the Universidad Politécnica de Madrid (Programa Propio I+D) and VELUX.

## Scientific Objectives

On 16 September 2020, the President of the European Commission presented the New European Bauhaus, an ambitious ecological, economic and cultural project. The initiative will implement the EU's future goals based on three fundamental pillars: sustainability, inclusiveness, and aesthetics. Designed with an eminently practical sense, the New European Bauhaus will show how sustainable innovation can be turned into concrete and positive everyday experiences.[1] In other words, its purpose is to make certain ideals physical, an issue that has been emphasized in all the documents produced so far.[2] It will thus address a problem of spatial design.[3] People and architecture are at the heart of the NEB.

Far from sticking to abstract goals, the EU continues to delve deeper into the meaning of its proposal. In particular, the New European Bauhaus Prizes have been awarded annually to the design of building materials, furniture projects, housing projects, or the restoration of cultural heritage and urban interventions. The winners share a desire to promote social cohesion and respect for our shared environment, regardless of the scale of their designs. Above all, however, they all avoid ambiguous reflection and focus the debate on concrete designs. In short, they show what the EU is trying to do: to recover the potential of design to participate in the political programmes of a society.

The designs that emerge from the New Bauhaus must not simply be a reflection of culture, but an active instrument in its formation. Hence the choice of the Bauhaus as an insignia of contemporaneity. It seeks to rescue an idea of critical production and the ability to think through design. The name also gives the project a pedagogical orientation. In the original Bauhaus, it was important to make, but also to teach how to do it. The Bauhaus, the old and the new, represents the link between production and

[1] European Union. 2021. *New European Bauhaus: new actions and funding to link sustainability to style and inclusion.*

[2] European Union. 2021. *New European Bauhaus Prizes.*

[3] EIT Urban Mobility. 2021. *What is the New European Bauhaus?* Interview with Xavier Troussard, EU Policy Lab Director.

experimentation, a model that can cover all scales: from the detail to the building, and from the building to urban and territorial planning.

As a result, the New Bauhaus challenges both active designers and schools of architecture. More than ever, the skills of the future designer must be based on authentic experience. Speculation on intellectual positions must give way to reflection on action: the focus is now on the work itself and the value of its design process. After years of debate, the EU has set out a clear path. Sustainability, social value and aesthetics are the critical positions. The challenge for architectural education now is how to teach future designers to translate these values into livable spaces, how to make the New Bauhaus a reality.

In this context, EAAE 2022, entitled "Towards a New European Bauhaus: Challenges in Architectural Education and Research," organized three tracks representing aspects that the New Bauhaus must face today: contemporary design, design education and critical positions. The final papers published in these proceedings are organized around the same tracks. The specific sessions during the conference were as follows:

Paper Track I. NEB and Contemporary Design

Session 1. Participatory Designs
Session 2. Confronting the Large
Session 3. Social and Environmental Sustainability

Paper Track II. NEB and Critical Positions

Session 4. Equality through design
Session 5. Old lessons and new challenges
Session 6. New Frontiers for the New Bauhaus

Paper Track III. NEB and Design Education

Session 7. Learning from practice
Session 8. Challenges in design curriculum
Session 9. New conceptual frameworks

These sessions represented the current challenges faced by all professionals and schools of architecture. In this time of uncertainty and instability, it is more necessary than ever to defend design as a problem-solving enterprise. It is also necessary to make room for unconventional practices and the new ways opened up by social networks and online and digital learning.

The Madrid Conference's commitment to the New European Bauhaus was a turning point for our teaching. The future is there for us to build: a new, beautiful, inclusive and sustainable reality.

Madrid, Spain — Manuel Blanco Lage
Winterthur, Switzerland — Oya Atalay Franck
Madrid, Spain — Nicolas Marine
Madrid, Spain — Manuel Rodrigo de la O Cabrera

# Organization

## EAAE 2022 Local Organizing Committee

Blanco Lage, Manuel, Dean of ETSAM-UPM.
De la O Cabrera, Manuel Rodrigo, ETSAM-UPM.
Marine, Nicolas, ETSAM-UPM.
Muñoz, Marta, ESTAM-UPM.
Pieltain, Alberto, ETSAM-UPM.

## Scientific Committee

Atalay Franck, Oya, ZHAW School of Architecture, Design and Civil Engineering.
Blanco Lage, Manuel, ETSAM-UPM.
Boutsen, Dag, KU Leven.
Cavallo, Roberto, Delft University of Technology.
De la O Cabrera, Manuel Rodrigo, ETSAM-UPM.
Flynn, Patrick, TU Dublin.
Marine, Nicolas, ETSAM-UPM.
Roth-Cerina, Mia, Faculty of Architecture, University of Zagreb.
Zupancic, Tadeja, Fakulteta za arhitekturo, Univerza v Ljubljani.

## Organized by

**European Association for Architectural Education**
Association Européenne pour l'Enseignement de l'Architecture

ETS de Arquitectura de Madrid (ETSAM)
**Universidad Politécnica de Madrid** (UPM)

## Supported by

GOBIERNO DE ESPAÑA
MINISTERIO DE TRANSPORTES, MOVILIDAD Y AGENDA URBANA

# Contents

# NEB and Design Education

# Envisioning a New European Bauhaus Stage—A Call for More Embodied Research in Architecture Education

Lisa Beißwanger(✉)

University of Koblenz, Koblenz, Germany
beisswanger@uni-koblenz.de

**Abstract.** The *New European Bauhaus* (NEB) set out to create an updated and enhanced version of its historical namesake. So far, this article observes, the initiative has somewhat avoided the fields of performance and stage work. Yet, these areas were not only of great importance for the original Bauhaus idea. Embodied knowledge and an experiential understanding of space also appear to be key competencies for the NEB's intended goals: shaping a positive, inclusive, and sustainable future. Drawing on the historic Bauhaus stage workshop, the article invites to consider the potentials that the fields of dance and theater offer for future design and architectural education and practice. More precisely, it proposes to envision a *New European Bauhaus Stage*, connecting the fields of architecture and the performing arts. Structured into two sections, the article first provides historical insights into the work and organization of the historical Bauhaus Stage. In the second part, it outlines some more recent developments in the field, introducing a selection of projects working at the intersections of architecture and performance. Reflecting on both, the older and the newer ideas and initiatives, and their inspirational potential for architecture and architecture education, it then goes on to outline what a *New European Bauhaus Stage* might look like. It proposes a decentral pan-European think-and-do-tank, aiming to foster embodied knowledge and research in architecture. Rather than offering a recipe or concrete didactic advice, the manifesto-like article aims to inspire its readers to develop new projects and collaborations.

**Keywords:** Architecture education · Dance · Choreography · Stage design · Bauhaus stage · Embodied knowledge

## 1 Introduction

According to the New European Bauhaus' (NEB) mission, the EU-initiative fosters design that creates positive "experiences", and aims to help create a "beautiful" future [1]. Beautiful, according to the NEB, are "places, practices, and experiences that are enriching, sustainable and inclusive" [1]. With this definition, the policy makers intentionally deviate from ideas associated with the technology-driven and functionalist modernism of the historical Bauhaus. However, considering the NEB's funding schemes and


M. Blanco Lage et al. (eds.), *Towards a New European Bauhaus—Challenges in Design Education*,
https://doi.org/10.1007/978-3-031-40188-6_1

projects to date, it still seems to prioritize technological innovation and the engineering disciplines over the arts, humanities, and social sciences. This article aims to challenge this general tendency.

Having previously worked on the intersections of performance studies and architecture theory including the historical Bauhaus Stage, I find that the NEB falls particularly short in this area. While the project does make use of dance and performance to present itself publicly as a vibrant and living institution, for example, during its first European Bauhaus Festival in Brussels in summer 2022 [2], the interdisciplinary "High-Level Roundtable", which is the project's expert steering committee, currently includes architects, natural scientist, investors and activists, but not a single performing arts specialist [3]. I am convinced that when it comes to the aspired inclusion, dialogue and environmental awareness, embodied knowledge is a key competence. And it is exactly in this area that the disciplines of dance and theater hold great potential for architecture practice and education. Therefore, I propose to envision a *New European Bauhaus Stage*, which will help to connect the fields of architecture and the performing arts and create a space for transdisciplinary learning and inspiration.

Taking a cue from the historical Bauhaus and its Stage Workshop, the article invites to think about the potentials that dance and theater practice and theory hold for design and architecture education, and in consequence, for future design and architecture practice. The article is divided into two sections. The first section will provide historical insights into the work and structure of the historical Bauhaus Stage. Taking up the NEB's reference to the historical institution, this will provide the point of departure as well as friction for the following discussion. The second section will focus on more recent developments in the field, introducing a selection of projects working at the intersections of architecture and performance. Reflecting on both, the old and the new ideas and initiatives, I will sketch out what a *New European Bauhaus Stage* might look like and conclude with some critical reflections on such a project's challenges and potentials.

## 2 Notes on the Historical Bauhaus Stage

The historical Bauhaus was the first—and to my knowledge the only—design and architecture school ever to include an actual performance department [4]. This department, the so-called Stage Workshop, was initiated in 1921, shortly after the Bauhaus' founding and it ceased only with the school's closing in 1932/33 [5]. In the early Weimar years the Stage Workshop had no dedicated space. Nevertheless, it formed an integral part of the school's holistic educational concept adding to the overall goal, as Walter Gropius put it, "to find a new and powerful working correlation of all the processes of artistic creation to culminate finally in a new cultural equilibrium of our visual environment" [6]. With the move to Dessau in 1925, a small professional stage was installed. Despite the size it held a prominent place within Gropius' building design, as it was located at the heart of the building between the school's auditorium and canteen.

The figure most closely associated with the Bauhaus Stage is the multidisciplinary artist Oskar Schlemmer, a painter, sculptor, stage designer, as well as choreographer, and performer. Schlemmer was the workshop's "Meister" (master) from 1923 to 1929, and responsible for most of its varied activities. These activities included the creation of semi-professional stage productions, for example a *Mechanisches Ballett* (*Mechanical Ballet*)

(1923) by Kurt Schmidt or the so-called *Bauhaus Tänze* (*Bauhaus Dances*) conceived between 1926 and 1929 by Schlemmer himself [7]. Among the common traits of these works were the de-individualization of performers, for example with masks or suits, the extension and/or alteration of their bodies, for example with props or costumes, and a general use of geometric shapes, primary colors and machine-like movement.

Contrary to first impressions, Schlemmer was not interested in creating robots or man-machines, but rather to study the relationship of the human body and the laws of geometrical space. In his essay *Man and Art Figure* (1924) [8] he explains how a dancing body in space can bridge the gap between what he conceived as geometrical space—for him represented by a Euclidean box-like concept of space—and the human body as representative of a natural and organic order (Fig. 1).

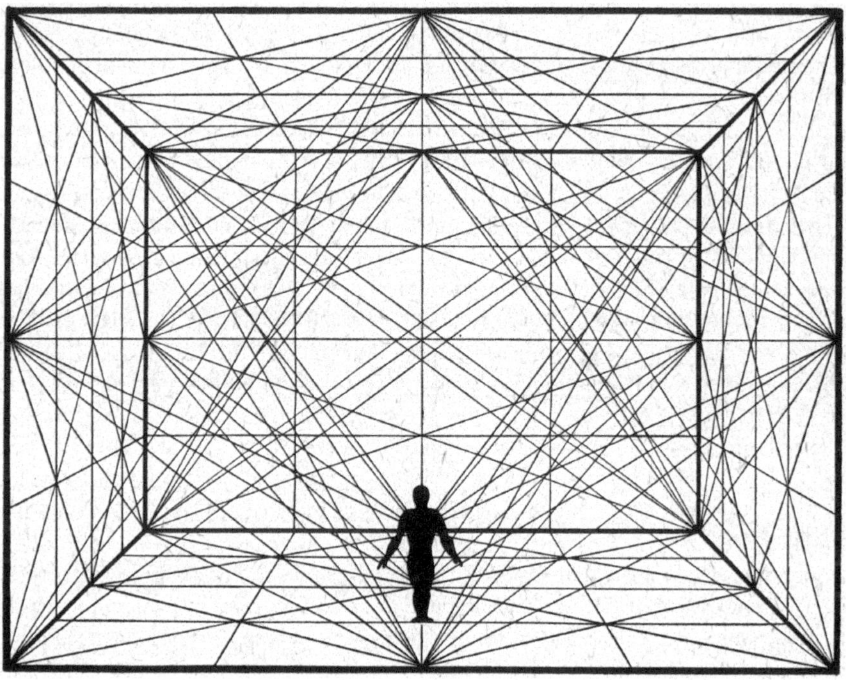

**Fig. 1** Oskar Schlemmer, *Drawing of the Abstract Stage.* In: Schlemmer O (1965 [1925]): Mensch und Kunstfigur. In: Schlemmer O et al.: Die Bühne im Bauhaus, Mainz: Kupferberg, p. 13

In the eyes of Gropius Schlemmer's stage work represented a transformation of "dancers and actors into moving architecture" [9]. More recently, theater scholar Melissa Trimingham noted that Schlemmer was "decisively countering disembodied approaches to design, and idealist notions of form detached from living", and instead highlighted the importance of "the body and the human being as central to any humanist quest" [10]. Indeed, Schlemmer's approach was deeply human-centered, which not only becomes manifest in his stage work, but also in his work as a visual artist and in his late theoretical Bauhaus introductory class "Der Mensch" (the human), in which he aimed to reach an understanding of the—in his view—universal character or essence of "the human" as a psychophysical being [11].

The Bauhaus Stage took on various functions. As a theater laboratory and spatial testing site, it allowed to investigate questions of movement in space and its living, organic, and somatic qualities, rather than posing constructional questions, or testing for material strength or functional performance. At the same time, and this seems very important, it was an intrinsically pedagogical and social space [12]. Its social function becomes particularly clear when considering the Stage Workshop's role in curating the legendary Bauhaus Parties such as the lantern (1922), white (1926), and metal (1929) parties. Schlemmer and his Workshop took the main role in planning these events, and they created costumes, props and performances for them [13].

Thus, the activities of the historical Stage Workshop went well beyond mere stage design. It was an open field for experimentation, adding body, space, and movement studies to the design curriculum. At the same time, it offered lessons in interdisciplinary collaboration, participation and group dynamics. Due to the stage's central location and social function for the Bauhaus as a whole, performance scholar RoseLee Goldberg called it the "beating heart of the Bauhaus" [14]. However, the workshop was never equal in status to the other workshops. Formally, it was labelled an "extraordinary" workshop, which means it was not permitted to grant graduation certificates, and it had no dedicated teachers besides Schlemmer [15]. It also had no full-time students but "shared" its participants with other workshops which also led to internal frictions. At the same time, the workshop was notoriously underfunded, and frequently endangered by budget cuts in the name of economic efficiency, especially in the later Dessau years. Even though Gropius and others were full of praise of the stage's innovative work, Schlemmer often complained that it was no more than a "flower in the button hole" to the overall institution [16], expressing his feeling that the stage was misused for entertainment and to impress officials, but otherwise neglected.

If suggesting to revive the idea of a Bauhaus Stage, it is important to note that from today's perspective Schlemmer's concepts and approaches would certainly need an update. This applies to his strictly mathematical and somewhat "neutral" Euclidian space-model, as well as to his rather anthropocentric and essentialist conception of "the human". Historical research has revealed a nationalist tendency in Schlemmer's work, as he was particularly interested in finding new ways for the German theater tradition [17]. Finally, the workshop was by no means an inclusive or democratic space in that it was organized hierarchically, with a sole "master" as its head, and with only a small number of students who participated in its activities, which certainly also limited its overall impact on the Bauhaus curriculum.

## 3 Towards a New European Bauhaus Stage

Now that I have outlined some of the innovations and challenges of the historical Bauhaus Stage, I will introduce some later projects that have continued the dialogue between architecture and the performing arts. With this I hope (1) to give some concrete inspiration as a starting point for anyone interested in taking up the subject, and (2) to show how a *New European Bauhaus Stage* can build on a range of past and existing initiatives, some with direct reference to the historical Bauhaus, others more independent from it. The second part of this section offers some suggestions how diverse projects such as these may be bound together to an overarching initiative for more embodied research in architecture education. I would like to stress, that my project selection is personal and somewhat subjective in scope and direction. There is an emphasis on the German speaking and European contexts as well as on dance rather than theater or stage design. Therefore, I would like to invite readers to continue and expand this selection in every direction possible [18].

### 3.1 Who Can We Learn From?

Immediately after the closure of the Bauhaus, the institution's transdisciplinary spirit was taken up in various places around the world, including the United States. With respect to stage work the most notable example may be Black Mountain College (1933–1957) in North Carolina, a new type of liberal arts college that merged elements of Bauhaus teaching with the pedagogical ideas of John Dewey who stressed the importance of art and creativity for every kind of learning process. A former Bauhaus Stage member, Xanti Schawinsky, was appointed to teach "Stage Studies" [19] and later, there were summer schools leading to collaborations between artists as diverse as John Cage, Merce Cunningham, Richard Buckminster Fuller, and Elaine and Willem De Kooning [20]. Although this school was rather short-lived, it remains an important reference point for radical transdisciplinary learning concepts with a strong focus on the performing arts.

A few years later, in the 1950s, dance pioneer Anna Halprin and her partner, landscape architect Lawrence Halprin, were inspired by Bauhaus pedagogy to give joined workshops for architects and dancers. They experimented with exploring space through dance and choreography, while focusing—much more so than was the case with the historical predecessor—on participatory and collaborative practices [21]. They also developed an innovative notation scheme called "motation" to capture motion through space, thus, finding a common language for dancers and architects [22].

More recently, in Germany, the legacy of the Bauhaus Stage was taken up once more. On the occasion of the Bauhaus anniversary in 2019, the Bauhaus Dessau—today a research facility and museum—has revisited the historical Bauhaus Stage with a project called *Bauhaus open Stage—Performative Architectonics* led by Torsten Blume. In the 2-year course of this initiative, students from seven international universities developed research projects based on historical Bauhaus Stage experiments leading up to final presentations in the context of the anniversary celebrations [22]. In 2020, the Bauhaus University Weimar offered a performative introductory course in architecture and urban studies titled *Dis-Tanz* (transl.: Dis-Dance) [23]. The performance took place in a large public square outside the school. A video documentation shows how participants, all clad in black and attached to each other with long white strips of fabric, experiment with moving in space in relation to each other. Happening at the height of the Covid pandemic and social distancing measures, dance and performance offered a way to experiment with the very pressing questions of closeness and distance.

The recent decades saw a growing number of interesting projects at universities and art schools. There have been several master and doctoral theses dedicated to aspects of dance and architecture, often using methods of practice-led research. At the Technical

University of Munich for example, researcher Katharina Voigt is working to develop architectural design methods, that include bodily forms of knowledge. Together with a team of dance practitioners, she develops architecture-related performances as well as workshops for designers [24]. One outcome of this work is an issue of the architecture journal *Dimensions* titled *Moving Experience* which contains several articles on dance and body-practices in design education [25]. Another approach to integrate dance into architecture studio teaching was taken by Zehra Ersoy at the Dokuz Eylül University in Izmir. Her students worked with two professional dancers to complete a number of tasks, such as using their bodies as architecture or developing a dance performance based on the study of iconic buildings. In one of the few scholarly articles on the topic, Ersoy reflects her approach critically, and offers some valuable hints for anyone who might wish to undertake a similar project [26].

Reflecting the digital turn, a project hosted at the Vienna University of Technology named *DANCR* currently investigates human–robot-interaction with the help of dancers and an AI-guided robot. The project's aim is to confront artificial intelligence with the human ability to improvise and perhaps enhance it as a result [27]. A researcher in the field Digital Design at the Technical University of Darmstadt, Samim Medizadeh, has recently worked with students on the physical animation of robotically molded concrete elements [28]. Collaborating with choreographers and dancers to activate these kinetic objects, the team explored the creative potentials of "living architecture" and human-architecture interactions (Fig. 2).

**Fig. 2** Susanne Grau (left) and Johana Kasperowitsch (right) in *Animate concrete*. Design studio and general concept: Samim Mehdizadeh, Oliver Tessmann (DDU/TU Darmstadt), choreography: Ashkan Afsharian with Johana Kasperowitsch and Susanne Grau

Beyond academia, the German Goethe Institute hosts and supports the program *Performing Architecture* which has initiated several events and exhibitions during Venice Architecture Biennials. As a final example I would like to mention *Theatrum Mundi*, a semi-commercial initiative based in London and Paris. Working at the intersections of performance and urban design, its members are conducting academic research but also offer workshops and courses to design schools [29].

This brief overview may show that even though there are no architecture schools with dedicated theater departments, there is a broad range of dispersed knowledge already in existence. However, accessing this knowledge remains challenging, since there is not yet a larger network in place to connect the various actors working in the field. A *New European Bauhaus Stage* would provide the opportunity to connect these actors and help bundle collective knowledge, make it more visible and thus, inspire architecture practice and education on a broader basis.

### 3.2 A Stage but Not a Place, a Workshop but No Master

So how could this *New European Bauhaus Stage* look like? Certainly, a mere resurrection of the historical Stage Workshop, led by a "master" and guided by modernist western, anthropocentric, and essentialist world views would be untimely, and not desirable. In the light of limited time and financial resources, it also does not seem feasible to install dedicated performance departments within existing architecture faculties or adding dance training to the already dense architecture curricula. In other words, architects don't need to become dancers. At the same time there is a lot for architecture—as an intrinsically interdisciplinary practice—to learn from performance theory and practice, especially when it comes to questions of embodied knowledge, and participatory and inclusive approaches.

Rather than building physical stages, my suggestion would be to stick with the New European Bauhaus' decentral concept. What I am thinking of is a pan-European, transdisciplinary think-and-do-tank, connecting actors and initiatives working on the intersections of design, engineering and the performing arts. This network could be guided by a core group or board of experts, which develops research and funding schemes, gathers and makes available ideas and resources through publications or meeting opportunities. The Activities would not necessarily be limited to the education sector, yet education would be a strong focal point. Institutions like the European Association of Architecture Education may be well suited to co-host an educational sub-network, offering concrete methodological guidance for educators interested in expanding their teaching and including performance-based practices.

The potential outcomes of such an initiative are manifold. They reach from innovations in stage design and actual stage productions to more abstract fields, such as analog and digital movement notation, motion tracking, or human–robot-interaction. A central concern will certainly be the strengthening of practice-led research in the field of architecture in general and the development of body-based design methods in particular to further integrate bodily and sensory knowledge into the architecture and design curriculum. A particularly socially relevant outcome may be a heightened sensitivity for what I like to call architecture's politics of space. In other words, how the built environment is shaping and at the same time shaped by human and non-human interactions and power dynamics

[30]. Here, the study of choreography can offer a deepened understanding of such spatially determined power dynamics and can help to develop design-strategies that navigate individual and collective experiences and foreground communication, collaboration and exchange.

## 4 Conclusion

Given the growing demand for an inclusive architecture with positive, experiential dimensions, it seems more important than ever to unlock embodied knowledge and foster participatory approaches towards beautiful design solutions as defined by the NEB. A New European Bauhaus Stage could operate at the very heart of this endeavor, transforming architecture education and in consequence, architecture as a whole.

Certainly, such a project holds challenges and I will outline three of them here.

(1) Re-evaluating the *historical* Bauhaus Stage made clear, that its *new* iteration will have to tackle master-narratives, and anthropocentric essentialism—as traits of Western modernism still not overcome—to open the way for design practices that are truly inclusive, and sustainable.
(2) A *New European Bauhaus Stage* has to be more than a "flower in the buttonhole" of the overall initiative. Sufficient funding and a thoughtful levelling of artistic, economic, and ethical aims will be key to create a basis for serious investigation, and to avoid exploitative and immaterial labor still too present in the creative sector.
(3) A special challenge will be to allow for theater's subversive potentials and to offer a space for open ended experimentation beyond neoliberal experience economy and the dominant logics of utility and cost efficiency. With historical hindsight, it seems evident that economically efficient solutions are by no means the most humane and technical advance does not equal social progress.

A *New European Bauhaus Stage* will be a forum to explore such issues. It has the potential to act as a catalyst for embodied knowledge and practices in design education, and help define what abstract buzzwords like "humane" and "sustainable" really mean to diverse individuals in varying contexts.

Finally, we have seen that many interesting initiatives already exist, albeit still under the radar of mainstream architectural discourse. A *New European Bauhaus Stage* has the potential to change this. It can help transform architecture from a profession centered on master architects putting their ingeniously designed boxes into space, to a collaborative and inclusive practice that creates relational and inhabitable spaces. This transformation is urgently needed and, at least partially, already on its way.

## References

1. New European Bauhaus Homepage. https://europa.eu/new-european-bauhaus/index_en. Accessed 28 Nov 2022
2. Festival Program, New European Bauhaus Homepage. https://new-european-bauhaus.europa.eu/get-involved/festival_en. Accessed 28 Nov 2022
3. The only partial exception being the architect and stage designer Pia Maier Schriever

4. The Architecture Association (2022) London offers the MA/MFA program "Spatial Performance and Design (AAIS)". Architecture Association Homepage. https://www.aaschool.ac.uk/academicprogrammes/postgraduate/spatial-performance-design. Accessed 28 Nov 2022
5. Scheper D (1988) Oskar Schlemmer. Das triadische Ballett und die Bauhausbühne. Akademie der Künste, Berlin
6. Gropius W, Wensinger AS (eds) (1987) The theater of the Bauhaus, 3rd edn. Wesleyan University Press, Middletown, p 7
7. Kaldrack I (2011) Die Bauhaustänze Oskar Schlemmers. Maske und Kothurn 1–2(57):123–140
8. Schlemmer O (2014) Man and art figure [1924]. In: Gropius/Wensinger (1987), pp 15–46
9. Gropius W (1987) Introduction [1961]. In: Gropius/Wensinger (1987), p 9
10. Trimingham M (2021) Gesamtkunstwerk, gestaltung and the bauhaus stage. In: Imhoof D, Menninger ME, Steinhoff AJ (eds) The total work of art. foundations, articulations, inspirations. Berghahn Books, New York, pp 95–114
11. Schlemmer O (1969) Der Mensch: Unterricht am Bauhaus, nachgelassene Aufzeichnungen. Florian Kupferberg Verlag, Mainz
12. Schlemmer writes: "Theater […] must—if only it would succeed—offer 'enchanting pedagogy' and be founded on Schillerian principles." Quoted in Trimingham M (2016) The theatre of the Bauhaus: the modern and postmodern stage of Oskar Schlemmer. Routledge, New York; London, p 110
13. Droste M (1993) Fest—Arbeit—Spiel. In: Droste M (ed) Bauhaus: 1919–1933. Taschen, Cologne, pp 50–57
14. RoseLee Goldberg quoted in Trimingham (2016), p 3
15. Scheper (1988), p 119
16. Oskar Schlemmer quoted in Scheper (1988), p 124
17. Schlemmer was also among those German artists who actively tried to comply with the Nationalist Socialist regieme. Droste M (2015) Ambitionen und Ambivalenzen: Oskar Schlemmer 1933/34. In: Ruppert W (ed) Künstler im Nationalsozialismus. Die »deutsche Kunst«, die Kunstpolitik und die Berliner Kunsthochschule. Böhlau Verlag, Göttingen, pp 177–202
18. For further historical, and more recent projects see Aubin C, Goldberg R, Minguez Carrasco C (eds) (2020) Bodybuilding: architecture and performance. Performa, New York
19. Schawinsky taught at Black Mountain Collage from 1936 to 1938. Laabs A, Blume T (eds) (2021) Xanti Schawinsky: from the Bauhaus into the world. Deutscher Kunstverlag, Berlin; Munich, pp 38–44
20. A notable collaboration between these artists was the play *Ruse of Medusa* (1948). Merce Cunningham trust homepage. https://www.mercecunningham.org/the-work/choreography/ruse-of-medusa-the/. Accessed 28 Nov 2022
21. Wasserman J (2012) A world in motion: the creative synergy of Lawrence and Anna Halprin. Landscape J 1/2(31):33–52
22. *Bauhaus Open Stage.* Bauhaus Dessau Homepage. https://www.bauhaus-dessau.de/de/online-magazin/im-gespraech/bauhaus-open-stage.html. Accessed 28 Nov 2022
23. DIS TANZ (2022) Introduction week 2020. Faculty of Architecture and Urbanism, Vimeo. https://vimeo.com/487238064. Accessed 28 Nov 2022
24. *Experiencing Sense(s)*, terra. Trienal de Lisboa Homepage. https://2022.trienaldelisboa.com/en/evento/experiencing-senses-2/. Accessed 28 Nov 2022; *ThinkBig! Festival Munich*, July 2022, Vimeo. https://vimeo.com/750743937/501e1c7866. Accessed 28 Nov 2022
25. Roy V, Voigt K (2021) Dimensions. J Archit Knowl 2(1):7–10. Spatial dimensions of moving experience. Bielefeld, transcript
26. Ersoy Z (2011) 'Building dancing': dance within the context of architectural design pedagogy. JADE 1(30):123–132

27. H.A.U.S. Humanoids in Architecture and Urban Space Homepage. https://h-a-u-s.org/index.php/2022/01/01/funded-to-develop-dancr-our-ai-dance-tool/. Accessed 28 Nov 2022
28. Animate Concrete DDU with TANZPOL, Vimeo. https://www.youtube.com/watch?v=QaQYl1uy3Ls. Accessed 28 Nov 2022
29. Theatrum Mundi Homepage. https://theatrum-mundi.org/. Accessed 28 Nov 2022
30. Beisswanger L (2021) Architecture and/as choreography: concepts of movement and the politics of space. Dimensions. J Archit Knowl 2(1):23–50

# Participatory Methods in the Education of Architects

Kristina Careva(✉) and Rene Lisac

Faculty of Architecture, University of Zagreb, Kačićeva 26, 10000 Zagreb, Croatia
kristina.careva@arhitekt.hr

**Abstract.** To understand the context, from the physical through the sociological to the economic, the architect must conduct field research. For a long time that primarily meant a tour of the location but recently, the importance of collecting the opinions of all actors of spatial development is becoming highly recognized. Moderating the conversation between all those actors' representatives is a demanding yet rewarding task. Since the architect already considers and tries to reconcile the views of various professions in his daily work, he could be a good moderator. Therefore, practical teaching of this nature is advisable to be included in architectural education. The theoretical foundations need to be supplemented by active work on a specific task. The elective course "Participatory Design of Space" at the Faculty of Architecture, University of Zagreb, practically introduces students to the importance and application modes of participation of all relevant stakeholders in creating a cognitive fund for solving a design or planning task. The five iterations of course held so far provide interesting insights into specific situations, point out the importance of comprehensive research fieldwork as well as indicate certain challenges within introducing the participatory framework to students.

**Keywords:** Participatory methods · Education of architects · City acupuncture · Participatory design of space

## 1 Introduction

Cities that have a high quality of life are cities that have a "human dimension" [1] that is, they consider the needs of citizens in designing, planning, transforming, and developing. The New Leipzig Charter states that the transformative power of European cities lies in the fact that they are places of "pluralism, creativity, and solidarity", and at the same time laboratories for new solutions and spaces that provide the possibility of social innovation [2].

To make architecture students aware of this important topic, it is good to familiarize them with the practical application of participatory methods in design and planning. Participation is certainly one of the topics that can be practically experienced during studies. When we talk about the participatory ladder as imagined by Arnstein, the aim in educating students is to practically examine the middle of the ladder meaning interviews, focus groups and co-creative workshops [3]. Only declarative, lower levels are tried to


M. Blanco Lage et al. (eds.), *Towards a New European Bauhaus—Challenges in Design Education*,
https://doi.org/10.1007/978-3-031-40188-6_2

be avoided and those empowering upper levels are hard to be reached since they demand level of awareness that is still not reached within Croatian society. The reason for this lies in the fact that, unlike countries that have been functioning as democratic societies for a long time, Croatia belongs to the circle of countries that emerged from socialism about 30 years ago in which a culture of participation was not built due to general social ownership. After the absence, we are witnessing a power and decision-making distribution with negative effect on participation where business and private capital in conjunction with government sector actively participate in decisions, while citizens and academia are left on the margins from where they are eventually included declaratively. Some of the conducted research carried out for some parts of Croatia confirms a low level of citizen participation and the absence of a statistically significant difference in citizen participation in the process of adoption of spatial plans and their dependence on the development of the citizenry [4].

Such a situation leads to a negative perception, mistrust, and even aversion of the Croatian public, as evidenced by some of the conducted research. In her work, Mišetić explain current Croatian context through elaboration of several conducted research: two of them made by the Institute of Social Sciences Ivo Pilar in Zagreb on a representative sample in 2008 and 2014, as well as the book Vital cities: a view from the perspective of experts in city administrations. According to the results of these surveys, only about a quarter of citizens actively participate in activities important to the local community, while almost half of the professionals employed in city administrations encountered resistance from the local community when it comes to project implementation. Furthermore, Mišetić concludes that there is a gap between the desired and actual influence of certain social actors, where the desired influence of experts and citizens is weak, and the influence of politicians and city authorities is strong [5].

As the public, led by NGOs and the academic community, is increasingly vocal in its efforts to actively engage in debates, this trend has been in a slow process of change over the last ten years. The importance of citizen participation as a guarantor of socially sustainable development was incorporated into the Spatial Development Strategy of the Republic of Croatia from 2017. Citizens themselves are becoming more and more active in signing petitions, filing complaints, participating in protests and public debates, and through contact with officials. The topic of participation and the importance of citizen involvement is increasingly discussed through numerous workshops and panels. The current position of Croatian society is characterized by a low level of political culture, and since it takes about 60 years to achieve a civil society, Croatia is currently halfway there [6].

As great potential in raising awareness of the need and benefits of citizen participation lies in the academic community, certain efforts in this direction have recently been made in the development of the University of Zagreb curriculum. In the domain of the Faculty of Architecture, graduate students can attend a spatial planning course that is conducted in cooperation with teachers and students of sociology, as well as enroll in an elective course that will be presented in this paper.

# 2 The Elective Course "Participatory Design of Space"

## 2.1 Pedagogical Methodology

The implementation of participatory knowledge in architecture curricula at the Faculty of Architecture of the University of Zagreb is acquired through the elective course "Participatory Space Design" at the graduate level. The intention was to offer an optional course dealing with the topic of participatory creation for master's students. The course is based on the City Acupuncture initiative, which has been active since 2009, and whose authors are teachers at the Faculty of Architecture of the University of Zagreb (Fig. 1). By researching participatory methods, but also by conducting numerous interdisciplinary and participatory workshops, the Acupuncture of the City method was developed for inclusive planning of common spaces [7]. The method aims to improve the quality of life in the city through the regeneration of public space. The method has 3 main principles: small and precise interventions in the urban fabric can significantly improve urban life; interventions should be designed in an interdisciplinary discourse; the process should involve all relevant stakeholders [8]. After 8 years of participatory action, the transfer of knowledge gathered through the process has matured to become a proposal for a new elective course in graduate studies.

**Fig. 1** Workshops (in the top row) and results (in the bottom row) of selected City Acupuncture activities

The course takes place in the summer semester as an intensive active work consisting of an intro, a field-based weekend workshop and a presentation. Besides students in Architecture, the course is open to students from other Faculties and Universities. The topic of each years' task is formed based on the needs of a community and its readiness for participatory activities that can range from information through surveys and interviews to involvement in implementation. The task that is thematized change for each academic year, but it always comes from a real need for some form of participatory design driven proposals.

The aim of the course is to point out to students the important facts related to participation of all relevant groups or group representatives. Roughly speaking, these are four groups important for the design process of socially relevant public (or semi-public)

space: the local government members; various expert groups representatives; representatives of companies interested in investment and development; and specific or future potential users. Students are explained and introduced to the quadruple helix model, which recognizes exactly these four main groups of actors in the innovation system [9].

The outcomes of the course that students acquire are:

- To identify and critically evaluate participatory content;
- To generate the conceptual level of planning—the intent;
- To discuss the intent with users;
- To translate it into a viable solution;
- To participate in realization, if possible.

All these outcomes are achieved through the students' creative work on the assigned task, where, in addition to listening to users' opinions, they should also respect each other while designing their project in groups.

## 2.2 Practical Teaching

The first iteration of the course took place in the academic year 2017/2018 when students tried to modernize and refine the common areas of the Faculty of Geotechnical Engineering in Varaždin as a remote part of the University of Zagreb. The total of twelve students, six of which from the Faculty of Architecture and six from the Faculty of Geotechnical Engineering, were divided into three groups. Employees' representatives of the thematized Faculty also took part and actively engaged in the conversations. The outcome was proposals for new identity for outdoor and indoor spaces. Through conversations with employees and students, it was observed that the faculty is changing its position, opening new study areas, and is looking for, among other things, a new visual identity, and a redefinition of common spaces so that they are less formal. The students tried to respond to the perceived needs by proposing a redefinition of the existing spaces without changing the structure and functional logic of the building. The interdisciplinary teams functioned very well with architecture students possessing the knowledge and skills for design response and geotechnical engineering students as users of the space being thematized.

The most complete participatory process that has been managed to achieve in this academic practice, followed in the academic year 2018/2019 when the students tried to improve the entrance space of the Ivan Gundulic's elementary school in Zagreb through design. 19 students, all of them from the Faculty of Architecture, were divided into five groups. Representatives of different groups (pupils, school employees, parents) took part in numerous conversations with students. The proposed solutions that students designed were explained to the school community (Fig. 2) which ended with selection of proposals and their realization at the end (Fig. 3). A seating area for students and guests of the school and a storage area next to the dining room have been realized. It is important to emphasize that the whole process took not more than a month [10].

Next year's task was not realized due to the Covid-19 pandemic, which put all academic activities in an online format. The task was to redesign the public space of the former barracks converted into a student campus in the city of Koprivnica. Instead, the teaching took place only through an online format that only managed to fulfill the

**Fig. 2** Student presenting their design solutions to the Ivan Gundulic elementary school community in the academic year 2018/2019

**Fig. 3** School staff, supervisors, students and parents working together on realization at the Ivan Gundulic elementary school in April 2019

theoretical knowledge, but made it impossible to carry out practical teaching, which is of utmost importance for achieving learning outcomes.

Although the Covid-19 pandemic was still present, classes were held in the 2020/2021 academic year. This was strongly influenced by the fact that Croatia was hit by several earthquakes during 2020, some in Zagreb, and some in the Banovina area, where the town of Petrinja was particularly affected. The will and need to help the inhabitants of the earthquake-affected areas conditioned outdoor activities with a high level of awareness of physical distance. The work of arranging the container settlement "Center of new life" in Petrinja was initiated by the company that participated in the construction of the settlement. A total of eight students, seven from the Faculty of Architecture and one from the Academy of Fine Arts, formed three groups and after discussions with the contractors, representatives of the authorities and some residents made proposals for the aesthetic and functional improvement of the settlement (Fig. 4). Unfortunately, implementation failed due to apparent administrative sluggishness.

**Fig. 4** Students talking with residents of the container settlement "Centre of new life" in Petrinja in April 2021

In the past academic year 2021/2022, the task was again related to the city of Petrinja, only this time the mayor opened the topic of the Main City Square. Five students of the Faculty of Architecture worked with government representatives and citizens to create a comprehensive map of possibilities, limitations and intentions. The collection of residents' opinions included survey results and on-site interviews, and ended with an extensive database of participatory responses that will be used in the near future to create the terms of reference for the architectural competition. In this task, the students were well acquainted and actively involved in one of the possible ways of implementing participation, but it was difficult for them to understand the real role of participation in the process of designing the space, since the creative part of designing in response to the collected needs of the users was absent.

## 3 The Importance of Teaching Participation to Students in Architecture

### 3.1 Positive Impact

After five years of implementation, the lessons conducted so far provide interesting insights into concrete situations and point to the importance of introducing students to the possibilities of complete field work and research to solve real problems, such as the actual demands of users, management structures, various experts as well as, of course, financial limitations. Most students are very satisfied with the course, and in surveys they say that they learned a significant amount of useful new knowledge during the course. Of course, it should be considered that the course is elective and only students who are interested in the theme of participation enroll in it. With the growing awareness of the importance of including participatory methods in design and planning, it is to be expected that a greater number of compulsory courses will introduce these methods into teaching. This is desirable, but it should be emphasized that even then there is a need for such practical and concise experiences such as the elective course Participatory design of space in which students try to get as much as possible from participation in a short period of time.

The students' creative response to the set topics and problems that the users presented in the conversations is a contribution to the community. Even mere conceptual proposals enrich the fund of knowledge, but it is best if they become a realization, which of course depends on financial parameters that are outside the scope of the educational paradigm.

### 3.2 Challenges and Solving Possibilities

Along with stating the importance and benefits of comprehensive participatory research field work in the education of architects, certain challenges that this process entails can also be indicated.

**Broader context**. The first one is understanding the broader context of inclusive planning and one's own design role. The large number of actors involved and interested in planning and the change in their importance in relation to a specific task, makes this challenge constantly present even though the accumulated experience reduces the time needed for adaptation to a specific task. Students who choose this elective course often encounter the concept of inclusive planning for the first time, still the breadth of the context does not frighten them but opens their curiosity. However, it is necessary to return them to their task so that they do not wander into areas whose methods are not taught at the faculty of architecture.

**Balance**. The second challenge is maintaining a balance between one's own professional creativity and the collected participatory information. Often, students in the graduation phase of their education are still examining the scope of their profession and are not yet sure of its boundaries. As participation presupposes a juggling act between understanding others and security in one's own creation, it is not easy for students (and architects in general) to maintain this balance. The final presentations and the user's response to the offered design proved to be a very valuable part of the course, in which each student gets a clear reaction to his proposal but can also see the reaction to the design proposals of his colleagues. Experience has shown that the best solutions are those that represent a clear design concept that incorporates an understanding of the user's needs.

**Objectivity**. The third challenge is the objective filtering of information. How to understand what is important, and what is just random (clumsy, insecure, provoked, angry) attitude of an individual? Not every participatory conversation presents these problems, but when they do, they need to be overcome with mentoring and discussions within the student group to work through them. This is particularly important because it is often precisely the attitudes expressed in affect that remain the most in the memory and it is more interesting for students to build their concepts on them. This, of course, can lead in a completely wrong direction and ultimately leave both users and students dissatisfied—users with unsuccessful solutions, and students with the belief that participation only creates problems in the design process. The joint discussion between mentors and students ensures objectivity in filtering the collected information and teaches how to form a collective intention about what the project should address.

**Participatory interactions**. Finally, approaching the interlocutor and communicating with him is the fourth challenge. This is where the personality of each student comes to the fore, and they themselves realize that not everyone is equally good at approaching, conducting conversations, listening, and encouraging interlocutors. The way to solve this unevenness is to divide the students into smaller groups (of two or three), where one student leads the conversation and the other takes notes.

Communication through graphic material is also important, whereby we distinguish between collecting user information as input data for the project response and obtaining the user's response to proposed project solutions. When collecting input data for greater coverage, a map on which the exact positions to which the user's comments refer are located through the dialogue, in order to avoid misinterpretations, proved to be very useful. When collecting users' opinions on spatial solutions proposals, especially when it involved a larger number of respondents, in addition to direct communication, leaflets were successfully used on which the project proposals were presented graphically, simply and clearly, with enough space on the leaflets for users to write or draw their comments and grades.

## 4 Conclusion

It is undeniable that today participatory methods are seen as important ways of designing and planning. Therefore, it is desirable to include the teaching of these methods in the curriculum of students in architecture. At the Faculty of Architecture of the University of Zagreb, a subject entirely dedicated to participation, Participatory Design of Space, is taught as an elective course at the graduate level. It consists of an introduction, a field-based weekend workshop and presentation and is open to students of other faculties alike. The topics change each year depending to real need of users. The aim is to familiarize the students with the quadruple helix model and help them to produce creative solution to the given task in cooperative surrounding. The course outcomes help the students to identify and critically evaluate participatory content, to generate the conceptual level of planning—the intent, to discuss the intent with users, to translate it into a viable solution, and to participate in realization (if possible).

Five iterations of classes held so far had different tasks, different number and structure of participants, and different outputs. Conclusions that could be drown out from this experience show importance of introducing students to participatory methods and the positive influence of their creative solutions to the community. During and after the class, several challenges that need to be faced were also observed, but the possibilities of solving them were also indicated. Understanding the broader context of inclusive planning and one's own design role can take students beyond their profession, so they need to be mentored. Maintaining a balance between one's own professional creativity and collected participatory information is always challenging, especially for students, but they can get a clear idea of how successful they are through joint presentations of ideas to the target audience. Objective filtering of information and separation of essential from non-essential information proved to be the easiest through joint mentored discussions of students. Finally, approaching the interlocutor and successfully conducting the conversation proved to be best when two or three students work together, as well as when graphic materials are properly used in communication.

In the future, based on previous practical experience, it is planned to dimension the tasks through obtaining information from users and providing a creative response that can be realized through existing funding. The current (modest) funds that enable NGOs to realize the program represent the potential for the realization of student design projects. Also, NGOs are more agile in the implementation of smaller interventions due to a simpler decision-making apparatus and clear deadlines in which they need to spend funds. Therefore, in the coming years, it is planned to focus on topics of interest to those associations that need help with spatial planning.

As Croatia shares its social context with countries that experienced the transition from a socialist to a democratic society at the end of the twentieth century, it is to be believed that the experiences gathered at the Faculty of Architecture in Zagreb can be applied, adapted, and developed at the architectural faculties of other countries in transition.

## References

1. Gehl J (2010) Cities for people. Island Press, Washington, DC
2. Euro Cities Homepage. https://eurocities.eu/wp-content/uploads/2020/12/New-leipzig-charter_2020.pdf. Accessed 28 Nov 2022
3. Duraiappah AK, Roddy P, Parry J-E (2005) Have participatory approaches increased capabilities? The International Institute for Sustainable Development. https://www.iisd.org/system/files/publications/economics_participatory_approaches.pdf. Accessed 18 Nov 2022
4. Radman Z (2014) Prirodna i kulturna baština u prostornim planovima. i sudjelovanje građana u javnopolitičkom procesu. Yearbook Titius: yearbook for interdisciplinary research of the Krka basin, vol 6–7
5. NGO Odraz Homepage. https://www.odraz.hr/novosti/odraz-ove-vijesti/participacija-nije-samo-javna-rasprava-vec-drustveni-proces-koji-ukljucuje-jako-puno-dionika/. Accessed 15 March 2023
6. Mišetić A (2022) Participacija građana u procesu urbane regeneracije: sociološka perspektiva. https://hrcak.srce.hr/file/378614. Accessed 15 March 2023

7. Careva K (ed) (2014) Improving the city life quality through small and precise interventions in urban structures. Zagreb Society of Architects, Zagreb
8. Careva K, Lisac R, Pletenac T, Vukić J (2017) Akupunktura grada kao participativni alat za revitalizaciju javnog prostora. Prostor 25(54):190–199
9. Schütz F, Heidingsfelder ML, Schraudner M (2019) Co-shaping the future in quadruple helix innovation systems: uncovering public preferences toward participatory research and innovation. She Ji J Des Econ Innov 5(2):128–146
10. Careva K, Lisac R (2019) Participatory design of space (course in academic year 2018/2019). In: Roth-Čerina M, Cavallo R (eds) The hidden school. TU Delft Open, Delft, pp 192–197

# Bauhaus Revisited: The Human Body as a Tool in Architectural Education

Carla Collevecchio[1(✉)], Sonia Vázquez-Díaz Author[2], and Zaida Garcia-Requejo[2]

[1] Umeå School of Architecture, Umeå University, 903 33 Umeå, Sweden
carla.collevecchio@umu.se

[2] Escola Técnica Superior de Arquitectura, Universidade da Coruña, 15008 A Coruña, Spain

**Abstract.** The current paper intends to reveal the importance of the human body as a tool for spatial production in architectural education, more specifically during the first year of the initiation process. The exercise Spatial Bodies, developed at Umeå School of Architecture in Sweden, aims to explore the phenomenon of space, through the lens of moving bodies, to discover what is the effect that space can have on movement and what is the effect movement can have on the creation of space itself. Through an experimental pedagogical methodology, that combines mapping techniques, choreography, material experimentation, design methods and performative instruments, students are asked to analyze, compose, build and perform a series of architectural artifacts that have the capacity to activate awareness of the interaction between body and space. This process enables the discovery and awareness of concepts such as geometry, structure, articulation, movement, material, surface, choreography and place, which go far beyond their surface meanings.

**Keywords:** Human body · Bauhaus · Architectural education · Spatial bodies · Movement

## 1 Introduction

The professional practice in architecture is becoming increasingly heterogeneous, specialised and strongly dominated by digitalisation. However, architecture teaching programmes are progressively concentrated in time. Therefore, it is necessary to consider rigorously on the teaching content and to determine which disciplinary bases need to be maintained or recovered.

The relationship between body and space, in which both interrelate and co-create, is fundamental. Space favours or restricts movements and actions, so it must be designed with full awareness of its consequences on living. To teach these concepts effectively, we turn our gaze to the Bauhaus, where apprentice architects and artists experimented with the human body: from the Vorkurs by Johannes Itten to the Theatre Workshop, firstly under Lothar Schreyer and later under Oskar Schlemmer, the aim is that students reflect on their own body occupying a space.


M. Blanco Lage et al. (eds.), *Towards a New European Bauhaus—Challenges in Design Education*,
https://doi.org/10.1007/978-3-031-40188-6_3

We want them to discover, among others the concepts of *affordance* [1] i.e. how objects invite us to interact depending on their shape and perceptual characteristics; and *inhabitability* [2], the capacity of a space to allow the possibility of performing a determinate action as a meaningful experience for us. To this end, we introduce a first-year exercise developed at Umeå School of architecture in Sweden.

The aim of this paper is to demonstrate the role of the human body as a tool for spatial production, and the use of choreography—movement practice—for the development of spatial design in architectural education. The intention is to explore the phenomenon of space, through the lens of moving bodies, in order to discover what is the effect that space (architecture) can have on movement and what is the effect movement can have on the creation of space itself.

## 2 Historical Context

One of the aims of the Bauhaus School directed under Gropius was to bring together all visual arts, including architecture and stage work. Proof of this is the existence of the Theater Workshop, something unprecedented in an arts school. This Workshop was directed by Lothar Schreyer first, and by Oskar Schlemmer from 1923 [3].

For Schlemmer dance could be related to space in two ways: using the geometry of the space (cubicle space) in relation to human biology, and on the other hand using the invisible functions of inner self in relation to space, where movement is determined mechanically and rationally, creating knowledge through experience [4]. Taking this into account, we can state that dance and architecture have common shared elements: body and space, and choreography is the tool to explore the relationship between space and movement, directly grounded in perception, body movement, and experience of physical or social nature. There exists an emotional relationship to space, an embodied experience, so the movement of the body through space (or dance) could be a way to find a common spatial experience.

The statement by Lászlo Moholy-Nagy, another Bauhaus Master in charge of the Basic course from 1923 to 1928, that spatial design, as "the design of relations between body positions (volume)", must be confronted with organic experience and that "space can be experienced most directly through movement, at a higher level through dance" [4]. But this hardly leads to architects or designers trying out whether they can "condense space" and "structure it" themselves by dancing. Bauhaus Dances, such as the ones created and directed by Schlemmer, used choreography and geometry to materialize this intrinsic and invisible relation between body and space, extending the body through materials, revealing the space in construction activated by the body [5].

According to Torsten Blume the man was explored as a dancer and abstractly geometricized as an "artificial figure"; this dancerman was less an expressive human being but rather a transformation of the human body as an architectural or "architecturized" apparatus for elementary artistic spatial productions.

## 3 First-Year Exercise at Umeå School of Architecture

The exercise, Spatial Bodies, is an experimental workshop that proposes the exploration of space through choreography, using the moving body as moving space, as a qualitative tool beyond quantitative aspects to create abstract systems and spatial forms (Fig. 1).

**Fig. 1** Spatial Bodies 2017. Umeå School of Architecture. Photo by Lucy Landaeta

The students, in small groups, must design and build a series of architectural artifacts that have the capacity to activate awareness of the interaction between body and space. The method used is Arts-Based Research, a participatory research practice that connects embodied, visual literacy to more traditional research practices [6]. Art is used to capture meanings that otherwise would be ineffable. Both the artist-architect/researcher/teacher and students use art disciplines to explore corporeal, unsaid and sensible experiences in a systematic and methodical way, such as the multidimensional relationship between body and space. The body is analyzed and explored as a geometric, mechanical, and performative instrument, to activate the awareness of the interaction between body and space (Fig. 2).

It is an interdisciplinary exploration that reveals how body, space and movement relate to each other; how movement induces tension and looseness, how space affects motion and velocity, and how body creates space within space. The process is structured in four interconnected parts: the toolbox, the documentation, the analysis and the synthesis.

This first part of the exercise, the Toolbox (or introductory tools), is a quick hands-on initiation that introduces movement and material, conceptually and instrumentally, through two small workshops. The first moment is about thinking through the body,

**Fig. 2** Spatial Bodies 2018. Umeå School of Architecture. Photo by Lucy Landaeta

becoming aware of our body: in movement, in pause, isolated, and in relation to space, surroundings and others. The content is based on the fundamental principles of dance: space, time, effort, form/shape, as a tool to activate our bodies in space, by exploring movement, reflecting on experience, perception and communication. The second moment is about thinking through the hands, bringing consciousness to the way we operate the materials. Crafting and making determine our way of translating abstract ideas in concrete objects. We manipulate multiple materials, linear and planar, rigid and flexible, understanding their potential and their limitations, by exploring systems, techniques, logics and patterns without scale by manipulating, cutting, bending, connecting, making small objects and joints.

The second phase, the Documentation, consists of collecting data and samples of the body(s) in space in different conditions. It goes deep into the exploration of body movements—as a series of actions that belong to a sequence—under three different physical conditions: site free, site specific and site tracing. In this case, simple movements are consciously selected, tested and defined as choreography. The body in motion is documented thoroughly using photography as a tool to decompose, and to register the successive phases of the sequence, like an analogue stop-motion system [7]. In this phase simple movements are consciously selected, tested and defined as a choreography, increasing the awareness of trivial situations of the body in space and its environment. This process is repeated several times according to each condition and number of bodies in the scene: one body, two bodies, three bodies and more. The variation in context allows to change the focus from a decontextualized body to a more concrete location, and therefore the choreography needs to consider how surroundings might influence the type of movements.

Some of the reflections we ask the students during this phase are:

- How can one body relate to another body? How might they affect or influence each other?
- How different spatial conditions such as the site allow or constrain for certain movements (i.e. a stair vs. a flat floor; a double height vs. a compressed room; a narrow corridor vs. an open platform; etc.)?
- What kind of boundaries do bodies create in space? What is the shape, dimension of it? How do they transform over time, or depending on the number of bodies in space?

The third part, the Analysis, is based on the collected data, and using geometry as a tool for abstraction, the analytical process begins. The aim is to blur the body figure by exploring the hidden: invisible forces, alignments, connections and interactions between bodies in space. Each frame is overlaid with tracing paper and analyzed through geometrical drawing to discover different elements: axis, boundaries, directions, extensions, rotations, articulations, projections, etc. The idea is to analyze progressively each condition to discover similarities or variations from one another. Students also produce a catalog of geometrical diagrams as the basis for the construction of the idea of the project (Fig. 3).

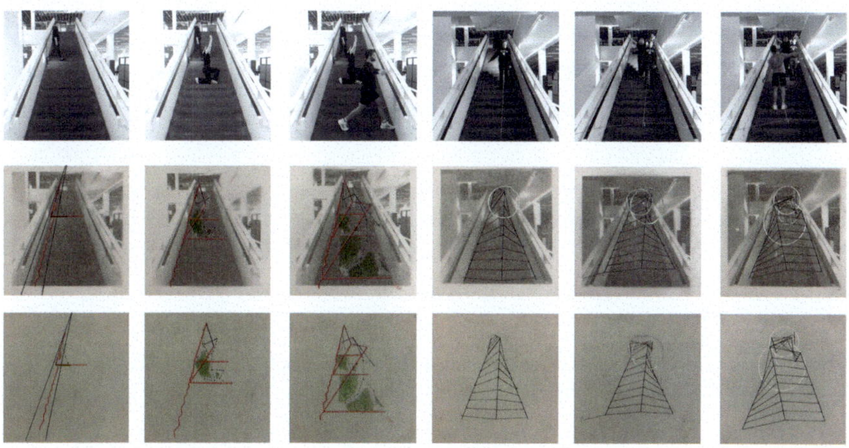

**Fig. 3** Analytical series. Spatial Bodies 2021. Umeå School of Architecture

The last phase, the synthesis, focuses on designing, prototyping, building and performing an architectural artifact in relation to the body that reactivates the awareness of the interaction between body and space. This phase is divided into 2 steps: models in scale (wire and dolls); and 1:1 artifact (designing, building and performing). After a process of total abstraction, we go back to the body again, not only in terms of dimension but also in terms of what the body produces in space. Based on the wireframe volumes, and using a doll that simulates a human body, we introduce scale, materiality, and constructive thinking. The wire is translated into paper straws or wooden sticks (linear elements); staples, clips, elastic bands and other materials are used to test the connection between elements and the mobility and transformation capacity.

This model is used to test notions of system, modularity, transformation, tectonic thinking (linear components) articulation and joinery, stability, rigidity, flexibility, and scale. Based on the mock-up doll model in scale 1:5, students extrapolate the learnings and define a strategy for making the structure in scale 1:1 using pvc pipes (light structures, hollowed, bendable, and stable—for the multiplicity of joints different materials are used, depending on the type of mechanical connection, scissor structure, rotation, sliding, expanding, growing, etc.) so fabric, elastic bands, velcro, bolts, screws, plastic and wooden elements can be used.

A progressive, iterative and consecutive design process is generated by combining different tools such as mapping body movement, wireframing in three-dimensions, prototyping mechanical systems, executing and performing live. The relation between the steps is not a literal translation, rather a flexible methodology open for interpretation (Fig. 4).

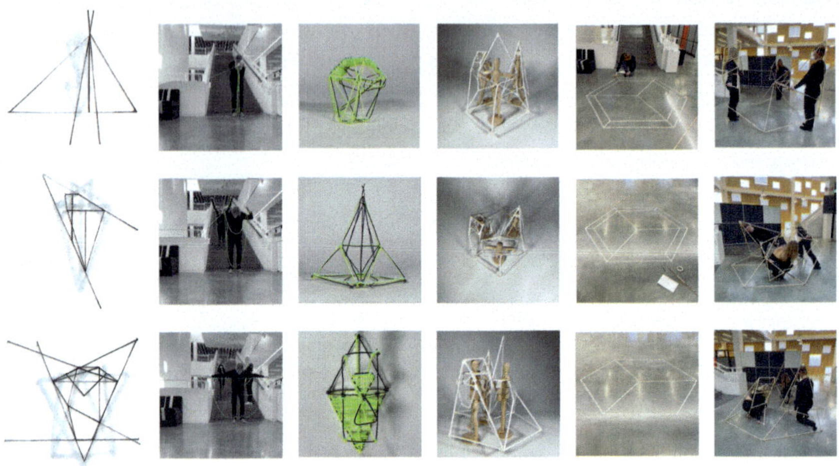

**Fig. 4** Design process. Spatial Bodies 2021. Umeå School of Architecture

Performance plays a fundamental role throughout the whole process: it is not a result, nor a goal, rather a design tool that informs the process, validating and challenging the strategy (Fig. 5). Students wear, inhabit, hold and connect their bodies to their artifacts during the design and construction phase, modifying, and adjusting the structure as well as the movements, depending on the different articulations and unexpected discoveries from the initial choreography. These architectural artifacts depend on the performance: without bodies no artifacts could perform on their own; bodies activate the artifacts and artifacts allow, restrict or invite bodies to move in a certain way, transforming each other constantly, revealing infinite types of compositions, geometries, forms, volumes, boundaries, and spaces, like an organism, like a creature [8].

To conclude the exercise, a collective and public performance takes place using the architecture school as an experimental stage. Any performance, as neurophysiologists confirmed, is able to activate the mirror neurons in the audience, so the discoveries made

**Fig. 5** Spatial Bodies 2018. Umeå School of Architecture. Photo by Lucy Landaeta

by the students are transferred in an emotional, cognitive and somatic way to the spectators. Therefore, the use of performance and dance as the core of the Arts-Based Research Method enhances the use of the body as a lens to capture the phenomena for the participants, and it is especially appropriate for educational purposes. This public performance has been explored in different ways: with no stage and simultaneous acts, where public and artifacts move together; with fixed stage and progressive acts with clear boundaries between performers and spectators; blending inside and outside spaces; inverting the stage space and the audience space; and with multiple stages, building a collective promenade between performers and spectators (Fig. 6). Beside the performance of the artifacts, the project challenges how to engage with the spectators, choreographing potential relations, but leaving space for playing with the unexpected, the unpredictable and the surprise effect.

## 4 Conclusions

The human body is used as a tool to introduce first-year students into the architectural discipline in Umeå School of Architecture, and it is an ongoing project since 2017 that continues to raise new questions, and to present new challenges. This exercise, following Schlemmer's approach and experimentation, highlights the possibilities of the body as a personalized tool, as a flexible instrument to perceive and inform spatial possibilities, as a tool for architectural design: the design studio becomes the laboratory to investigate, design, build and experiment these architectural artifacts. Results have been collected for six years and they will be used in the near future to conduct a systematic study that

**Fig. 6** Spatial Bodies 2017. Umeå School of Architecture. Photo by Lucy Landaeta

will lead to find patterns of learning outcomes and conceptualize the phenomenon of interaction with space (Fig. 7).

**Fig. 7** Spatial Bodies 2019. Umeå School of Architecture. Photo by Samuel Pettersson

Students are able to visualize and materialize the inseparable relation between body and space, and also to demonstrate that architectural form is linked to the human body at multiple scales. They learn through their own experience, not through a dictionary concept, achieving multiple outcomes like: strong abstract-thinking approach; sense of awareness of body in space by grasping architectural concepts (scale and distance, affordance and inhabitability); geometry as a system of order and a tool for abstraction; structure as the materialization of a tectonic system, as a shelter; articulation as the connection between components to enable transformation; **movement** as temporality and as a design component; material as a physical means for translating ideas into reality; choreography as a performative act, as an ephemeral interaction; surface as a tactile, permeable and malleable envelope that blurs and frames; place as the construction of an experience, a memory, a meaning.

This exercise moves in between architecture, theater, sculpture, dance and performance. Students are invited to learn from architecture in a new way; we teach them how to think through the body, how to think through the hands, how to be conscious and aware of corporeal needs in architecture, of embodiment as an attitude towards the discipline, as an instrumental approach that goes beyond skills. The architectural artifacts can be recognized as a primary architecture, a type of architecture that emerges from the body into the space, a space that celebrates the body, providing new meanings in relation to this obvious but radical way of thinking—activating the awareness of the body in space (Fig. 8).

**Fig. 8** Spatial Bodies 2021. Umeå School of Architecture. Photo by Samuel Pettersson

## References


1. Gibson JJ (1977) The theory of affordances. In: Shaw RE, Bransford J (eds) Perceiving, acting and knowing: toward an ecological psychology. Lawrence Erlbaum Associates, Hillsdale, pp 67–82
2. Franco A (2019) Our everyday aesthetic evaluations of architecture. Br J Aesthet 59(4):393–412
3. Droste M (2019) Bauhaus. TASCHEN España S.A Unipersonal, Berlin
4. Nikolic S (2014) The Bauhaus theater: Oskar Schlemmer's design-in motion concept. SAJ Serb Architect J 6:43–62
5. Blume T (2015) Dance the Bauhaus. E.A. Seemann, Leipzig
6. Pentassuglia M, Boylan M (2017) "The Art(ist) is present": arts-based research perspective in educational research. Cogent Educ 4(1):1301011
7. Braun M (1994) Picturing time: the work of Etienne-Jules Marey. University of Chicago press, Chicago
8. Herzog L (2021) Strandbeest: the dream machines of Theo Jansen. TASCHEN, Köln

# 2 Method of the Study

## 2.1 Context and Case Study

Our study took place over four months within the first year of the master's program in civil engineering and architecture within the design studio at the University of Liège and its connection with the course on Sustainable Environmental Design.

In a context similar to an architectural design competition, 21 students are challenged to work together in 6 teams through a collaborative learning process. They are called to design a contemporary building while respecting complex programmatic requirements, form, function, structural systems, spatial qualities…

Sustainability issues in this pedagogical context are represented in terms of the High Environmental Quality initiative for Building (HEQ), based on 14 targets and divided into 4 themes: energy, environment, health, and comfort. For some pedagogical choices made by the teaching staff, only 6 targets are considered in this experience as follows: physical relationship of the buildings with their immediate environment, integrated choice of construction processes and products, energy, water, and waste management, and hydrothermal comfort with its various parameters.

The particularity of this experience is that through this architectural design process, and besides the interconnection between the design studio and the theoretical course on sustainability, students benefit once or twice a month, depending on the evolution of the design process, from the collaboration with different professionals and experts. Therefore, multidisciplinary contributions were provided from several fields including architecture, building envelope, environmental quality, structure, fire safety, accessibility standards, heating, and ventilation to best respond to their architectural choices.

## 2.2 Methodology and Observation

Throughout the whole design process, the methodology involves qualitative methods that include examining learning activities, teaching methods, professionals' inputs as well as students' outcomes and interviewing them to know their feedback at the end of the design process as shown in Fig. 1.

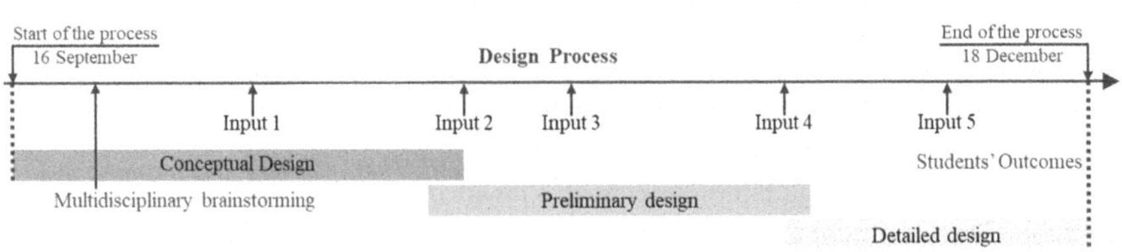


**Fig. 1** Professionals' inputs within the design process

## 3 Multidisciplinary Design Process

Starting with multidisciplinary brainstorming, the design process becomes more heterogeneous, with several diverse experts involved as it can be seen in Table 1.

**Table 1** Multidisciplinary contributions of different professionals

| | Time frame | Professionals | Design studio layout | Inputs |
|---|---|---|---|---|
| Design process (within the design studio) | Multidisciplinary brainstorming starting with all disciplines | | | |
| | 14 October |  | 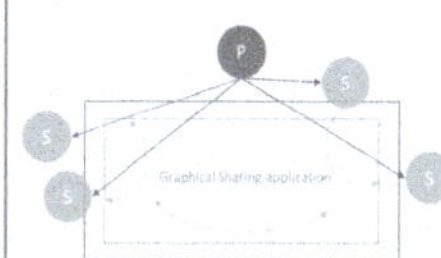 Architect 1 | Research phase of architectural choices<br>Functional programming |
| | 24 October |  | 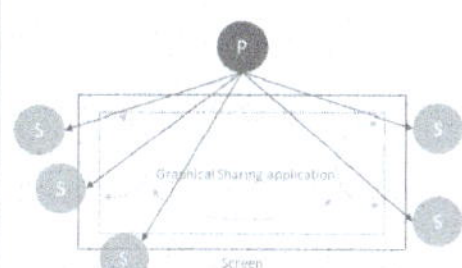 Architect 2 | Architectural composition<br>Formal concept understanding of the urban and architectural context |
| | 6 November |  | 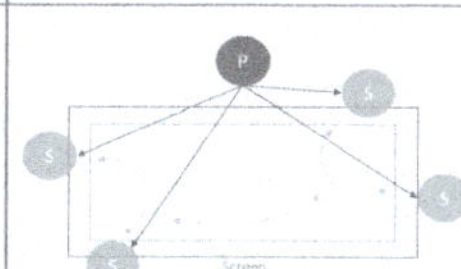 Engineer (Envelope) | Sustainable materials<br>Envelope<br>Energy efficiency |

(*continued*)

**Table 1** (*continued*)

| | Time frame | Professionals | Design studio layout | Inputs |
|---|---|---|---|---|
| | 21 November |  | 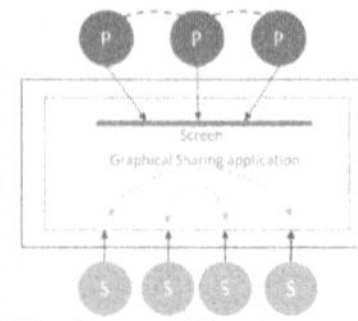 Engineers (Structure…) | Architectural technology Structural system and regulations |
| | 28 November |  | 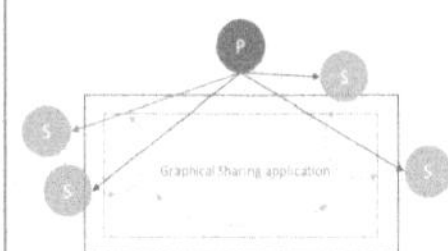 Architect 1 | Mastery phase of the design, considering functional, structural and environmental aspects |

Collaborative learning and teamwork, Multidisciplinary interactions and inputs

This leads us to develop an integrated approach that combines the different aspects of the building design of the architectural composition, structural engineering, architectural technology, and environmental aspects (HEQ Targets).

## 4 Results

Through the intervention of different professionals, students acquire and put into practice an interdisciplinary approach using a collaborative, learner-centered approach as opposed to the more traditional teacher-centered approach in the architectural design studio. These contributions lead us to the concept of the integrated design as a teaching method adapted to the pedagogical context. It is defined as an interdisciplinary conceptual approach based on collaborative work. Teaching methods should serve as the bridge to industry, helping students to contextualize their class work in the light of their projections for future roles [10]. Starting from the very beginning of the design process, the goal is to bridge the gap between professional practice and teaching. Their interventions foster the integration of sustainability criteria through different design process phases within an interdisciplinary approach.

The Fig. 2 shows the mobilization of sustainability criteria along the observed design process. It can be concluded that, from the early phases of the conceptual process, most of the sustainability issues are discussed coherently in the various aspects of the design. This paradigm is developed in a holistic way of design thinking that goes hand in hand with all the functional, aesthetic, and technological aspects of the design process while respecting the requirements of the design process.

Based on the concepts of synergies and interconnectedness, this approach is based on the concept of the Whole Building Design and involves two components: an integrated

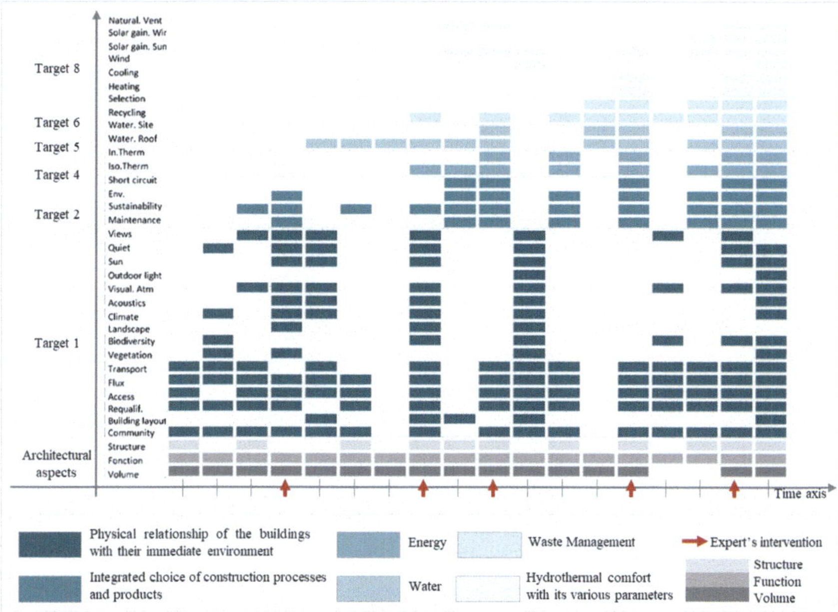


**Fig. 2** Visualization of sustainability criteria within the design process (1 team)

design approach and an integrated team process. Through the integrated design approach, all building stakeholders are encouraged to work together throughout project phases and to consider materials, systems, and assemblies from various angles [11]. In contrast to the typical design process, in which specialists with specialized knowledge work isolated from one another, this approach draws on the expertise of a team of specialists who are experienced in their own specialties.

Since the beginning of the design process, students are guided through the conceptual research phase and site analysis to define design solutions that can both optimize natural conditions and create spatial design challenges and opportunities. Then during the preliminary phase and with the intervention of architects and different engineers, students use compositional aspects and volumetric design to propose an architectural choices that can represent a design language for sustainability.

At the end, they have developed construction details and displayed environmental technological solutions as a Mastery phase.

Although results are fully satisfactory, directing students into sustainability criteria turns out relatively difficult, at least at the beginning of the process.

However, with time, it appears that dealing with specific technical environmental and engineering issues in many cases requires corrections to the initial schemes and sketches of the project, with the final benefit for the overall proposal.

This is how the Bottom-Up approach is reached in the design process which is substantially the opposite of the Top-Down approach, where first technical problems are solved and then the architectural composition of the design is created [12].

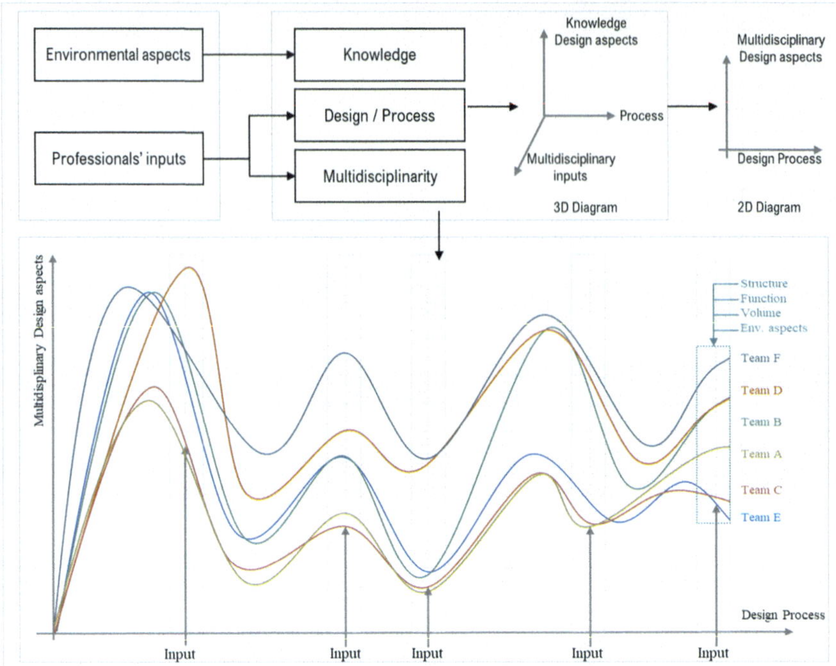


**Fig. 3** Illustration of the impact of professionals' inputs

The multidisciplinary aspect of the design starts from the brainstorming stage so that various disciplines intervene from the beginning of the learning process.

Based on the analysis of the data collected during the observation, and on learning outcomes, the study reveals that, after professionals' reviews, each team path is influenced by their mindset and feedback, like shown in Fig. 3.

Their impact differs from one group of students to another. It is the case for all the teams but not with the same level. It depends on different parameters including the complexity of the project, the chosen strategy and the design process.

Dealing with collaboration between students itself needs different levels of knowledge, skills and abilities to acquire to best carry out the design process.

On the other hand, students' answers on different questions show positive feedback towards this experience as we can witness from their comments as follow.

"Sustainable and environmental quality was integrated entirely all along the project and not just at the end of the process" (team B).

"It was an enriching learning experience in terms of the multidisciplinary intervention of several experts from different fields. There was a completely different way of thinking about the design project" (team F).

"The project is more viable than a classic architectural object… Not only have we got an architectural design that is beautiful, inclusive, and sustainable… Beyond the sustainability, but there is also an entire consideration of the architectural design as a whole building design" (team A).

## 5 Discussion and Conclusion

From the perspective of this study, the projects developed by the students reveal the benefits of professionals' interventions and the collaborative learning that enhance the integration of sustainability within the different phases of the design process.

The biggest challenge faced in this experience is this multidisciplinary character of the profession of architecture but also considering the constraints generated from the early phases of the design process, and the conditions of possible integration of sustainability issues.

In our case, some of the constraints detected are the design process, complexities of the architectural project, the multilayered collaboration, the students' skills (group management skills; inquiry skills; different levels of ability, diverse backgrounds, experiences, and ideas, team strategy…).

Therefore, many learning tools and strategies integrating disciplinary-specific knowledge should be used to achieve an efficient collaborative experience. Formal lectures and theoretical courses are the most common however these teaching approaches are still traditional. We can focus on "hands-on" experiential learning and try more case studies during lectures. Enhancing the collaboration on a macro scale between schools of architecture and engineering may raise awareness among teaching staff and educators. A close architect-engineer collaboration within teaching is also needed with an early consideration of sustainability issues within the design process.

**Acknowledgements.** We would like to address our deepest thanks and gratefulness to the teaching staff, students and professionals within the master's program in civil engineering and architecture at the University of Liège for their contributions. This article was previously presented during the EAAE Annual Conference and General Assembly 2022, Madrid, Spain, Towards a new European Bauhaus: Challenges in Design Education and Research. This article presents the results of a study conducted within the first semester of a master's degree program in civil engineering and architecture at the University of Liège, with the approval of all the participants during this study.

## References

1. Rosado-García M, Kubus R, Argüelles-Bustillo R, García-García M (2021) A new European Bauhaus for a culture of transversality and sustainability. Sustainability 13:11844. https://doi.org/10.3390/su132111844
2. Salama A (2015) Spatial design education: new directions for pedagogy in architecture and beyond. https://doi.org/10.4324/9781315610276
3. Why the building sector? Architecture 2030. https://architecture2030.org/why-the-building-sector
4. Earth overshoot day home—#MoveTheDate. https://www.overshootday.org
5. Salama AM (2002) Environmental knowledge and paradigm shifts: sustainability and architectural pedagogy in Africa and the Middle East. In: Salama AM, O'Reilly W, Noschis K (ed) Architectural education today. Architecture and Behaviour Colloquia, Lausanne, pp 51–59
6. Altomonte S et al (2012) Educate! Sustainable environmental design in architectural education and practice
7. New European Bauhaus: beautiful, sustainable, together. https://new-european-bauhaus.europa.eu

8. A new European Bauhaus: op-ed by Ursula von der Leyen. European Commission - European Commission. https://ec.europa.eu
9. Nyka L (2019) Bridging the gap between architectural and environmental engineering education in the context of climate change. World Trans Eng Technol Educ 17:204–209
10. Fowler T (2009) Collaborative integrative-interdisciplinary digital-design studio (CIDS)
11. Fathi A (2014) Integrated building design studio: a cumulative methodology to accommodate and apply different design approaches in architectural education stage. Int J Innov Res Sci Eng Technol 3:8951
12. Sgambi L (2017) Collaborative design in a multidisciplinary architectural design approach. In: Présenté à IABSE symposium report, pp 224–225. https://doi.org/10.2749/222137817821233198

# The Great Game: Teaching Architectural Design as Collective Practice

Valeria Federighi[1(✉)], Camilla Forina[1], Tommaso Listo[1], Edoardo Bruno[1], and Sofia Leoni[2]

[1] DAD Department of Architecture and Design, Politecnico di Torino, Viale Mattioli 39, 10125 Turin, Italy
valeria.federighi@polito.it

[2] DIST Interuniversity Department of Regional and Urban Studies and Planning, Politecnico di Torino, Viale Mattioli 39, 10125 Turin, Italy

**Abstract.** Architects appear torn between pursuing an irreducible specificity of their profession and the need to constantly negotiate that specificity with a variety of other actors, from the politician to the technician to the citizens' association. Architectural design is a collective and political practice, but the model that is transmitted and reproduced in the design studio is often that of architects who follow their own unique, creative idea. Students learn that design is about ideas, professionals know that ideas—even good ones—are to be negotiated, modified and legitimated through continuous and energy-intensive exchanges with the other agents that participate in the transformation of space. Can this particular competence be taught, or should students just wait and learn it through trial and error, after graduating? Acknowledging the social and political complexity of the contemporary world, the pedagogical experiment of the Great Game tries to address the need to move from individual creativity to collective practice. The players' primary objective is that of defining viable strategies and associating the right set of actors to carry them through, thus developing an understanding of design practice as collective practice, and of the mechanisms that make it possible to socialize ideas into the real world of political exchanges. The game has been tested in two successive academic years: this paper will trace its theoretical positioning, the way it has unfolded and its possibilities for application.

**Keywords:** Design strategy · Game · Design practice · Design theory · Design pedagogy

## 1 A Pedagogical Experiment

In the design studios of architecture schools everywhere, a simulated environment is constructed through a system of constraints aimed at guiding a linear progression from design brief to final design proposal. The tutors are the makers and guarantors of this system, acting as sole interpreters of the world and as sole interlocutors for the work of the students. In this model, a perspective made of absolute values is transmitted, which is intended to give students the confidence, once they become professionals operating in the real world of exchanges, to stand by their ideas, and see them through.


M. Blanco Lage et al. (eds.), *Towards a New European Bauhaus—Challenges in Design Education*,
https://doi.org/10.1007/978-3-031-40188-6_5

Meanwhile, in the profession, architects appear torn between pursuing an irreducible specificity of their competence and the need to constantly negotiate that specificity with a variety of other actors, from the politician to the technician to the citizens' association. Architectural design is increasingly acknowledged as a collective [1] and political practice [2] Therefore students learn that design is about ideas, but professionals know that ideas—even good ones—are to be negotiated, modified and legitimated through continuous and energy-intensive exchanges with the other agents that participate in the transformation of space [3]. Can this particular competence be taught, or should students just wait and learn it through trial and error, after graduating?

"Architects do not make buildings; they make drawings of buildings:" in 1989 Robin Evans [4] famously stated a fundamental distinction between design as the product of the work of architects and design as a tool for the construction of artifacts, in contrast to the authorial positions that largely promote the immediate connection between architect and built object. Such distinction allows to pragmatically isolate the one product on which architects should focus to reclaim agency: not buildings, but drawings—or, rather, documents [5].

If we separate the production of documents from the production of buildings, it is much easier to understand design practice as a social, other than technical, practice. Documents can be and are exchanged, modified, negotiated, revised, and exchanged again. Documents exist as objects in the world (a digital file, a printed sheet of paper, an image that is circulating on many platforms), as well as projections of other objects yet to be built. While students are usually taught how to produce design documents, they are rarely exposed to the need to circulate them, to dedicate time and energy to the construction of platforms for exchange, to carefully measure the actions that would allow for the most effective result with the least expenditure of energy. In short, they are taught how to prefigure future worlds, but not how to prefigure a course of action that can make the development of such worlds a possibility in the specific time and place in which they are acting.

To make measurable the network of relations that exists and on which architectural design can claim agency, we propose a game of strategies as pedagogical experiment. In the game, actions, negotiations, and exchanges, rather then built objects, are the focus of attention. But how can such a game work?

## 2 The Game

The experimental design studio here presented has been conceptualized and then implemented on these theoretical premises. The main hypothesis is that design is a collective act rather than an authorial one, and as such depends on exchange and negotiation as much as (and probably more than) physical objects. If it is so, then:

1. Architectural design research should be concerned with defining ways for design to claim agency in the world by developing tools to increase the effectiveness of exchange and negotiation;
2. Architectural pedagogy should be concerned with the problem of making such complexity communicable, and replicable.

Acknowledging the social and political complexity of the contemporary world, the pedagogical experiment of the Great Game tries to address the need to move from individual creativity to collective practice. Within a game of strategies, each group of practitioners has a different client, clients have a story, their intentions are not always clear, they don't know exactly what they want, they change their mind; norms and regulations can be negotiated, they go undetected for a time; economic interests change, developers withdraw support, flashy narratives turn out to be more effective than well thought-out volume compositions. The practitioners' primary objective is that of defining viable strategies and associating the right set of actors to carry them through, thus developing an understanding of design practice as collective practice, and of the mechanisms that make it possible to socialize ideas into the real world of political exchanges.

A specific Detroit neighborhood is offered to students as "board of the game", animated by opposing interests and controversies, an entanglement in which it is not possible to make easy distinctions between analysis and synthesis: a place of Latourian matters of concern rather than matters of fact [6]. Students design in the same time and in the same space in which other students work; different groups' projects areas intentionally overlap to encourage interaction.

Students find themselves interacting within a complex and contradictory environment in the role of architects that deal not only with the client, but also with other actors who they wish to associate to carry out their design strategy.

Actors, in turn, support, negotiate or oppose the proposals from their own perspective, whether it is a public or a private one. Actors are either interpreted by tutors, or by external experts who simulate a specific role in the game—for example, a history professor from Detroit (where the Great Game has been staged for a couple of years) was invited to interpret the role of the historical commissioner.

As a result, groups' design proposals overlap throughout the game, both with other groups' proposals as well as with ongoing transformations in the city. Different groups with different objectives negotiate projects and strategies together in order to carry their strategies further and enlarge the collective that revolves around their own project. That is, the objective is to progressively increase the number of people, institutions, norms, that support the proposed transformation through cycles of production, exchange and revision of the design project.

## 3 A Pedagogical Experiment Actions, Reactions: Making a Move in the Game

At the start of the game, students are thrown into a world that is far from a blank canvas: it has existing actors, it has rules, it has power relations, and it has perspectives. In the simulation of the game, it is a world in motion, that has its own memory and that exists independently from the students' work. It is the same world for everyone, and it is a world that can be modified: one in which every action done by the students can have an effect. This world, which we represent as the board of the game, displays a specific state of affairs at the beginning of the experiment, and will necessarily display a different one at the end of it.

For ten weeks, different groups of architects struggle frantically over a few blocks of the city, devising strategies and attempting different actions to leave their mark on its physical configuration. The rules are simple: they can do everything that architects can do in the real world, provided it is something that leaves traces; they can simulate to take a walk, for instance, but it has no bearing on the game; if they simulate to take a walk with a group of residents and a journalist writes an article about it, that has bearing on the game. In this way, after two and a half editions of the game, we have a large repository of actions of many types. We will not attempt to give an exhaustive account of every one of them—something that would need far more space than what is available here—nor a sound categorization of the different types, but rather trace a few of them as they unfolded in the game.

- *Choosing what to draw, and how*: sketching, drawing, tracing lines is the first thing that every good architect does right after being contacted by a client, visiting the site for the first time, being challenged with a specific spatial problem. The act of drawing can have different social implications: it can remain confined within the intimacy of one's own office, or it can be a tool to associate numerous actors. *The architects who work for the owner of the casino mark the property around a new compound of luxury residences with a continuous double line. The municipality, however, is critical of what appears to be a project for a gated community in downtown Detroit. The architects then transform the boundary line into a softer line representing a system of water channels and discontinuous walls.*
- *Copying* is often more effective than drawing; particularly when the objective is not to produce a final, good design, but rather to give a non-technical audience a general idea of what the future world we propose could be like. *The architects working for the mayoral candidate, in a matter of days, hand their client a rich, seductive proposal for a new commercial boulevard using existing images of other projects with minimum modifications.*
- *Picking the right norms*: norms and rules are carefully selected and appropriated so as to support certain design choices through the construction of a very specific normative framework. *The architects working for the big farmer employ the normative framework allowing their client to make an offer for a number of adjacent plots of land before they are sold at auction, to define the boundaries of a large project site.*
- *Meddling with power relations*: design documents can be exchanged to attempt modifications of existing power relations. *In an unexpected move, the architects working for the big developer offer the manager of the other developer access to a prime plot in the project in order to have her support in the negotiations that would follow.*
- *Opening up the bottega* [7]: design production processes are not all the same, and more and more often the boundaries of the traditional studio (the *bottega*) become porous and networked. *The architects working for the private developer succeed in associating the residents' association by organizing two days of co-design, during which their initial plan is taken apart, and reworked by a host of different hands.*
- *Producing extra documents, just to be sure*: architects often have one shot at associating a certain actor, or set of actors; anticipating their doubts, limitations, and aspirations is part of their job. *The architects working for the farming cooperative prepare four design scenarios. They start presenting the one that they know will not*

*be accepted by their client because of the high cost of construction of one specific building, that they have recently added; since their clients do not rule out the possibility of adding the building but only object to the cost, the architects take out a specific one of the three spare scenarios that they had prepared. The other two are never shown, and are immediately stored away in the archive folder.*

- *Archiving*: architects organize thick digital folders on their offices' servers to safely archive every document related to a specific project. *The architects working for the private developer appoint one member of the team to organize the archive and keep it updated.*
- *Negotiating*: project proposals affecting the city are characterized by temporal coexistence, spatial contiguity and scarcity of resources. *Two groups of architects working in collaboration with two public agencies develop diverging proposals for the same neighborhood: one involves the development of a 'smart community', the other of a grassroots urban farming community. The former is far more expensive but can count on private funding, while the latter would have the support of local associations. The two groups eventually work together to develop a physical proposal that can carry forward both narratives.*

## 4 Actions, Projections, Retrospections

The game is a simulation: as such, any action that is undertaken by participating groups has no absolute value to it, i.e. it cannot be intended to give students the practical experience they would get from participating in real-world exchanges. The objective of the game is, instead, to learn to manage the development of a design strategy as it evolves, incrementally: this means managing the time of action (the present move); the time of retrospection (past moves); and the time of projection (future moves). Every week, students are asked to represent, in a diachronic diagram, the current move (i.e. the action they are undertaking, the documents they are producing and the actors they intend to associate), their past moves (which actions they have undertaken up to that moment, which deviations the project has undergone, which actors it has succeeded in associating); and the future moves (which actions they will undertake, to associate which actors, and which possible deviations they anticipate).

The ability to formulate strategic hypotheses, that is, to orient one's actions towards the future, is the main skill that the pedagogical experiment is intended to develop, as students get deeper into the game. Initially, the moves of the students/architects are proposed with the aim of immediate feedback; but it is only after a few weeks that they have the opportunity, by tracing the progress of the project along its process, to understand how the same process is subject to the projective manipulation characteristic of project drawings. The prefiguration can thus be implemented with drawings, but also through overall strategies, made up of communication, processes, and subdivision of the proposals into temporal phases.

On the other hand, the retrospective work involved in the construction of the process diagram leads to a reconceptualization of what an architectural project is: the aim is to bring out the awareness that the project outcome does not stop at the materialization of an authorial intention, but has been produced through interaction over time with the multiple actors and agents (Figs. 1 and 2).

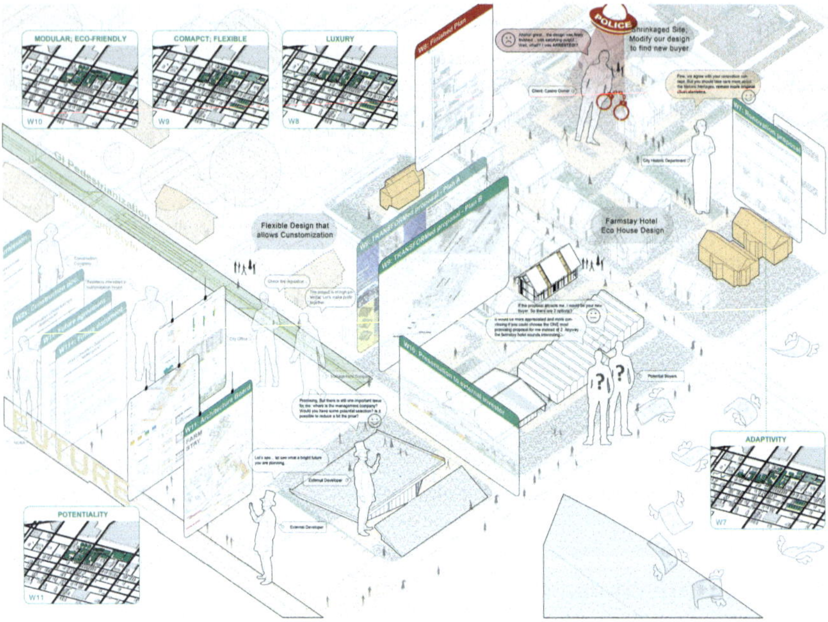


**Fig. 1** Retrospective diagram showing the main actions of the architects working for the casino owner. *Source* Studio's final panel © Parisi Giordana, Yu Ziyue, Zhong Jihao, Phan Tran Khue Tu, Zhang Xian

Through weekly reiteration of the diachronic representation of a process, which traces and projects their own strategy in its past, present and future implications and in its interactions with the other strategies of other agents, by the end of the ten weeks students learn to use the project not only as a technical tool—to represent ideas that are in their heads—but also as a social tool—to associate actors by successive cycles of production, exchange and revision of design documents (Figs. 3 and 4).

## 5 Conclusions

Where does the Great Game stop and why? How far can its simulation go? Although the exercise carried out by the students simulates that of professional practices, is the meta-work of retrospection, action and projection that stands at the core of the experiment, having as its objective the conceptualization of design activity. This meta-level holds it firmly within the didactic and academic field; it is on this level, in fact, that simulations can be managed and designed to not respond deterministically to dictates and conventions of the world, but rather to facilitate the exploration of different possibilities and experimentation.

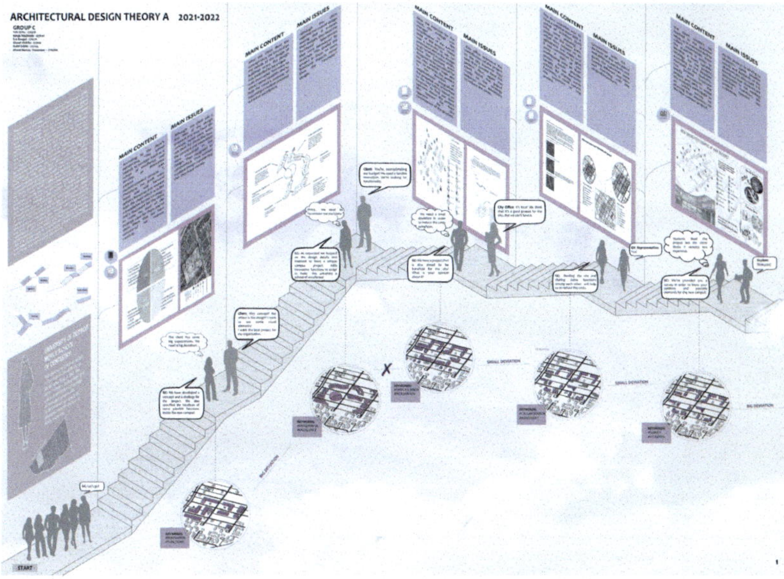


**Fig. 2** Retrospective diagram showing the main actions of the architects working for the University of Detroit Mercy School of Dentistry. *Source* Studio's final panel © Nasibzade Nargiz, Shakiba Ghazal, Zabihi Sadaf, Erinc Yeliz, Ozogul Ece, Tinaztepe Ahmet Berkay

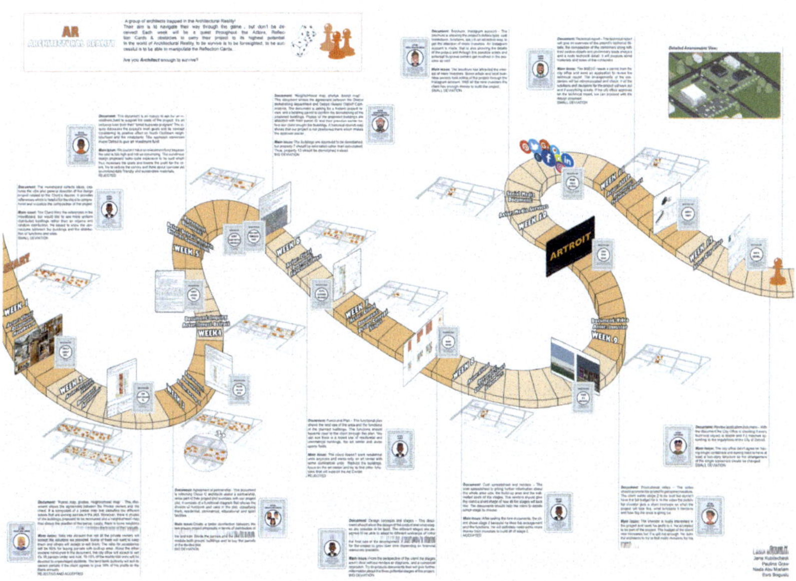


**Fig. 3** Retrospective diagram showing the main actions of the architects working for the small developer. *Source* Studio's final panel © Graw Paulina, Kubitscheck Jana, Abumariam Nada Bassam Ibrahim, Boguslu Esra, Buyukuzun Laden

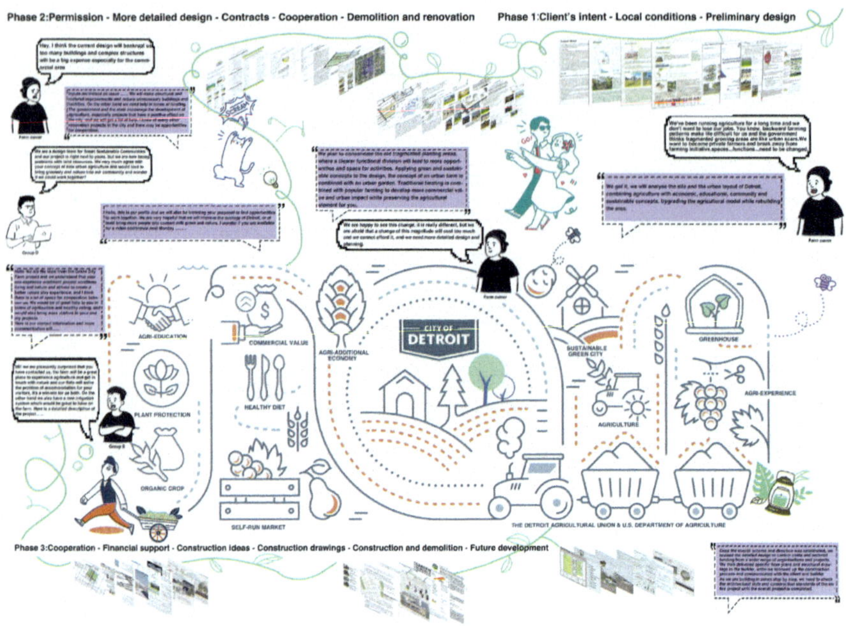


**Fig. 4** Retrospective diagram showing the main actions of the architects working for the farmers' cooperative. *Source* Studio's final panel © Li Mengyao, Liu Beicheng, Niu Zitong, Erdene, Ochir Sevjiddulam, Zhou Zhihao

The Great Game presupposes that students have already learnt the main tools of design; it does not, in any way, wish to substitute the role of traditional design studios and the compositional competence that they allow students to develop; rather, it pushes for their progressive integration with another set of skills, which are probably not more important in contemporary practice, but hold a large potential in the search for a renewed relevance of the discipline.

## References


1. Loukissas YA (2012) Co-designer, cultures of computer simulation in architecture. Routledge, New York
2. Awan N, Schneider T, Till J (2011) Spatial agency: other ways of doing architecture. Routledge, London
3. Yaneva A (2016) Mapping controversies in architecture. Routledge, London
4. Evans R (1989) Architectural projection. In: Blau E, Kaufman E (eds) Architecture and its image. MIT Press, Cambridge, pp 19–36
5. Ferraris M (2012) Documentality: why it is necessary to leave traces. Fordham University Press, New York
6. Latour B (2004) Why has critique run out of steam? From matters of fact to matters of concern. Crit Inquiry 30:237–241
7. Latour B (2010) Cogitamus. Six lettres sur les humanités scientifiques, La découverte, Paris

# The Frankfurt Kitchen: A Case Study for a New Education

Mayka García-Hípola(✉)

Navarra University, 28027 Madrid, Spain
mghipola@post.harvard.edu

**Abstract.** This article presents a proposal for a new pedagogical model that uses a case study as a new way to teach design studio based on the method employed by business schools. In the same way the Bauhaus looked for a new way do architecture, this case uses the Frankfurt Kitchen as an example to teach architecture in a different way and debate about concepts as important nowadays as prefabrication, flexibility and the role of sociology and gender in spatial design. The project assumes a problem that is the capability of attention of the students nowadays. Education has to compete with the cellular and apps as Spotify or YouTube. This case looks for an answer to this problem and uses a catchy tune by a modern pop singer to attract the students' interest and think about the possibilities of the proposed case. The research question of the case is how could be the kitchen of the future and its aim is to make students think about the concepts enunciated thought a new pedagogical model that revisits the 1920s project in the 2020s. The project is address to 5th year Architectural Studio based in Housing and this exercise concentrates in the design of the kitchen. The task after capturing their attention thought the video clip is to study other cases as well as making their own proposal. The outcome of the project will be a kitchen design that answers to the proposed questions in the twenty-first century. The potential to be replicated is enormous not only though other spaces of the house but also to other typologies related to education, sports or public institutions so it could be transformed in a teaching method with different case studies.

**Keywords:** New pedagogy · Kitchen · Case study · Margarete Schütte-Lihotzky

## 1 Introduction

This case study presents a new way to teach design studio based on the case study method employed by business schools in Anglo-Saxons as Harvard Business School. Other countries as Spain has imported it to prestigious schools as the IESE, with which the author of this article and School of Architecture from the Universidad de Navarra have collaborated [1]. These type of schools use the method to look for an answer to a given problem. They use the case methodology in their pedagogy, where a teacher presents the case to the students for their resolution. The objective of the case presented in this article is a practical proposal to employ this methodology so that can be used in the field of architectural studio and in workshops for other subjects of a creative or technical nature.


M. Blanco Lage et al. (eds.), *Towards a New European Bauhaus—Challenges in Design Education*,
https://doi.org/10.1007/978-3-031-40188-6_6

In the same way, the Bauhaus looked for a new way to teach and do architecture, this case uses the Frankfurt Kitchen as an example to debate things as important nowadays as prefabrication or the role of sociology and gender in architecture. Even though the protagonist of the case, Margarete Schütte-Lihotzky, was not part of Bauhaus, she was connected to it through the new ideas for a new life, a new man, and a new house and proposed how these are reflected in design.

## 2 Proposed Case of Study

The Frankfurt Kitchen is an example of the application of industrialization and prefabrication to the domestic space that took place in Germany in the 1920s. The Frankfurt Kitchen illustrates key principles of this time: objectivity, functionalism, and above all, standardization [2]. The concept of standardization was connected not only with production techniques, but also reflected the ideological position of the Bauhaus and Werkbund activists, who saw the uniform design of everyday objects as a contribution towards leveling the differences between classes. This example serves as an object of discussion of the concepts of flexibility and repetition applied to design. It also discusses how the shape of the space affects its use in terms of movement.

The case shows how this example is part of a series of other projects within the trajectory of its author that reflect her ideals and a personal and architectural coherence, which at the same time has influenced other contemporary domestic designs and architects. During the case, different questions are raised, such as whether this space works from the point of view of the user and their social relationship. The relationship between cuisine, women and architecture is an interesting topic that arises.

This case clarifies for students aspects to consider when designing, combining apparently different aspects such as flexibility and repetition, taking into account the relationship between form, use and distribution. After discussing the case, students will be able to work with these tools and be more sensitive towards the domestic space and its users, and propose their own design for the twenty-first century kitchen: how, for and by whom should it be designed.

## 3 The Trigger

Rotifer is a pop singer that makes this interesting song about an architect that was an old friend of his grandmother.

The track is taken off his 2008 "Coach Number 12 of 11". In his video clip, he uses a catchy tune as well as interesting and informative diagrams and visual resources related to architecture to explain the virtues of a new kitchen, the Frankfurt Kitchen. As can be read in its lyrics and understood from its diagrams. This space is the perfect fusion between form and function, but it also points the fact that the woman is isolated as a worker of the kitchen from the rest of the house.

The students are asked to look for his videoclip so they can see the different diagrams in YouTube and download the catchy tune in Spotify so they can listen it in their way to school or at any free time the can have. The project assumes a problem that is the capability of attention of the students nowadays. Education has to compete with the cellular and apps. This case looks for an answer to this problem and uses a catchy tune by a modern pop singer to attract the students' interest and think about the possibilities of the proposed case. The research question of the case is how could be the kitchen of the future and its aim is to make students think about the concepts enunciated thought a new pedagogical model that revisits the 1920s project in the 2020s. The project is address to 5th year Architectural Studio based in Housing and this exercise concentrates in the design of the kitchen.

## 4 The Protagonist

Then the protagonist of the case is presented. The mother of the built-in kitchen, Margarete Schütte-Lihotzky (1897–2000), or "Grete" as she was known to her friends was the first female architect in Austria.

Grete graduated in 1918 and worked from 1921 at the Viennese Department of Housing, with Adolf Loos as chief architect, designing houses and domestic equipment. Both shared the political vision and interest in economizing rationalization strategies, such as the reduction of housing. From its beginnings, Schütte-Lihotzky devoted great effort to improving the spaces for domestic work, from both theory, politics and the project, considering that they enslaved women.

At this time, Germany stood out for its industrialization process and this architect applied it to the domestic space, using Taylorist criteria that encouraged prefabrication and the reduction of actions to achieve maximum productivity. This concept, which in the United States was also applied to industry by Henry Ford and to the kitchen by other notable and unknown female architects, had numerous socio-cultural implications when applying business criteria to the domestic context, an area in which the protagonist at that time was women, who took care of the kitchen and the family.

Margarete Schütte-Lihotzky and Ernst May were the ones who spread the most about the principles of new buildings. In the magazine Das Neue Frankfurt, founded by May, and in international magazines and conferences, Schütte-Lihotzky argued that: in Frankfurt it was people's income and not formal ideals that determined projects. Rationalization and mass production were necessary to have more time for the education of children, culture, free time and sports [3].

She received numerous awards: in 1978 for merit for the liberation of Austria; in 1980 the prize for architecture of the city of Vienna; in 1987 the Prechtl medal from the Vienna Polytechnic and in 1997 the Gold Cross of Honor, Austria's highest award. (…). She had a committed position from the beginning of his career regarding society and the responsibility of people who practice architecture, a position that will mark both his professional and personal life.

She was also in the resistance fighting the Nazis with Rotifer's grandmother. She was arrested by the Gestapo in 1941 and still managed to survive. Both had joined the underground Austrian Communist Party, and when the war was over, they were involved in the foundation of the party's feminist organization, where they acted as presidents. She also did also some wonderful designs that can been seen in the video and that deal with flexibility and movement as:

## 5 The Damned Kitchen

The Frankfurt Kitchen was a project realized in Germany in the time of industrialization that took place in the 1920s and reflects how the industrial development affected the domestic and functional space of the kitchen.

At the end of the World War I, kitchen design, which had remained stable for a century, began to attract the attention of designers, manufacturers and suppliers. Many companies, especially in the United States, established modern test kitchens employing professional home economist to collaborate with industrial designers on innovative products for expanding markets.

In 1925 Ernst May proposed her to work in Frankfurt so that she could incorporate her own ideas and experiences into its popular housing program. In the first projects, which began to be built in 1926, they installed what will be known as the "Frankfurt kitchen". The kitchen transformed the modern home. It was designed based on efficiency in daily work, with easy-to-clean metallic and tiled surfaces, the modular relationship of the parts and technological advances. The kitchen was a modular system, the first in history, to lower costs with mass-produced element systems and that allowed its shape and size to be varied depending on the type of home in which it was placed. Therefore, it was flexible.

Schütte-Lihotzk's Frankfurt Kitchen illustrates key principles of the 1920s: objectivity, functionalism, and above all, standardization. The concept of standardization was connected not only with production techniques, but also reflected the ideological position of the Bauhaus and Werkbund activists, who saw the uniform design of everyday objects as a contribution towards leveling the differences between classes.

As it is mentioned in the videoclip and reflected in its diagrams, it presents novelties as the shortest route from larder to worktop, to cupboard, to cutlery drawer, to worktop, to bin. This also educates the Spanish students in the use of facilities vocabulary. Students. Is also presents new features as light-weight aluminium drawers, worktop, cupboards, easy-action drawers, overhead storage, two-sided bin. However, it also offers space for some criticism as the work in isolation. It incorporated flexibility and movement to the space of the kitchen and offers new ways of using this space of the kitchen taking the advantages of industrialization and repetition.

It is interesting how we can see for the first time the use of the woman as a modulor. It is so important than in 2016 Moma in New York devoted an exhibition to this Kitchen. As said in this exposition, as heightened tensions developed into Word World II, the New Kitchen's emphasis on economic use of resources, hygiene, and health was crucial to the home front on both sides of the conflict.

## 6 Problems. Goals to Solve

The pros and cons of this type of kitchen are various and debatable. The kitchen worker and his/her isolation also is an issue to be considered. The organization of the different appliances in the longitudinal space of this kitchen conditions its use in a certain way following Taylorist ideas related to work and industrialization versus the leisure of cooking. The Frankfurt Kitchen was one of the models propagated for the "new life" of the "new man". This case actualizes the concept to the 2020s and asks the student about how the "new kitchen" should be nowadays. The have to complete a table with the pros and cons of this design. Some of the possible answer are:

| **Pros:** |
| --- |
| – Covers spatial and economic criteria |
| – The organization of the different appliances in the longitudinal space of this kitchen conditions its use in a certain way |
| **Cons:** |
| – The kitchen worker and his/her isolation |
| – It eliminates the leisure please of cooking |

## 7 Pedagogical Results

This pedagogical project is address to 5th year Architectural Studio based in Housing and this exercise concentrates in the design of the kitchen. The task after capturing their attention thought the video clip is to study other cases as well as making their own proposal. The outcome of the project will be a kitchen design that answers to the proposed questions in the twenty-first century. The potential to be replicated is enormous not only though other spaces of the house but also to other typologies related to education, sports or public institutions so it could be transformed in a teaching method with different case studies. This project carries Schütte-Lihotzky's design into the 2020s and asks students how an updated, "new kitchen" should look and be designed today.

Some of the research question that have been asked to the students are: Is nowadays the woman the only worker in the kitchen? Do you know other woman architects? Is the gender problem in this case the isolation of the woman in the kitchen or the isolation of woman in architecture something to deal with nowadays? Is this example based on prefabrication, efficiency and Taylorist criteria good from the point of view of the twenty-first century user? Is the kitchen considered a social space, a place to eat what you order at home, a place to cook as a master chef, a laboratory where things are done with the precision and speed of a factory?

During the study of the case, students analyses the role of women in architecture and in the kitchen, investigate the relationship between use and form and spatial distribution, and use the diagram as a tool for analysis and representation. Some of the conclusions achieved for these particular case of studies that are learning outcomes that the students take home, are related to the importance of the kitchen as a social space and the relation between "server spaces" and "served space". They understand that the kitchen can be a new place of application of the new technologies typical of the current context: IT (Information Technology), computer-aided design or programming as a tool that amplifies design through the understanding of complex and multivariable scenarios. After a final debate about how the kitchen space of the future should be like, students can assess, as key topics, concepts such as flexibility and repetition, which are apparently opposed or antagonistic and may be related. They can propose industrial development to the domestic and functional space of the kitchen, because times of crisis are opportunities for new developments and designs.

## Bibliography


1. García Hípola M (2019) Proyecto de innovación docente. Aplicación del método del caso a talleres de asignaturas con carácter creativo y técnico". In: CUICIID 2019: Congreso universitario internacional sobre la comunicación en la profesión y en la Universidad de hoy IX: Contenidos, investigación, innovación y docencia: 23 y 24 octubre 2019, p 273. ISBN 978-84-09-17043-2
2. Dreysse DW (1988) Ernst May Housing Estates: architectural guide to eight new Frankfort Estates, 1926–1930. Fricke Verlag, Frankfurt
3. https://collections.vam.ac.uk/item/O121079/frankfurt-kitchen-kitchen-schutte-lihotzky-margarete/frankfurt-kitchen-kitchen-sch%C3%BCtte-lihotzky-margarete/
4. Bullock N (1988) First the Kitchen—then the Façade. J Des Hist 1(3/4):177–192
5. Godau M (2016) On the organization of products in German Design. DESIGNABILITIES: Design Res J (11) 2016. ISSN 2511–6274
6. Henderson S (1996) A revolution in the woman's sphere: Grete Lihotzky and the Frankfurt Kitchen. In: Coleman D, Danze E, Henderson C (eds) Architecture and feminism. Princeton Architectural Press, New York, pp 221–248
7. Markgraf M (ed) (2018) Bauhaus World Heritage Site (English edition). Spector Books, Leipzig
8. Melching K (2006) Frankfurt Kitchen: patina follows function. Conserv J 53
9. MoMA|Counter Space: the Frankfurt kitchen. www.moma.org. Retrieved 25 Aug 2020
10. Noever P (ed) (1996) Margarete Schütte-Lihotzky. Soziale Architektur—Zeitzeugin eines Jahrhunderts. Böhlau, Vienna

# Architectural Analysis and 'Living Archives': The Norman Foster Foundation Archive as a Pedagogical Tool at ETSAM-UPM

Gabriel Hernández(✉)

Universidad Politécnica de Madrid, Madrid, Spain
g.hernandez@upm.es

**Abstract.** The creation of the New European Bauhaus indicates how crucial architects and designers are to take meaningful action in an era marked by the UN's Sustainable Development Goals for 2030. In this direction, architectural archives provide an excellent field for learning from particular interpretations of the built environment. Despite its potential as a valuable tool for research, archives are often disregarded in architectural education, vacating their understanding and interpretation to art and architecture historians. But what is the role of architectural archives in the contemporary teaching of architecture? What can younger generations learn from the lessons from the past? Can archives be not only a tool for research but also a tool for practitioners? This contribution aims to share the experience of bringing together institutions that have a role in designing the built environment through an architecture educational program, enhancing a multi-directional knowledge transfer in the same spirit of the New European Bauhaus agenda. On one side, undergraduate students and professors of Architectural Analysis at the ETSAM Madrid School of Architecture (Universidad Politécnica de Madrid), and on the other, the Norman Foster Foundation (NFF) through its Archive, also located in Madrid. The resulting pedagogical experiment encourages acquiring research tools that motivate new readings and understanding of archives as a proactive agency in producing a more sustainable future and society through design.

**Keywords:** Archival research · Architecture analysis · Norman Foster Foundation · Design ethnography

## 1 Introduction: Milestones and Context

The creation of the New European Bauhaus indicates how crucial architects and designers are to take meaningful action in an era marked by the United Nations' Sustainable Development Goals for 2030. To achieve such complex goals, architectural educators must create alliances among various stakeholders to amplify their impact. The European Association of Architecture Education (AEEA) sets the tone in this regard. Following the same spirit, this paper informs how architectural archives provide an excellent field to learn from particular interpretations on how to relate to the built environment and the agency that can be acquired by archival research, dealing simultaneously with heritage


M. Blanco Lage et al. (eds.), *Towards a New European Bauhaus—Challenges in Design Education*,
https://doi.org/10.1007/978-3-031-40188-6_7

and future thinking. The present context informs that, despite its potential as a valuable tool for research, archives often need to be considered in architectural education, vacating their understanding and interpretation to art and architecture historians. In the 1990s, philosopher Jacques Derrida identified an *archive fever* that subsequently situated the *archival turn,* arriving in the last decade in the architectural field through the work of design ethnographers. In this context, architectural researchers and educators should identify their agency in this field. What is the role of architectural archives in the contemporary teaching of architecture? What can younger generations learn from the lessons from the past? Can archives be not only a tool for research but also a tool for practitioners?

This paper shares the experience of a pedagogical experiment that involved institutions with an active role in designing the built environment through an educational program, enhancing a multi-directional knowledge transfer in the same spirit of the New European Bauhaus agenda. On one side, undergraduate students and professors of the Architectural Analysis seminar at ETSAM Madrid School of Architecture (Universidad Politécnica de Madrid), and on the other, the Norman Foster Foundation (NFF) through its Archive, also in Madrid.

This text is structured into three sections to unfold the programme's complexity. The first section provides a theoretical framework which addresses how archival research can be beneficial in opening the scope of critical thinking in architecture students through access to the Foundation's Archive as a case study. The second addresses the syllabus of the Architectural Analysis seminar crafted around the combination of Norman Foster's architectural thinking through the study of archival content at his Foundation's Archive. The third section concludes with an analysis of the program's outcomes—such as an exhibition of student work-connecting to what the New European Bauhaus identifies as an 'inspiring projects and ideas' project typology. Lastly, it is relevant to point out the context of this contribution, the European Association of Architectural Education (EAAE) summit in Madrid, as it allowed the opportunity to open a debate through feedback amongst colleagues, educators and design thinkers.

## 2 Introducing Archival Research

In 2017, the British architect Norman Foster established his Foundation's headquarters in Madrid.[1] The Norman Foster Foundation headquarters,[2] a former three-storey residential mansion, houses the architect's professional and personal archives, comprising a growing collection of at least 120,000 items.[3] The Foundation's Archive hosts Norman

[1] 1999 Pritzker prize winner Foster is the Executive Director of Foster + Partners and the President of the Norman Foster Foundation. He is also Doctor Honoris Causa of the Universidad Politécnica de Madrid.

[2] The Norman Foster Foundation was founded in London in 1998. In 2015, its headquarters were transferred to Madrid, motivating the creation of an education and research program around the Foundation's Archive. The author of this text was part of the founding team members of the Foundation, firstly as the Archive and Projects Coordinator (2015–2017), and lastly, as Head of Education and Research Units (2017–2019).

[3] Its last inventory identified 116,302 items in 2019, with an ever-growing collection due to other fond acquisitions and Foster's ongoing work and active role in his practice.

Foster's impressive drawings, plans and models collection. Most importantly, its archival contents suggest an early understanding of sustainability and multidisciplinary thinking as essential architectural thinking and practice tools. As the program's backbone is to enhance archival research within architectural education programs, locally accessing the archives of one of the most influential global architects has been an excellent opportunity to strengthen archival research for undergraduate architecture students.

The strong connection between archival research and the skills students can acquire from consulting architectural archives implies an extra effort from educators to make this apparent to the student body. Archives provide an opportunity to acknowledge how practitioners have related to the environment through built or unbuilt initiatives that offer a broad scope of data and experiences, including failure or success. From a historical perspective, we can also learn how architects have provided an array of multidisciplinary approaches to sustainability through changing technology, with an increasing sensibility that has varied through time.

Naturally, the epistemological approach for this experimental syllabus is based on the *archival turn*. Derrida's *Mal d'Archive*, a seminal piece in this field, connects archives and archiving with the creation of memory and narrative, combining archival science with Philosophy, History of Science and Anthropology. A recent update on archival theory can be found through anthropologist Albena Yaneva, considered one of Bruno Latour's most prominent disciples. *Crafting History: archiving, and the Quest for architecture legacy*, one of her last scholarly works, unfolds the relevance of the archive in contemporary architectural practice, underlining the processes of archival making and arguing the archive as a specific practice in itself.

Additionally, the *archival turn* can be combined with Latour's *ethnographical turn*, which provides a framework for knowledge transformation through research and design thinking, especially uniting design, photography, and information systems. Through ethnography in information systems, Latour describes how the progress of ethnography and design are combined: on one side, *research* through instances and patterns- and on the other, *design* through design concepts and prototypes.

The combination of both *turns* applied to architectural education provides the theoretical basis for the syllabus described in this text. In analogy to Latour's terminology, *research* will become *archival research*, based on Norman Foster's archival contents, and *design* will become *design education* through the syllabus of the Architectural Analysis seminar. Both institutions ally with the ultimate objective of enhancing critical design thinking in undergraduate architecture students, seeking to generate a debate based on additional questions: can archival research motivate students and inspire architectural education? What could architectural archives' role be in contemporary architecture pedagogy? How can the study of architectural archives become a valuable tool for design thinking and, most importantly, in the 2030 Agenda for Sustainable Development Goals?

## 3 Architecture Analysis Experiments

The Architectural Analysis seminar's syllabus seeks to answer the previous questions through an educational experiment by bringing together students and educators with archivists. During the Spring 2022 semester at ETSAM, 80 undergraduate students enrolled in the program to analyse Norman Foster's work under the guidance of professors and architects Javier Madera and Gabriel Hernández. These educators crafted the program to familiarise students with the previously described research methodologies connected to the consultation of architectural archives for critical architectural thinking, pivoting the experiment into an additional motivation through direct contact with Norman Foster and the contents of his impressive architectural collection and Archive.

Architectural Analysis is a second-year course seminar. With a combination of theory and practice, the syllabus' learning objectives include the following: analysis and theory of form and the laws of visual perception to foster the implementation of ideas of form's knowledge, the acknowledgement of composition and architectural types, and to study the performance of methods of the process of symbolisation, practical functions and ergonomics. The skills students acquire are as follows: spatial vision, aesthetic sensitivity, historical culture, critical reasoning and knowledge of other cultures and traditions. This subject enhances critical thinking and reasoning applied to architecture and the built environment.

The seminar's syllabus consists of a five-analysis sequence to guide students in dissecting -and therefore understanding- architecture: *parietal* (façades), *massive* (volumes), *spatial* (interior spaces), *functional* (systems) and *structural* (constructive) analysis. Separately developed, students are challenged to apply the masterclasses' theoretical contents to a selection of buildings which serve as a case study. On the 2022 spring semester, Professor Madera and Professor Hernández curated a selection of sixteen works where Foster's practice had addressed contemporary topics such as sustainability, resources and material economy. The program underlined the following Sustainable Development Goals (SDGs): #4: Quality Education, #9: Industry, Innovation and Infrastructure, #11: Sustainable Citizen Communities, #12: Responsible Conception of Production, and lastly, #17: Partnership for Goals.

As previously described, the partnership consisted of the Architecture School of Madrid (ETSAM/UPM), a public university, and the Norman Foster Foundation, a private Institution created in 2015. This combination seeks to provide a seductive framework in which the students can learn from both sides, from the institution (NFF) to the practice in design, the Architectural Analysis subject at the ETSAM—in a clear analogy to Bruno Latour's diagram on the ethnography of design.

The NFF's Archive comprises the early work of Norman and Wendy Foster, who jointly founded and codirected Foster Associates in 1967. The Archive gathers most of Team 4 and Fosters Associates' early work, providing a large set of archival items that generate an extensive 'archival framework' for this program. Impressively, at the onset of the 1970 decade at Foster Associates, it is possible to identify a significant enthusiasm for sustainability, material efficiency, green energy and ecology. Several projects where environmental explorations stand out serve as proto-examples of how to deal with the current ecological crisis. Two are the *Climatroffice* project and the urban planning for La Gomera island in the Canaries. Although both were unbuilt, these projects set the tone for one of Foster Associates' research lines: the quest for efficiency and sustainability.

Given this framework, Prof Madera and Prof. Hernández structured the Architectural Analysis program (five separate analyses and a final project) into two overlapping sections. The first one comprises a combination of theoretical and practical sessions based on curating archival contents extracted from the Foundation's Archive. This selection included original drawings, plans, slides, photographs, and artworks, providing a multidisciplinary ground on which to develop the previously described analysis variations (Fig. 1).

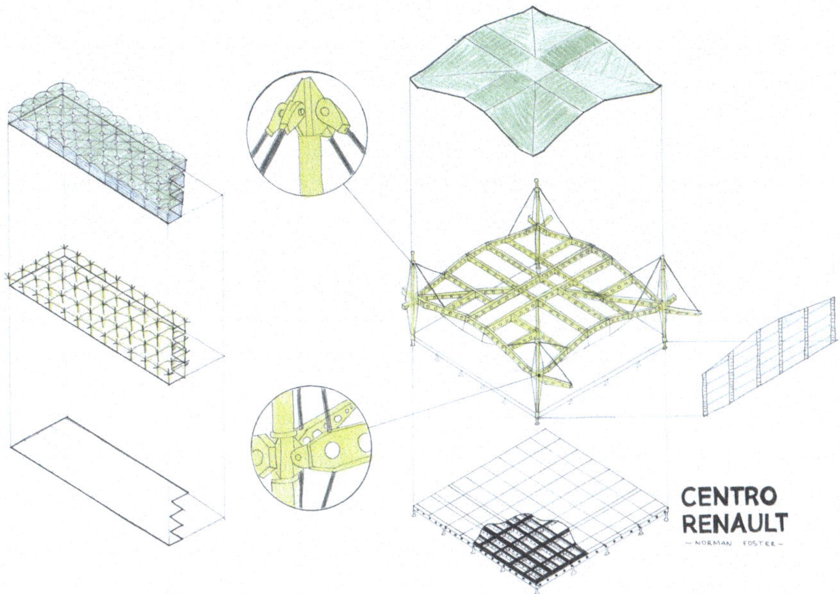


**Fig. 1** Structural analysis: sample of student work, May 2022 © Universidad Politécnica de Madrid

The second part implied a series of on-site visits to the Foundation's Archive. The students could consult original material and expand the selection with additional content distributed at the Foundation's Archive Galleries. A further aspect that enriched the collaboration between the ETSAM and the Foundation is that the NFF's Archive has developed an online and open-access database, allowing students to consult most of Foster's Collection remotely, enhancing curiosity within an expanded set of material. Acquainting students with archival research methodologies permitted them to produce a unique set of analytical drawings that confirmed that the program's learning objectives had been achieved. At the end of the semester, the last exercise is the Architectural Language analysis, tailored to allow students to apply all of the acquired skills in a self-assessment process (Fig. 2).

The architectural language analysis is an open-format exercise, allowing students to enjoy absolute freedom (and control) on its contents and results - in a process similar to gamification. The students showcased creativity and unprecedented mastery of archival research tools. After three intensive weeks of work, the students curated their archival collection through specific archive contents, such as plans, drawings and photographs. These were successfully arranged in utterly plural outcomes: from comic strips showcasing Foster's drawing evolution to a cookbook fanzine where buildings are described as

**Fig. 2** ETSAM students visiting the Norman Foster Foundation's Archive Galleries, February 2022 © Norman Foster Foundation

culinary masterpieces. Various board games included a *Foster Risk* -conquering regions where Foster has built-, a Foster Guess Who -tracing analogies within environmental policies within buildings- and *Memory* card games, and many more. Additional formats featured *Foster Metro* plans, musical-architectural scores and *Constructor* games (Fig. 3).

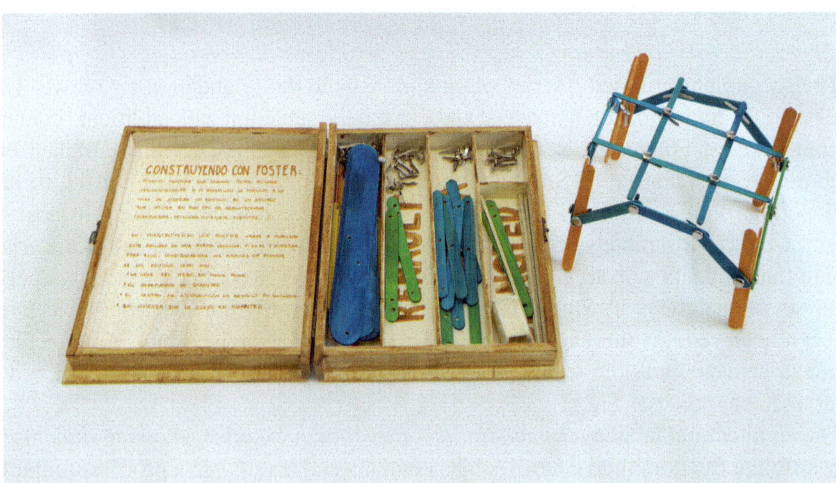


**Fig. 3** Architectural language analysis: sample of student work, May 2022 © Universidad Politécnica de Madrid

The semester's final presentations took place at the Foundation's Archive Galleries. Students presented the semester's outcomes to an audience of architects and archivists, who provided feedback, remarks and debate around the students' learning experience. Interestingly, surrounded by Foster's original archive contents, the students showcased their reinterpretations of Foster's archive through their Architectural Language analysis final exercise, resulting in a wide variety of formats ( drawings, books, fanzines, mock-ups, board games, etc.) that produced a debate between the original and informed archival material. The students' final exercises filled the Foundation's central Library oval desk in a display prepared for Norman Foster, who was visiting Madrid the following week. Foster's positive feedback -and pictures of him enjoying the games- motivated students to realise the importance of archival research. Most importantly, it leads to their empowerment by understanding how each is an agent for a positive change in the built environment (Fig. 4).

**Fig. 4** Student final presentations at the Norman Foster Foundation's Library, June 2022 © Norman Foster Foundation

## 4 Program Outcomes and Conclusions

The analytical work developed by the ETSAM students at the Architectural Analysis seminar in Spring 2022 allowed them to acquire archival research tools, engage in multidisciplinary thinking and collaborate with several institutions. The semester's syllabus focused on analysing a series of Norman Foster's works by accessing his Foundation's archival fonds to enhance students to acknowledge their agency for a positive change in the built environment.

The wide variety of outcomes at the last exercise demonstrated a highly motivated student body that even surprised Norman Foster, paving the way to addressing the learning goal of how archival research can inspire future generations. Moreover, the display of students' final work at the Norman Foundation main Library space motivated an unexpected result: an invitation to showcase the student's work at the Foundation during Architecture Week between 5–7 October 2022.[4] Located at the Norman Foster Foundation's Pavilion of Inspirations, this milestone was the closing event of the educational program in which ETSAM undergraduate students shared their newly acquired architecture analysis skills with a broader audience, receiving hundreds of visitors.

The resulting pedagogical experiment and collaboration between the Department of Architectural Composition at ETSAM, Universidad Politécnica de Madrid, and the Norman Foster Foundation encourages more knowledge transfers that motivate new generations of architects to acquire additional archival research skills and agency to produce a more sustainable future and society (Fig. 5).

[4] The author would like to acknowledge and thank the support from Prof. Javier Madera, the Departamento de Composición Arquitectónica (DCA) of the Universidad Politécnica de Madrid (UPM), the Norman Foster Foundation (NFF) team members and Fundación Arquitectura COAM. I would also like to extend a special gratitude for the engagement and enthusiasm of the following students who made the exhibition possible: Maider Lavín Cachorro, Jaime de Andrés Lerma, Víctor Manuel Martín Sánchez and Ignacio Masoliver Sánchez.

**Fig. 5** Volunteer students preparing the *Learning from Foster's Archive* exhibition, 5–7 October 2022 © Norman Foster Foundation

# Architectural Education: Between the Discipline and the Profession

Bilge İmamoğlu(✉)

TED University, Architecture, TED Üniversitesi, Mimarlık Bölümü, Ziya Gökalp Caddesi No.48 06420, 06420 Kolej Çankaya, Ankara, Turkey
bilge.imamoglu@tedu.edu.tr

**Abstract.** This paper will discuss the social role of the architectural education in the continuous reproduction of, not only the disciplinary knowledge, but also the ethos associated with the practices through which an individual utilizes their disciplinary training in placing their social standing. It will be argued that the need to re-evaluate the means and ends regarding the communication between the disciplinary knowledge and professional roles is especially compelling in the field of architecture, with the new agenda that the twenty-first century has introduced. In this frame, the significant historic contribution of the Bauhaus in the early twentieth century will be referred particularly in the critical form that this school had reformulated the triad of relationships taking place between the discipline, the professional and the society at large. Referring to the literature of sociology of professions as well as the Foucauldian concept of "discipline" and Bourdieu's concept of "field", it will be discussed that, just like social space itself, such relationships are also cohesive social products and are subject to constant change. It will be argued the knowledge content in a discipline may be translated into sustainable results on the professional's social role and function specifically if such reforms is coupled with reforms on practices of introducing the renewed knowledge content to the student of the field.

**Keywords:** Professionalism · Architectural discipline · Critical theory · Bauhaus

## 1 Introduction

If a very quick review of some recent debate on architectural education is allowed, at least one of the ways to attempt to define an overarching summary would be through the concept of "content update". One can exemplify for instance, as a rather extensive event, the Oxford Conference of 2008 titled "A Re-evaluation of Education in Architecture" and presented as "50 Years on - Resetting the Agenda for Architectural Education", being the follow-up of the similar conference in 1958. It is highly visible in this event that the emphasis is on the urgent and the difficult task of confronting emerging global challenges, where we "have to change the way we do things to adapt to rapidly evolving circumstances" [1]. There seems to be an ever-growing consensus, and with good reason, on that the twenty-first century is presenting the world new issues and conditions that were not present or as pressing in the previous century, which was when the


M. Blanco Lage et al. (eds.), *Towards a New European Bauhaus—Challenges in Design Education*,
https://doi.org/10.1007/978-3-031-40188-6_8

fundamentals of modern theory and practice on architectural education were basically framed. Such issues and conditions are generally observed to be on ecological problems as exemplified in climate change; on technological developments both in terms of design and production as exemplified in digital technologies; and new social and political challenges as exemplified in concepts of diversity and sustainability and issues of refugees as a result of exacerbated armed conflicts. Eventually, the debate on how to update the content of architectural education with reinforced knowledge and skillsets so that architectural students are equipped to encounter the challenges that the twenty-first century brings in their professional world is rightfully a hot one.

While strongly acknowledging that all the effort on such content update is necessary and urgent, this paper aims to contribute to a discussion which should be running parallel to the one on the knowledge content and the practical requirements of the profession: That is the discussion on the identity of the profession and its social function, and the role of the university as the institution assigned with the task of training future professionals. This paper will aim to remind that, when the modern theory and practice of architectural education was established in the early twentieth century with pioneering cases such as Bauhaus, it was not by a mindless and affirmative reflection of the requirements of the professional practice on the knowledge content of the discipline, but instead by a social critique of how the two relate to each other.

## 2 Critical Theories on Professions, Fields and Disciplines

For any discussion that sets out to frame architectural education as being "between the discipline and the profession"—as the title of this paper does, it seems quite unavoidable to refer to the popular dichotomy on the theory versus practice. A rather reductionist conception of this dichotomy assumes that theory is about 'knowing' without necessary practical results, while practice does acquire results without necessarily knowing 'why' [2]. However, critical theory is not represented in this formula, as it attempts to combine 'why's with 'how's; not only dwelling at 'why it is working or not', but at times extending to 'why it should *not*', or '*not* that way'; and at times making its assertions to claim simply '*that* is *not* why'.

This paper basically aims to remind in this sense that the modern university, as a social concept and a shared idea, as well as an institution for the organization of higher education, assumes a much more complex function in the modern society than merely being an instrument for the transmission of the practical and theoretical knowledge of a given discipline to future professionals. Besides, it also involves the transmission and incorporation—and hence inevitably reproduction, of the discursive and/or ideological set of premises on how the tripartite relationship of the professional subject, disciplinary knowledge and the society is organized and sustained. In the process of their education, students incorporate all of those at once: how to professionally construct the relationship of the professional subject to the disciplinary knowledge (i.e. how one thinks, acts and talks like an architect); the relationship of the disciplinary knowledge to the society (i.e. the social function of the architectural action and the assertion on why the society actually needs that); and the relationship of the professional subject to the society (i.e. the socially established identification of the architect subject and the status and privileges that comes within). Such constructions contain established forms of knowledge in epistemological, technical and ethical settings; yet they may also go beyond and utilize such forms in reproducing idealized self-reflections and social identifications.

It will be suggested here that there is set of a well-established critical theory, which all individually focus on one string in this three-partite relationships and can all be effectively brought together to re-evaluate the whole, and particularly the role of institutions of professional education in all. The architect's and scholar's relation to the field of knowledge can be analyzed in the studies of Bourdieu with the term "field", and especially—and critically—in the problematic relationship of the cultural and the economic fields of the architectural habitus to the university field of the scholar one. The field of knowledge's relationship to the society can be subject to a Foucauldian discourse analysis. And finally the architect's—or the student's, relationship to the society can be reassessed through ideological deconstruction through the critical readings of professionalism.

In order to briefly summarize how they all relate, we may begin with the last one. The reference here is to the scholars of sociology of professions who especially coin to the so-called 'market monopoly model' following Sarfatti-Larson's work [3]. This model reveals the determining role of the implicit expectation on economic and social privileges embedded within the manifest ethos built around the professional identity and professional act. The construction of the professional ethos is coupled with the search for "adequate ideological legitimations for the monopolistic exclusion of competitors" in the market for the profession's services [4]. According to Abbott, "it is the history of jurisdictional disputes that is the real, the determining history of the professions" [5]. This concludes that the social and cultural role attributed to architecture is the historicity of this dispute, and not necessarily how the knowledge content autonomously evolved.

Field, for Bourdieu, is a structure where all sorts of interrelations between the individuals, institutions and discourses are organized through rules and regularities that frame the given "space of play" to regulate the competition for the control of the resources and capitals specific to that field [6]. Individuals contest for the rewards presented by the field according to a set of rules, but the struggle is simultaneously and conditionally for the control over the rules. Following Bourdieu, Stevens points out that the practice of the production of the built environment actually relates to fields with many diverse and complex agencies, such as the economic field, where architects do not possess the tools to autonomously control [7]. Therefore they tend to inflate the knowledge content with references, which will push the play further in the cultural field where claims on autonomy becomes significantly more acceptable. That is quite parallel to Sarfatti-Larson and her discussion on architecture, where she similarly argues that, especially after the technological developments of the nineteenth century, architects could defend their occupational role only in "stylistic, symbolic and eminently theoretical terms", carrying their play into the cultural field [4].

Bourdieu's studies on the field are inspiring especially remembering that the argument here aims to be conclusive on architectural education and not professional practice. In that sense the issue at hand does not simply dwell into the dynamics of one particular field but attempts to problematize the relationship of two; the field of architecture on the one hand and the university field on the other. Bourdieu researched into the university field by focusing on the French academic circles of a particular period, yet his observations are exclusively teaching on how the university is indeed a field with the set of rules and regulations of itself for status-seeking and power-generating actions of the agents, such as practices of publication and student-supervisor relationships, who all adopt into the habitus of the field, which present them a system of shared social dispositions and cognitive structures generating perceptions, appreciations and actions [8]. This eventually means that knowledge production at the academia is not as self-dependent and self-motivated as pictured in the presentation of universal and objective the scientific methodologies.

The approach of critically problematizing the formation, production and distribution of knowledge is shared also in Foucault's conceptualization of "disciplines". For Foucault, disciplines "are at one and the same time social entities and generators of the very knowledge which they apply to society" and practices related to knowledge are always an exercise of power [9]. He focuses his studies on the institutional and structural (or rather discursive) attributes of its practices and essentially as experienced by who are subject to power rather then who exercises it. Yet his understanding of discipline is similar to Bourdieu's field in how they both differentiate from studies on professions; while sociologists of professions recognize the profession's knowledge basis as a given, they both give a central importance to its formation.

## 3 Bauhaus

In this frame, the significant historic contribution of the Bauhaus in the early twentieth century should be referred as not being confined in the content of the knowledge on design and architecture, but particularly in the critical form that this school had reformulated the triad of relationships taking place between the discipline, the professional and the society at large, opposing the professionalist norms which were in the process of being established in its time. It may be suggested that, besides the intellectual formation of the individuals that composed it, what rendered Bauhaus as a unique and precious case was how it responded to the context of the local and the historic developments, in a socially and critically engaged manner. As for such developments, one should immediately refer to the context of the German industrialization which eventually created the Deutsche Werkbund as a quite unique cooperation of the bureaucracy, industry, artists, designers and craftsmen (and which included a strong critical—and theoretical statement on the practice). Another important component of the historic context should be taken as the nineteenth century evolution of Humboldtian higher education with the co-existence and cooperation of theoretical research and professional training. It was this essentially German reinstitutionalization of the university, which universally established that it is the autonomous and self-motivated practices of knowledge that conditions the social claims of professional practice and not vice-versa [10]. And finally, and only after the other two, one can discuss how Bauhaus brought up an alternative design theory in an industrial age that builds upon and replaces the British Arts and Crafts Movement.

In that sense it should be argued that Bauhaus can historically be defined as a perfectly balanced combination of those three, which transfers a combined critical social theory into a theory on design education. In that respect, the Bauhaus Manifesto comes forth, with its references on the barrier between the artists, architects and craftsmen and the implications that such barrier is based on a professionalist and exclusivist ideology and its removal is a class issue [11]. Similarly, Moholy-Nagy should be reminded as he ridicules the concept of talent by saying that everybody is talented and attempts to replace the myth of talent in creative training by formulation of transparent and reasonable methodologies [12]. Finally, Itten can be added to the list, as he redefines artistic training as an intellectual self-realization and rediscovery [13].

The Bauhaus experience thus presents an educational reformulation that is based on the unity and balance between abstract knowledge and applied knowledge, a conciliation between the free individual in an intellectual self-realization and her social responsibilities, and cooperation between autonomous production of disciplinary knowledge and social legitimization of professional identities. It should be seen particularly marvellous that the greatest artefact to come out of all this theory was the "Vorkurs"; the basic design study of the first introductory semester.

## 4 Conclusion

It is not possible to claim that the social critique fully prevailed in what followed the Bauhaus Manifesto of 1919; modern design theory and architectural practice soon evolved to create its own version of the heroic professional as the privileged and creative individual, especially in its post-war versions and fuelled the "stardom profession" later. This paper is simply pointing out that with the new issues and challenges that the twenty-first century presents, it is urgent to reconsolidate the knowledge content and pedagogical practices in architecture in communication with a renewed critical theory on professional (and professionalist) practices and refreshed debate on the identity of the profession and its social function.

Methodological renovations in architectural education should not evolve into reclaiming the lost role of the modernist master-conductor in any shade, but rather should promote enhanced communicative practices facilitated for all actors and agencies involved in the production of the built environment. In our department of architecture, we attempt to hint at such a path. The initial assumption is that a portion of the training should be allocated at the equipment of our students as able and intellectual individuals and not exclusively as well-equipped professionals. In that end we have willingly gave up some departmental courses to make room for sets of cross-disciplinary elective courses that directly aim such reinforced communicational experiences and habits, as well as interactive societal roles (the course set on "global citizenship" is a rather good example). The departmental courses are also being redirected similarly on highlighting communication, interaction and diversity of roles, with the motive being promotion of increased communication and cooperation in between courses, including the studio.

As a similar reference, and an inspiration, Akcan's adaptation of the theory of translation in her historiography can be referred, where she could brilliantly overcome problematic relation of the "original" to "else" in her study of the architectural history of modern Turkey [14]. As she also acknowledges, her interpretation of the concept of translation can easily be extended into architectural pedagogy, in the ways that "it advocate[s] a commitment to a new culture of translatability from below and in multiple directions for truly cosmopolitan ethics and global justice" [15]. In our own pedagogical experiments, which we published in detail elsewhere [16], we attempted to apply similar methodologies of interactive communication not as tools of historiography but as design. The scope of this paper will not allow to further dwell into those examples, but nonetheless, they can be brought up to exemplify how the discussion above can modestly be echoed in educational studio practices simply by seeking out to replace the myth of the talented, original creation of the designer with methodological and conversed cooperation, a replacement which was argued here to be what the historic Bauhaus did best in its time.

## References


1. Roaf S (2008) Introduction. In: Roaf S, Bairstow A (eds) The Oxford conference: a re-evaluation of education in architecture. WIT Press, Southampton
2. In reference to the widely famous quote from Hermann Hesse (1877–1962), where he says: "Theory is knowledge that doesn't work. Practice is when everything works and you don't know why"
3. Sarfatti-Larson M (1977) The rise of professionalism. University of California Press, Berkeley
4. Sarfatti-Larson M (1983) Emblem and exception: the historical definition of the architect's professional role. In: Blau JR et al (eds) Professionals and urban form. SUNY Press. Albany
5. Abbott A (1988) The system of professions: an essay on the division of expert labor. The University of Chicago Press, Chicago
6. Bourdieu P (1992) An invitation to reflexive sociology. The University of Chicago Press, Chicago
7. Stevens G (1998) The favored circle: the social foundations of architectural distinction. MIT Press, Cambridge
8. Bourdieu P (1984) Homo Academicus. Stanford University Press, Stanford
9. Goldstein J (1984) Foucault among the Sociologists: the "disciplines" and the history of the professions. Hist Theory 23(2):170–192
10. Timur T (2000) Toplumsal Değişme ve Üniversiteler. İmge Yay, Ankara. Also see: Günther KH (1988) Profiles of educators: Wilhelm von Humboldt (1767–1835). Prospects 18(1)
11. Gropius W (1919) Programme of the Staatliches Bauhaus in Weimar. In: Conrads U (ed) (1970) Programs and manifestoes in 20th century architecture. The MIT Press, Cambridge, pp 49–53
12. Kostelanetz R (ed) (1991) Moholy-Nagy: an anthology. Da Capo Press, Michigan
13. Itten J (1975) Design and form, Revised Edition: the basic course at the Bauhaus and later. Wiley, New York
14. Akcan E (2012) Architecture in translation: Germany, Turkey and the modern house. Duke University Press, Durham
15. Akcan E (2018) Writing a global history through translation: an afterword on pedagogical perspectives. Art Transl 10(1):136–142
16. İnan AD, İmamoğlu B (2019) Housing reference, house in reference. In: Dostoğlu N (ed) Architectural episodes, international conference on educational pursuits and experiences. İstanbul Kültür Üniversitesi Publication, İstanbul

# Time as the Immaterial Structure of Architectural Education

Mert Zafer Kara(✉) and Sevgi Türkkan

Department of Architecture, Istanbul Technical University, Istanbul, Turkey
{karamert,turkkan}@itu.edu.tr

**Abstract.** Time is one of the lesser-addressed elements of architectural education, which fundamentally shapes its organizational structure, but also the values, culture, and motives stemming from the educational experience. Time has long been assigned meanings and roles in architectural education: as a framework, a limitation, a measurement, a commodity, a source, a route to follow, etc. The changing priorities, learning technologies, and philosophies exemplified by online architectural education and the increasing commodification of time in educational establishments have raised the importance of re-evaluating the often-invisible time-bound practices of architectural pedagogies. In this context, this paper aims to briefly review the roles of time in the explicit and hidden curriculum to structure and shape the culture of architectural learning via a range of formal and informal educational practices, from pre-institutional traditions to influential nineteenth and twentieth-century models, and lastly some twenty-first-century cases. The selected practices present an array of internalized temporal practices and organizations still visible in architectural education today. This brief review of architectural pedagogy from the lens of time puts into focus the immaterial structure of architectural education and time-bound elements that characterize the formation of the architect. Through examples, it is aimed to discuss these temporally inscribed pedagogical motives and conventions from the past and present to reflect on the potential of time as an active and critical agent of architectural education.

**Keywords:** Architectural education · Architectural learning · Temporality · Time · Architectural pedagogy · Curriculum

## 1 The Cultural Construction of Time in Architectural Education

The notion of time is a complex, innate, and yet highly potent element of architectural education that is often invisible and academically under-addressed. Time has long been a formative aspect of architectural pedagogy not only in its organization but also in instigating specific values, characters, cultures, and motives that serve the formation of "the architect" figure. It is culturally assigned varied meanings and roles as a framework, a limitation, a measure, a commodity, a source, a route to follow, etc. However, many of the time-bound norms and their pedagogic implications seen as timeless and universal facts in architectural education originate from the Western educational models formed


M. Blanco Lage et al. (eds.), *Towards a New European Bauhaus—Challenges in Design Education*,
https://doi.org/10.1007/978-3-031-40188-6_9

in the eighteenth century onwards, as Parcell states for the disciplinary conventions in architecture [1]. Today, the temporal organization of architectural education is shaped by the continuation of these historical traditions and values but also by external influences: the pressure from professional bodies, national policies, and the educational market. Crysler criticizes this situation by mentioning a double bind produced in the transmission model of education due to the unmoving, stagnant time of the canon and the pace of knowledge consumption [2].

Meanwhile, the need to re-evaluate architectural education's temporal structures and practices comes to the fore with changing priorities and urgencies in architectural education: climatic and social crises, new mediums and technologies of learning and information, the developments in learning-centered educational philosophies, etc. This calls for the often veiled but impactful use of time in architectural education to be addressed, questioned, and rethought both in the light of inherited norms and emerging urgencies. In this context, this paper aims to briefly review the role of time in shaping the cultural construction of architectural education. Initially, it discusses the roles of time in explicit and hidden curricula and proceeds by investigating the temporal organizations of a range of influential architectural learning models and the time-bound pedagogical practices formed by it. Hence, it aims to point out the importance of time as an agent of architectural learning and the necessity to reconsider its role in the future organization and practices of architectural education.

## 2 Time as an Element of the Explicit and Hidden Curriculum in Architectural Education

Time is often conceptualized as an event, an experience, and a measuring tool [3]. As an element of measurement, it allows (im)material things to be understood in terms of rate, frequency, duration, and intensity [4]. For educational organizations, it is primarily addressed and organized through the daily, weekly, or yearly schedules of the "curriculum", although it is not solely limited to that. A curriculum works as a structuring device of an educational organization on account of designating the length and the order of educational knowledge and experience in a specified period [5]. But in the meantime, it also transmits the scope, content, and values embedded in an educational model. The word "curriculum" is rooted in the verb *currere* in Latin, which means "to run, course of action or race" [6]. This etymological root implies a time-bound characteristic that ties it with the notions of speed and competition, while the etymological root of the word "school", *scholē* in Greek, refers to the notion of "leisure" [7]. The conflict between these two words' denoting different temporal modalities feeds the complexity of questions about time in education; one implies a racing effort, the other ease of leisure.

Education theorists mark the two types of curriculums that work reciprocally: explicit and hidden curriculums, which instrumentalize time in particular ways (Fig. 1). The explicit or formal curriculum is the official, presented, or published version that is usually taken as the one-and-only curriculum and is detailed in the official documentation of learning activities and events [8]. It reveals perhaps the most stated relationship in which time is given the role of framing educational activities; therefore, it directly describes the temporal structure of an institution. Through the explicit curriculum, time becomes a tool that draws the boundaries of an educational experience and marks the appropriate lengths, thresholds, orders, and durations of content and activities prescribed for one's journey to professional competence. The temporal inscription of the formal structure gives way to using time as a measuring tool, a limit to evaluate success, productivity, and even proficiency, translatable into grades, crediting systems, and assessment of professional qualifications.

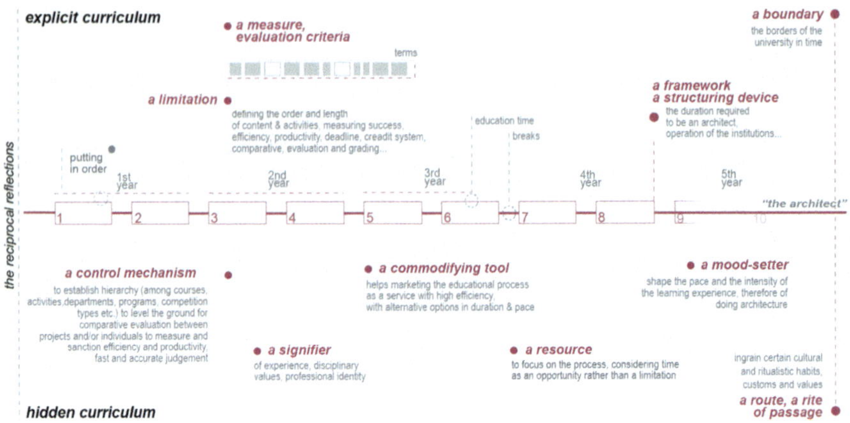


**Fig. 1** Time's roles in architectural education as an element of the explicit and hidden curriculum. Produced by the authors, Mert Zafer Kara and Sevgi Türkkan, 2022

Meanwhile, the hidden curriculum is constructed from the unstated values, attitudes, and norms that stem tacitly from the social relations of the school and classroom as well as the content of the course [9]. The hidden curriculum, recognized through prolonged exposure to contextual socialization and orientation, is significantly associated with the concept of time [10]. Through the hidden aspects of the curriculum, time gains varied functions to do with values, habits, modes, and hierarchies imbued in architectural culture. Considering the role of time as a limitation, it may be used to control and establish hierarchies among the courses, activities, programs, and even individuals or projects. The prescribed amount of time to be devoted to a project or course assumes its importance level in comparison to others while enabling competition. Time also functions as a mood-setter that shapes the pace and the intensity of the learning experience, therefore of doing architecture. It helps the educational process to be commodified as a service with high efficiency, speed, and alternative options in duration and pace to fit diverse target groups. Differently, time can appear as a resource that enables process-oriented, slow, and long-term learning.

All these diverse uses of time through the hidden and explicit curriculum appear to characterize the process as a route or a rite of passage that ingrains certain formal and cultural, professional and ritualistic behaviors, habits, customs, and values. Curriculums perform as transparent, equalizing structures that even out a learning process on an egalitarian temporal basis for different individuals going through it, which for centuries used to be an undefined and singular journey with no specific ending marker. Rethinking curriculums as mechanisms that materialize the otherwise immaterial structure of the learning experience may therefore enable us to see the roles of time as an informed agent in architectural education and culture.

## 3 Temporal Organization and Time-Bound Practices in Selected Architectural Pedagogies

This chapter addresses and discusses the time-bound practices of a number of architectural education practices by looking into their temporal structures and motives inherent in their pedagogies. The selected examples range from pre-institutional traditions to the influential nineteenth and twentieth-century models of Beaux-Arts and Bauhaus, and a number of twentieth and twenty-first-century institutions and organizations. Firstly, the explicit temporal organizations of these architectural education practices are visualized in diagrams for collective reading (Fig. 2). Secondly, for an understanding of the inherent pedagogic implications of time, as discussed in the hidden curriculum, some of these cases will be discussed through the time-bound themes, and motives they devised.

The diagrammatic visualization allows for a brief reading of the different temporal structures of architectural learning from various periods, geographies, and cultures. The initial look reveals the different uses of time-bound fragments, thresholds, or pedagogic motives either strictly or loosely, overlapping or isolated, repetitive or optional to demarcate a process/system. While some motives are unique to these examples, some are commonly inherited from earlier formats and precedents (e.g., apprenticeship durations, competitions, and competency exams). Moving towards the nineteenth century, the contingencies of learning characterized by open-ended processes gradually give way to regulated structures with predefined temporal thresholds and a variety of optional educational pathways. The overall duration to obtain the architect title seemingly shortens and crystallizes, though the requirements become more complex and intertwined. Despite the growing efforts to standardize and regulate architectural programs in the twentieth and twenty-first centuries, the temporal structures still seem to vary. The integration of self-initiated, asynchronous online models opens ways for even more diversified temporal reorganization of architectural education.

The diagram starts with pre-institutional architectural learning practices, commonly characterized as loosely defined, personal journeys based on emulation and first-hand experience, patronized and guarded by knowledge-practicing bodies. According to Parcell's definition of *Technê* in ancient Greece, as an earlier root of architecture, the education of pupils could take five to ten years to become proficient and learn the ancestral traditions of the craft and acquire sufficient experience for the practice. In the twelfth and thirteenth centuries in England and France, students were apprenticed for nearly 3 years in carpentry or masonry, and 3 years in geometry and building components before spending another 3 years gaining experience in varied kinds of jobs [11]. In both cases, learning traditional knowledge was occupying a large part of the pupil's life. Time, in this case, was the measurement of devotion as maybe one of the oldest canonical values that continues to impact architectural learning culture.

In one of the earliest institutional examples of architectural education, the seminal École des Beaux-Arts, education was still an open process in terms of the length, yet

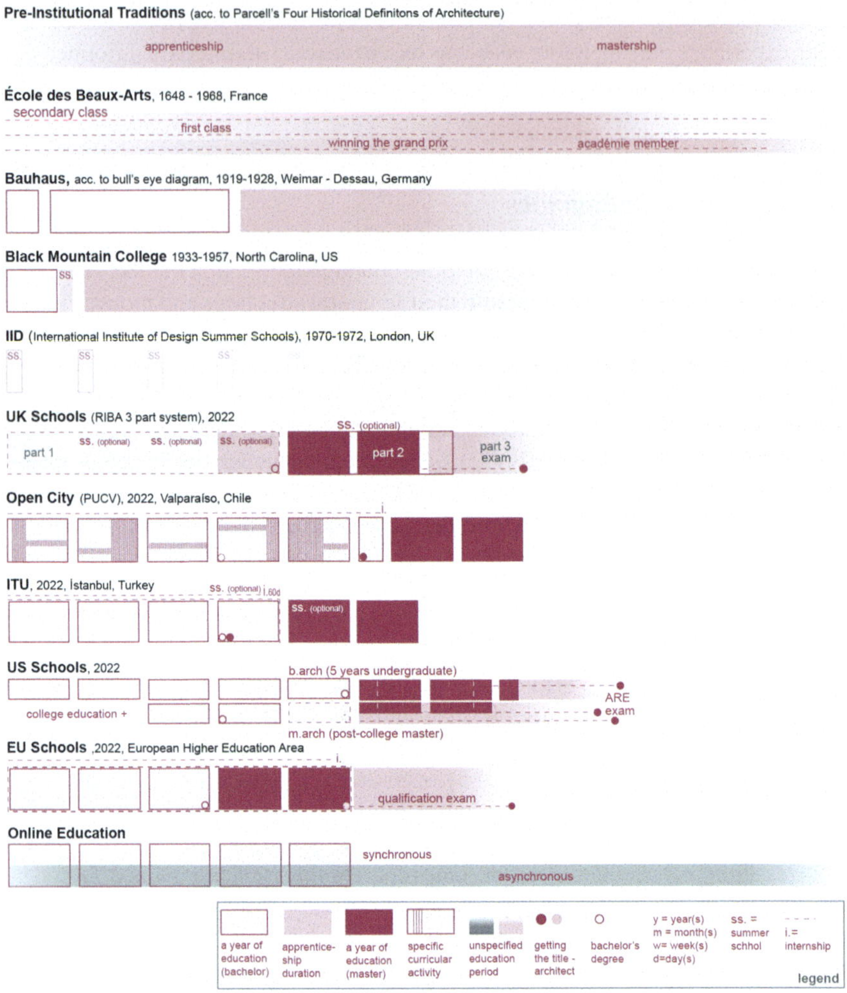


**Fig. 2** Visualization of the temporal structures of selected architectural education models. Produced by the authors, Mert Zafer Kara and Sevgi Türkkan, 2022

specific time-bound pedagogic elements were created to structure it (Fig. 3). The curriculum allowed students to go at their own pace [12]. However, the pedagogic motives that the system was built on, such as competitions, courses, exhibitions, etc. were predetermined and specified in duration in detail in the *Reglement* booklet published by the school for every educational year. Architectural learning was organized around a series of design competitions rather than formal classroom training [13]. Competitions were the central and foundational time-bound motive: ateliers and individuals competed against each other in given time-frames, but also against time itself. These competitive time spans would be spent in the atelier or loges [14], collectively or in isolation. The different lengths and values of competitions were formulated into a credit system to demarcate their prestige and importance. For example, the most important competition for a student's career as an architect, *the Grand Prix,* was the longest in duration. First, the competitors were subjected to a 24-h sketch exam. Those who passed this stage were given 3 months to complete their projects, working day and night. The winner was sent to Rome for a five-year experience [15]. As seen in this example, time was a hierarchy-building tool via pedagogic elements in the system, a limitation to prove one's judgemental and technical skills, and a measure for comparative evaluation and elimination. One's success was judged not by how long they studied but by how well they performed in predetermined time frames. Many of the characteristics of École des Beaux-Arts remained as conventions in architectural education today: charrettes exalting the importance of speed, "the architect as all-nighter" working towards a "deadline" or relatively short jury reviews located at the end of an intense production period.

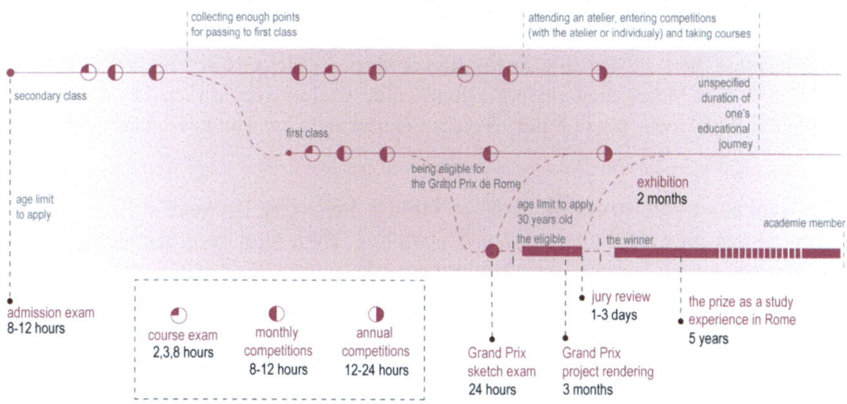


**Fig. 3** The education process and time-bound elements in École des Beaux-Arts. Produced by the authors, Mert Zafer Kara and Sevgi Türkkan, 2022

The Bauhaus also stands out as one of the most influential pedagogical models of the twentieth century in adapting new time-bound concepts such as industrialization, standardization as well as experimentality, experience, and free time in architectural learning. Bauhaus was the precursor of the modern architectural curriculum which was later adopted by numerous schools, but also embodied temporally loose and traditional learning models such as apprenticeship. In this model, students could still continue their education at their own pace. But for the first time, the immaterial notion of time was graphically inscribed as pedagogic phases of the overall structure. The famous bull's-eye diagram of Walter Gropius (Fig. 4a) stated the duration for each phase of the educational program in relation to each other and framed the first six months as a preparatory period, i.e., preliminary course. Hannes Mayer added to the visualization attempts of the Bauhaus education with his curricular diagram (Fig. 4b). In Gropius's curriculum, after the preliminary course, the students would continue their education in the workshops. After nearly 3 years of attending workshops, only the most talented students were admitted to the "architecture course" which until 1927 meant working in Gropius's office as an apprentice [16].

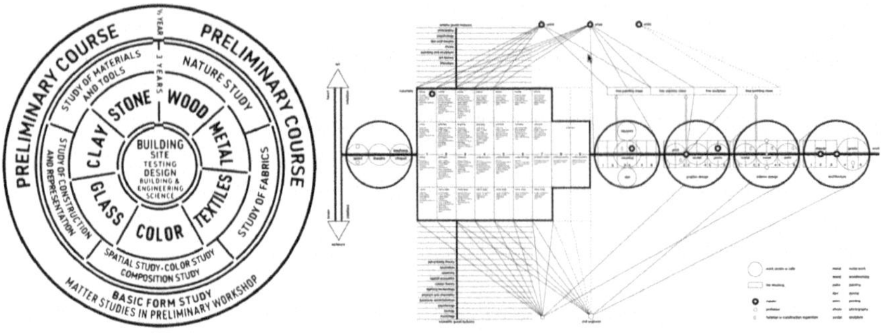


(a) (b)

**Fig. 4** **a** The Bauhaus curriculum (adapted), Walter Gropius, 1922. From Walter Gropius, Statutes of the State Bauhaus in Weimar, July 1922. Bauhaus Typography Collection, 1919–1937. The Getty Research Institute, 850513. © 2019 Artists Rights Society (ARS), New York/VG Bild-Kunst, Bonn [17], **b** Hannes Meyer, diagram for the Bauhaus curriculum, reconstructed by Klaus-Jürgen Winkler in the 1980s (courtesy of Bauhaus-Archiv Berlin and Foundation Bauhaus Dessau) [18]

The Bauhaus model used the time to build a hierarchy between different phases of design education with a clear visual declaration. Their signature pedagogic motive, the foundation course, demarcated 6 months as the necessary duration for students to obtain the primer skills prerequisite for design education. Although the forthcoming schools that adopted the Bauhaus model varied this duration up to 3 semesters, the idea of a preliminary foundational period became a common norm for twentieth-century architectural institutions.

Another significant pedagogic concept in Bauhaus was the notion of free time associated with experience, playfulness, and experimentality. Alongside the focus on hands-on learning, free time was also considered part of a continuous, uninterrupted educational experience. The Dessau campus was designed as a living community with dorms and masters houses on-site, which enabled 24-h exposure to the educational culture and environment beyond the course/workshop hours. Masters and students shared time in both extracurricular activities and their free time. Students lived and dined together, spent their free time playing sports, designing publications, organizing parties and festivals, and collaborating on art projects which required expertise from the workshops (Fig. 5a).

Simultaneously, industrialization and standardization were having significant impacts on the conception of time in architectural education by the late 1940s. A prominent example of the industrial notion of time becoming a design parameter in architectural education is the Bauhaus graduate Ernst Neufert's "Rapid Design course" where the students were using a supplementary 'open-source' design solutions catalog of ready-to-use samples to copy and assemble to design new architectural projects [19]. Neufert thought an architectural database for an efficient, time-saving, and rationalized design process was necessary to cope with the complications of designing from scratch [20]. Students of this course were compelled to design small yet complete housing projects in only three hours, using this open-source catalog at hand. This course led to the development of the

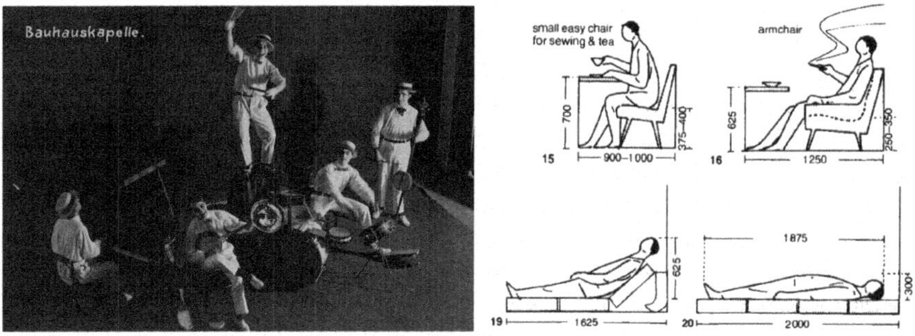


(a) (b)

**Fig. 5** **a** Bauhaus Band, Lucia Moholy, 1925/26 [19], **b** Dimensions and Space Requirements, Architect's Data, Ernst Neufert, 1936 [20]

global best-seller architecture book, *Architect's Data* (Fig. 5b), furthering the ethos of time efficiency and speed as required parameters of design and design learning.

Henceforth, alternative late twentieth-century pedagogies embraced slowness and process orientation to counter the competitive and efficiency-oriented notion of time in architectural pedagogy. Founded by a group of architects and poets in the 1970s as an experimental annex to The Universidad Catolica de Valparaiso (PUCV), the Open City (Ciudad Abierta) is continued by the generation of students that enroll in its design and construction [23]. Design-learning activities in the Open City are organized through loosely defined temporal modalities and in various time intervals, from semester projects to Wednesday meetings and opening day activities to decades-long engagements. The design-build projects are taken over by succeeding semesters or generations of students, as not all projects are aimed to be finalized. Time is embraced as design input, leaving its traces on projects along with other effects of natural forces [24]. Slowness is welcomed as an architectural experience. Time becomes an enabler, a resource, not an evaluation criterion or an end goal. Instead of fixation on a deadline or a final image, the focus is placed on the process.

Another temporal modality common to architectural education is the extra or non-curricular pedagogic structures such as workshops and summer schools. With the development of universitarian curriculum in the US and Europe after the 1950s, previously informal yet integral practices such as internships, excursions etc. were placed in the leftover time slots from the formal semester. The academic year is set to prioritize the formal over the informal time and curricular over extracurricular learning. However, some twentieth-century non-curricular pedagogies were as influential as the formal ones for their emancipatory learning and organizational modes. The short-lived but influential Black Mountain College's famous summer schools, inspired by Dewey's pedagogic theories and experiences from Bauhaus, were exemplary of informal learning settings with communal living, codesign, and outdoor activities during the summers of 1944 to 1953. International Institute of Design Summer Schools (IID), organized between 1970 and 1972 by Alvin Boyarsky, created an international, interdisciplinary "alternative ambiance" constructed of ongoing ideas, dialogue, and activity, a deviant learning experience, which later informed the unit system in Architectural Association [25].

Twenty-first-century pedagogical practices are marked by concepts that free learning from predetermined and fixed spatiotemporal frameworks. "Lifelong learning" defines a form of unconstrained, self-initiated education that is focused on personal development with an autonomous learning pace outside of a formal educational institute, such as a school, university, or corporate training [26]. The Asynchronous Model of Online Education (MOOC), an element of online education experienced globally by architectural schools during the 2019 pandemic, is thought to liberate the learning experience from spatiotemporal boundaries. One can learn at any time, at any pace, independently, or in connection with a formal institutional structure [27]. A recent study states that the future universities may be a life-long "open loop" where students can decide the time for learning in line with their own pace [28]. These flexible time-bound concepts have the potential to enrich the learning process in terms of autonomy, affordability, and inclusivity, which has been culturally problematic in architectural education.

Architectural institutions today are challenged in their temporal structures due to economic, systemic and cultural pressures: the educational market's demand for commodifiable programs for fast-track education, the high cost of architectural education, and regulatory norms from governmental and professional bodies. Meanwhile, new opportunities arise in reorganizing time–space and accessing information with changing learning mediums and technologies. Raising concerns about the well-being of students and teachers, as well as climatic and social urgencies give way to alternative conceptions of time as in slow learning and slow architecture. The amalgamation of these challenges and opportunities marks a new horizon of questions: Can architectural education be reorganized to liberate from its problematic spatiotemporal conventions, or will it adhere to its hands-on, collective, studio-based rituals, routines, and social culture?

## 4 Conclusions

This paper briefly examined time as an immaterial structure and pedagogic motive of architectural education through various cases in a historic array. Time appears as a crucial element not only for organizing architectural education but for imbuing certain values, habits and culture through experience. Being aware of the entanglement between the changing conception of time, learning models, and architectural agendas can help us frame new conceptual and practical organization models for architectural learning more in tune with temporal contingencies and urgencies. Therefore, the critical review and re-evaluation of the often-invisible and culturally consolidated time-bound pedagogical norms both in the hidden and explicit curriculums of architectural education appear to be essential. Hence, the critical agency of time as a pedagogic element could trigger other emancipatory formulations for architectural education as well as practice.

## References


1. Parcell S (2007) Four historical definitions of architecture. McGill-Queen's Press, Montreal
2. Crysler CG (1995) Critical pedagogy and architectural education. J Architect Educ 48 (4):208–217
3. Ressa T, Daniels D, Wells-Jensen S (2021) Time as a hidden curriculum: qualitative study of challenges faced by students with mobility, speech, and visual disabilities in P-12 settings. Int J Educ Res Rev 6(3):250–263
4. Le Poidevin R (2004) The experience and perception of time. In: Zalta EN (eds) The Stanford encyclopedia of philosophy. Last accessed 20 Aug 2022. https://plato.stanford.edu/cgi-bin/encyclopedia/archinfo.cgi?entry=time-experience
5. Kelly AV (2004) The curriculum: theory and practice, 5th edn. Sage, London
6. "Curriculum" In Merriam-Webster.com dictionary. Last accessed 22 Aug 2022
7. "School" In Merriam-Webster.com dictionary. Last accessed 22 Aug 202
8. Nelms TP (1991) Has the curriculum revolution revolutionized the definition of curriculum? J Nurs Educ 30(1):5–8
9. Giroux HA, Penna AN (1979) Social education in the classroom: the dynamics of the hidden curriculum. Theory Res Social Educ 7(1):21–42
10. Ressa, Daniels, Wells-Jensen: 252
11. Parcell: 110
12. Levine N (1982) The competition for the Grand Prix in 1824: a case study in architectural education at the École des Beaux-Arts. The Beaux-Arts and 19th-Century French Architecture. Thames & Hudson, London
13. Garric JP (2017) The French Beaux-Arts. In: Bressani M, Contandriopoulos C (eds) The companions to the history of architecture, vol III, Nineteenth-Century Architecture. Wiley, New York, pp. 1–15
14. Türkkan S (2023) The architecture of the making of the author. OASE J 113 (Authorships. Eds: Grafe C., Avermaete T, Davidovici I., Patteuw V., NAI010 Publishers)
15. Jacques A (1982) The programs for the architectural education of the École des Beaux-Arts, The Beaux-Arts and 19th-Century French Architecture. Thames & Hudson, London
16. Siebenbrodt M, Schöbe L (2009) Bauhaus: 1919–1933. Parkstone Press, New York, p 15
17. Diagram of the Bauhaus curriculum, Walter Gropius, 1922, https://www.getty.edu/research/exhibitions_events/exhibitions/bauhaus/new_artist/history/principles_curriculum/. Last accessed 22 Aug 2022
18. Lueder C (2013) Diagram Utopias: Rota and network as instrument and mirror of Utopia and Agronica. J Architect Educ 67(2):224–233
19. Neufert, E (1980) Architects' data. In: Jones V (ed) Architects' data. Blackwell, London
20. Weckherlin G (2007) Ernst Neufert's Architects' data: anxiety, creativity and authorial abdication. In: Anstey T, Grillner K, Hughes R (eds) Architecture and authorship. Black Dog Publishing, London, p 146
21. Siebenbrodt, Schöbe: 234
22. Neufert: 11
23. De Arce R, Oyarzún FP (2003) Valparaíso school: Open City Group (No. 7–12). McGill-Queen's Press, Montreal
24. Harden M (2013) Ciudad Abierta: the open city of Ritoque, Chile
25. Sunwoo I (2009) Pedagogy's progress: Alvin Boyarsky's International Institute of Design. Grey Room 34:28–57
26. Osborne M, Houston M, Toman N (2007) The pedagogy of lifelong learning. Routledge, Abingdon

27. Calise M, Kloos CD, Reich J, Ruiperez-Valiente JA, Wirsing M (eds) (2019) Digital education: at the MOOC crossroads where the interests of academia and business converge. In: 6th European MOOCs Stakeholders Summit, Naples, Proceedings. Springer
28. Massey J (2016) Changing course in architecture academia. https://www.architecturalrecord.com/articles/11858-changing-course-in-architecture-academia. Last accessed 20 Aug 2022

# Planning and Implementation of Project Management Processes in the Work of Architecture Students

Damir Mance(✉) and Dubravko Bačić

Faculty of Architecture, University of Zagreb, Fra Andrije Kačića Miošića 26, 10000 Zagreb, Croatia
dmance@arhitekt.hr

**Abstract.** Students usually describe the work process at the university as chaotic. There are many courses with different topics and many students with interests known or unknown to their peers and professors. All-nighters are still more than an exception, and often such practices continue even after graduation during work in the office. In the architectural profession, time planning stands out as one of the project planning components that significantly impact the project's (financial) outcome. The 5th-year Project Management (PM) course discussed in this article intends to acquaint students with PM processes (scope, schedule/time, cost, quality, resources, communication, and risks) through the application and development of awareness of their impact on scheduling and time management. By applying the PM process, the students initiate, plan, execute, control, supervise, close their work in the final semester of studies, and submit a seminar paper as a self-assessment of the execution. With thorough planning and execution of the plan, it is possible to improve efficiency and speed up the decision-making process without affecting quality. Decision-making is a disadvantage during students' work due to the lack of experience, resulting in uncertainty and indecisiveness that consume time. The introduction of PM courses in the earlier years of study and the application of the PM process can positively affect the quality of studies, student satisfaction, and the adoption of productive habits as a positive contribution to the progress and improvement of students' efficiency in later work.

**Keywords:** Project management · Time management · Architecture students · Students' efficiency · Quality of studies · Productive habits

## 1 Introduction

### 1.1 What Best Describes the Work of the Architecture Students?

In preparation for a professional career, we see many issues in the education of architecture students. The work process of most of the students is chaotic. Students tend to procrastinate and prolong their commitments. Most of the time, they are not motivated to do and present their tasks regularly. One of the main reasons could be that there are many courses with different topics, combined with lots of interests among the students, which makes loads of information to process.


M. Blanco Lage et al. (eds.), *Towards a New European Bauhaus—Challenges in Design Education*,
https://doi.org/10.1007/978-3-031-40188-6_10

Many myths that prevail within architectural schools were aptly detected and discussed in AIAS 2002 report (like "architectural education should require personal and physical sacrifice", "the best design ideas only come in the middle of the night", "creative energy only comes from the pressure of deadlines", "students do not have the power to make changes within architecture programs or the design studio" just to name few). The report stated that these myths can influence the mentality of students and promote certain patterns of behavior equally through the education and in future professional work [1].

Has anything changed since then? We believe not. Since then, students must deal with the procrastination issue emphasized by recent disruptive technologies (such as e-mails, instant messaging, and social platforms) and the enormous quantity of information—unstructured, hyperlinked, and available within a tap of a finger. Multitude of different curricular tasks in combination with disruptive social media technologies can lead to a state of persistent attention residue, where act of transitioning between tasks can have implications on students' engagement in a subsequent task [2].

Uneven workload distribution throughout a semester is a common phenomenon in architectural education. In the aforementioned AIAS report, time is pointed out as "more than a constantly endangered resource" since, in the professional world, time is a valuable resource and is closely related to the financial outcome of the project [1].

Based on the analysis of the 3rd semester of the Master in Architecture graduate program courses at the University of Zagreb Faculty of Architecture, in Fig. 1 we present the potential of obligatory and elective course integration as a part of studios. The other classes could and should be a supplement to the studios. However, there are issues in organization and communication between the studio and course instructors that could improve. Usually, instructors receive no training in teaching design, and they rely on what they have learned as students, through practice, on their interests, and their instinct [3]. The optimal goal for education in the ever-growing (inter)disciplinary knowledge base is to generate autonomous learners who can integrate the knowledge acquired in formal classes with real-life problems [4]. Being an independent learner is a skill that can also be taught to both the student and the instructor [5]. One can achieve that through workshops where studio instructors and academic researchers present their work to each other and discuss their insights [6].

## 1.2 Hypothesis

To easily process huge quantities of information, students should plan and monitor their work through execution. Each student has significant importance for the performance of a design team, so students need to be acquainted with the processes that lead to a quality project.

The introduction of PM courses in the earlier years of study and the application of the PM processes in the work of the students and the organization of the Studios can positively affect the quality of studies, students' satisfaction, and the adoption of productive habits, all that as a positive contribution to the progress and improvement of students' efficiency in later work.

To elaborate the hypothesis, we prepared a multi-criteria analysis based on the data collected through the survey and seminar papers.

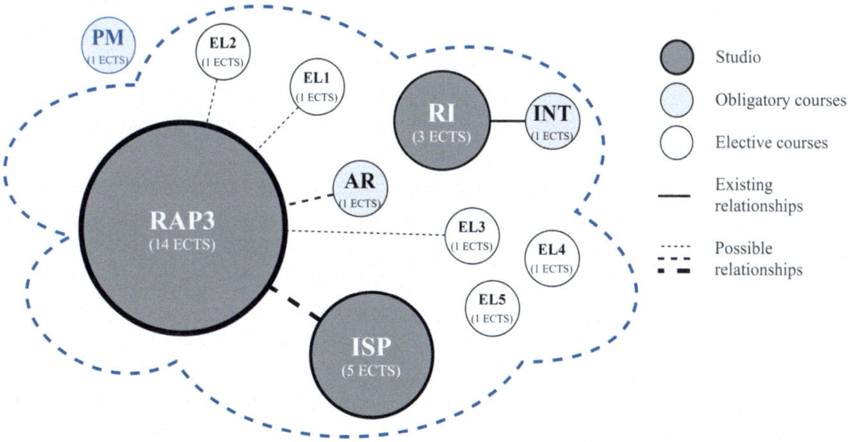


**Fig. 1** 3rd-semester structure at the Master of Architecture program, University of Zagreb Faculty of Architecture: studios (RAP3—architectural/urban design; ISP—research; RI—interior design), courses, and interconnections

## 2 Assignment

### 2.1 Project Lifecycle and PM Processes

The PM course discussed in this article is obligatory lecture-type course in the 5th year at University of Zagreb, Faculty of Architecture. The course intends to introduce PM processes [7] through the application and development of awareness of their interdependency with scheduling and time management.

As a part of the assignment, the student initiates a project including all courses and studios during the final semester of graduate study (see Fig. 1) to complete the semester with a positive outcome. By implementing the PM process groups, the students initiate, plan, execute, control, monitor, and close their work in the final semester of graduate studies (see Fig. 2). Due to the multitude of courses, massive quantities of information, and interdependencies between them, integration, with characteristics like unification, consolidation, communication, and interrelationship [7], becomes the most crucial process for the success of the project in all process groups.

In the end, each student submits a seminar paper that represents a self-assessment of the planning and execution process groups.

**Scope** represents the basics of the project. Students specify the deliverables for each studio and create a work breakdown structure (WBS) for each semestral task. Within the **schedule**, students plan the timeline and milestones for each studio. The milestones based on project scope serve as reference points for control of the assignment. Through **cost**, students estimate the financial resources required for the semestral task and define the budget. **Quality** is one of the assignment's most general and least specific processes. It is not easy to decide the quality standard for the studio deliverables as it often depends on the student's previous experiences and acquired knowledge, but most importantly instructor's opinion (as the studio instructor is, effectively, the person in charge of the quality control through milestones). Part of the task was to define **resources** needed for

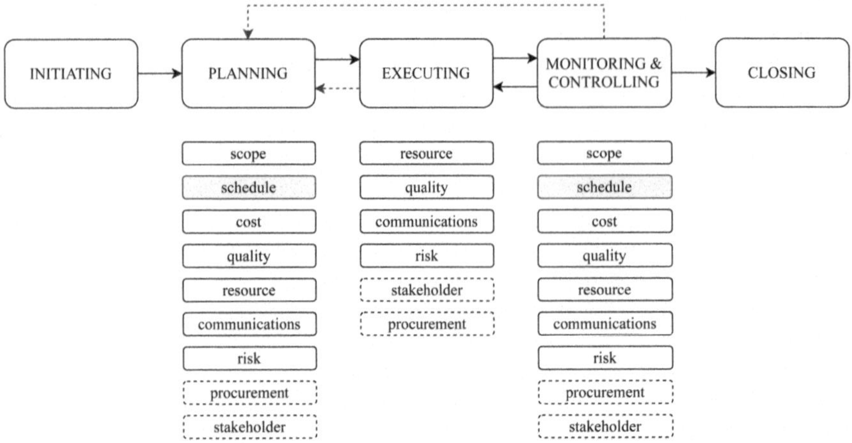


**Fig. 2** Students' assignment structure

the project completion and to plan **communication** protocols based on best practices in professional communication. The point is to make students aware of the importance of communicating the right information, on time, to the right person as the core of every successful project. Lastly, students identify the **risks** in order to plan reactions (should unexpected circumstances occur) with the aim of achieving the timely execution of the plan.

## 2.2 Planning

Planning includes all PM processes, and, through execution, monitoring, and controlling, each one of them can affect a plan. In architectural profession, time planning stands out as one with a significant impact on the financial outcome of a project. Schedule, as time management component of any plan, includes the processes required to manage the timely completion of the project [7] and is the most critical part of the student's seminar paper.

Initially, during the Bologna Process reform, the ECTS system was implemented under the assumption that one credit should be equivalent to approx. 25–30 working hours (including class time, individual study, exam preparation, practical work, etc.) which represents a typically required workload (10–12 working hours per workday) for the completion of a module within a study program [8].

Consequently, an equal distribution according to ECTS credits system was suggested as a basis for the time estimate in this assignment. It was encouraged to treat the studying as a full-time job, with 8 to 11 working hours per workday during the 15 weeks of the term (winter holidays and weekends were to be treated as non-working days).

In the workload distribution diagram (see Fig. 3), the left part of the diagram represents the studios, and the right part the lecture courses. Only the data for the three studios were processed due to their practical nature. Students are divided into several groups within each studio, with different instructors and specific assignments, so higher workload deviations were to be expected.

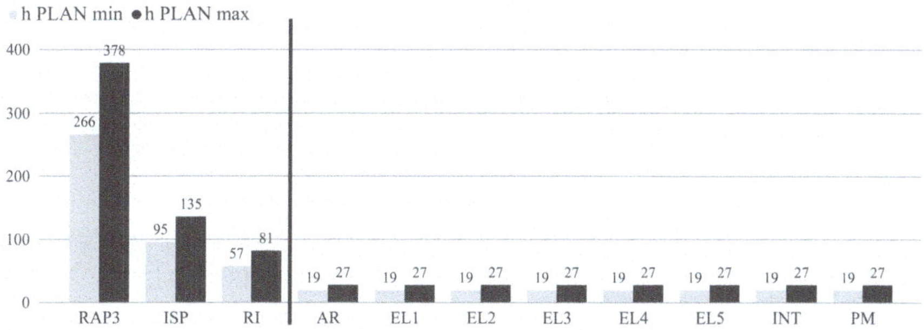


**Fig. 3** Ideal workload distributed through courses according to ECTS (h PLAN min = 8 work hours/day and h PLAN max = 11 work hours/day), 3rd semester of the Master in Architecture graduate program at the University of Zagreb Faculty of Architecture

In the planning phase, some of the students tend to present a diagram with a heavier load towards the end of the semester (see Fig. 4). We typically assign that to the acquired experience from previous years of study when the end of the semester was the most stressful.

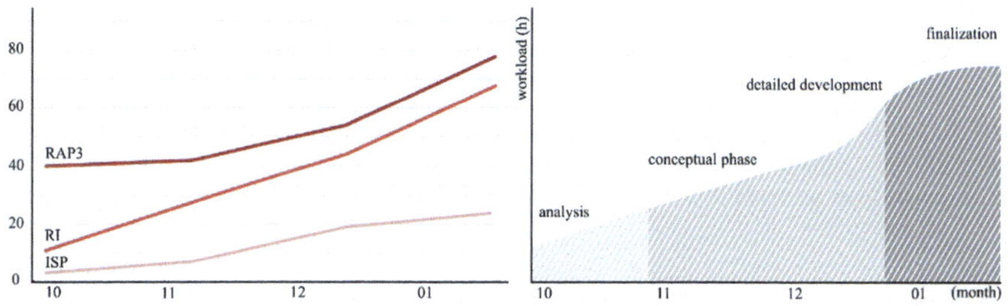


**Fig. 4** Examples of workload planned by a student as presented in seminar papers

## 2.3 Execution

During the execution of the plan, students monitor and control all mentioned PM processes and track their workload on the assignments. The collected data, combined with detailed reports in seminar papers serves as a basis for our analysis. The total of executed work hours was less than planned as presented in “Eqs. (1) and (2)”

$$\textit{average values}: \ 523 \ \textit{work hours executed} < 625 \ \textit{work hours planned} \tag{1}$$

$$\textit{median values}: \ 464 \ \textit{work hours executed} < 660 \ \textit{work hours planned} \tag{2}$$

In the students’ diagrams of the workload distribution through the semester (see Fig. 5), one can notice that the executed workload is lighter in the first weeks of the semester and that it gets heavier as the end of the semester approaches.

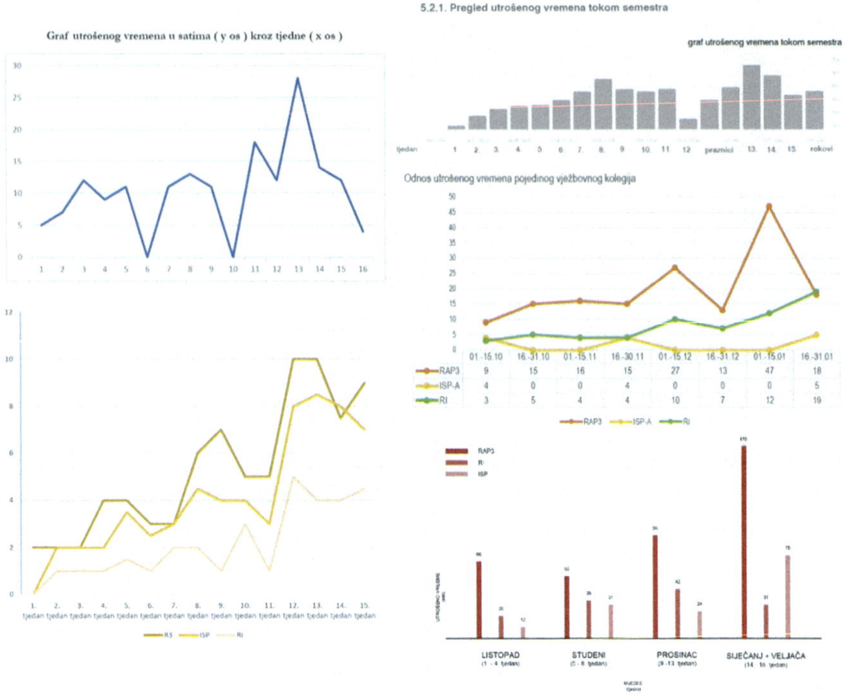


**Fig. 5** Examples of executed student workload as presented in students seminar papers (date/workload diagrams with date presented as the work week or time interval on the horizontal axis, and workload presented on the vertical axis)

In diagram (see Fig. 6), we compare workloads for each student where one can see that, for majority of the students, the executed value was smaller than the planned value. As expected, the workload distribution among the students is uneven, which depends on multiple factors such as previously acquired knowledge, students' attitudes and interest in the topic, diligence, and specifics of the assignments, to name just a few.

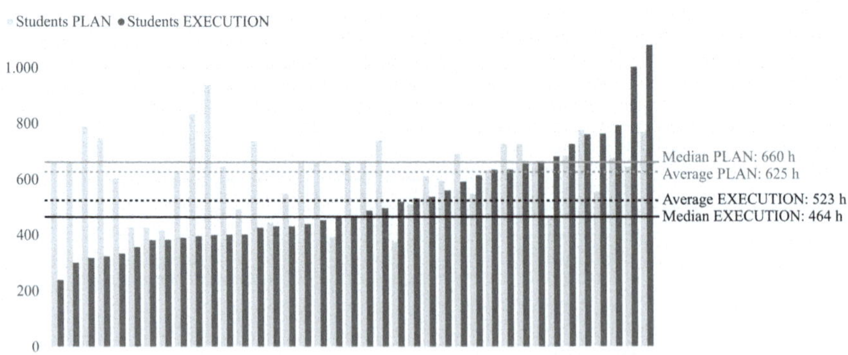


**Fig. 6** Planned and executed workload per student comparison

## 3 Survey

Following the analysis and our initial assumptions, the survey was conducted among students to detect the reasons for such uneven workload distribution. The survey served as the criteria for filtering the information collected through students' seminar papers. It consisted of fourteen (14) questions, and in this paper, we will focus on questions relevant to the hypothesis. Out of one hundred and twenty-four (124) students that enrolled in the class, sixty-two (62) answered the survey.

The first set of questions was intended to help confirm and filter the collected data's relevance. Those students who did not submit the seminar paper, did not enroll in all Studios, or did not track their workload during the semester were excluded from the analysis. In the end, a total of thirty-nine (39) complete survey responses (approximately one-third of the class) were selected as a relevant data pool and used for the analysis.

In the second set of questions, students evaluated their commitment to the assignment and the efficiency of planning the semester. As expected, a large majority of students (84%) intend to improve their organization in the future (see Fig. 7) and believe that procrastination is a big issue in their work habits (see Fig. 8).

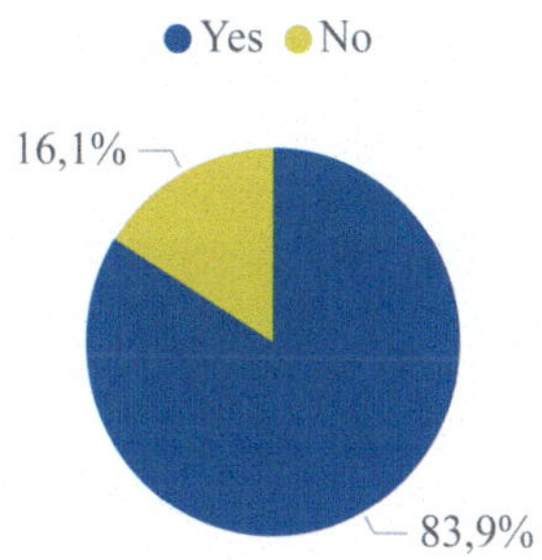


**Fig. 7** Question 5: do you intend to improve your own organization by creating and following (long-term) plans and continue to apply some of the project management processes in the future (yes/no)?

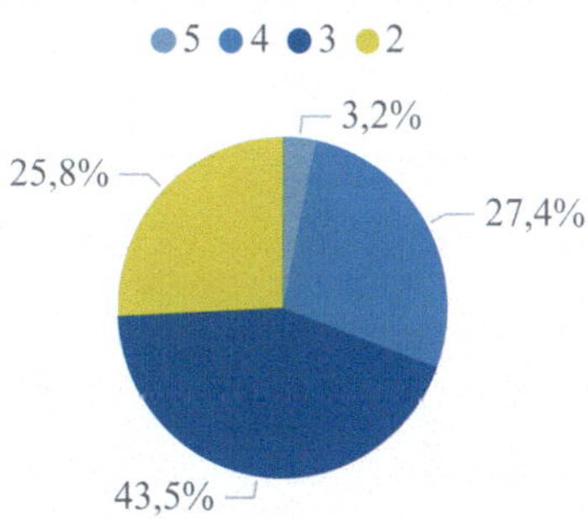


**Fig. 8** Question 6: do you resort to procrastination / postponing the fulfillment of obligations (5 = always to 1 = never)?

Students also found the course helpful in semestral tasks and useful for keeping the procrastination under control (see Fig. 9). They also believe that the data collected through the monitoring of the plan can help them to improve their self-organization in the future (see Fig. 10).

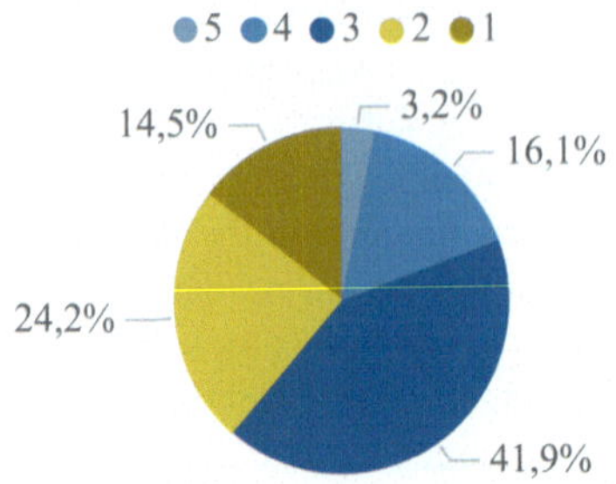


**Fig. 9** Question 7: how much did this assignment help you to complete the semester and control procrastination (5 = helped me a lot to 1 = didn't help me at all)?

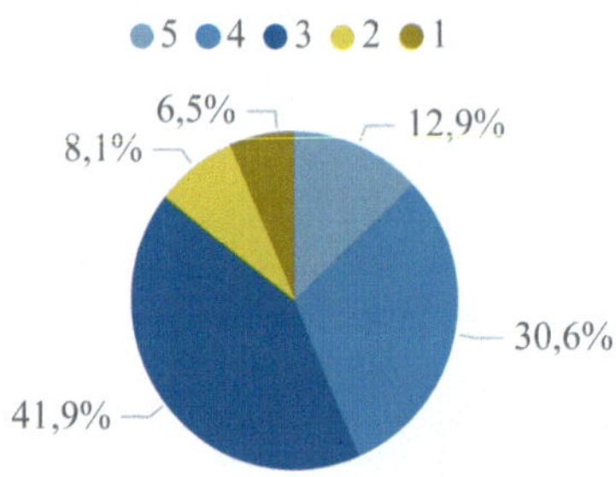


**Fig. 10** Question 8: did the collected data help you to detect ways to improve your organization and time planning (5 = helped me a lot to 1 = didn't help me at all)?

The majority of students (87%) point out the schedule as the most crucial process in PM (see Fig. 11). We should point out here that the PM processes are interdependent, and that the process importance depends solely on the characteristics of the specific project.

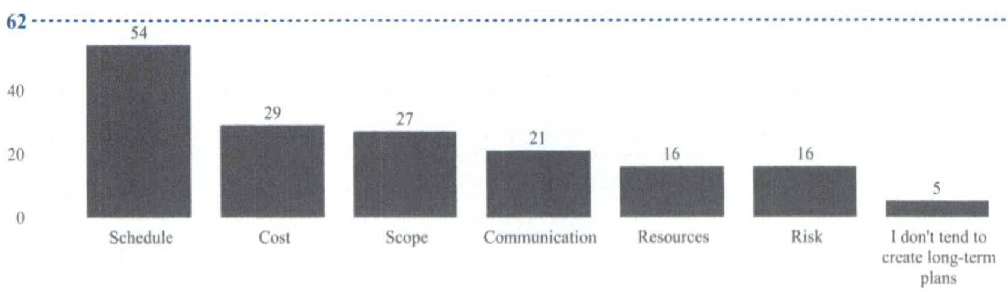


**Fig. 11** Question 9: please mark which processes you consider crucial for the successful planning of your activities (select multiple answers—presented in a diagram is the number of students who selected a specific process)

The last set of questions was about curriculum in general and workload distribution (see Figs. 12 and 13), wherein students also evaluated the statements that could help them achieve more evenly distributed workload throughout the semester (see Fig. 14).

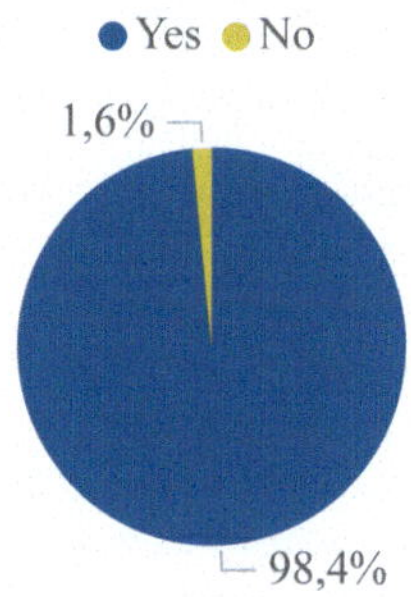


**Fig. 12** Question 10: should the distribution of students' workload and engagement in individual courses be proportional to ECTS credits?

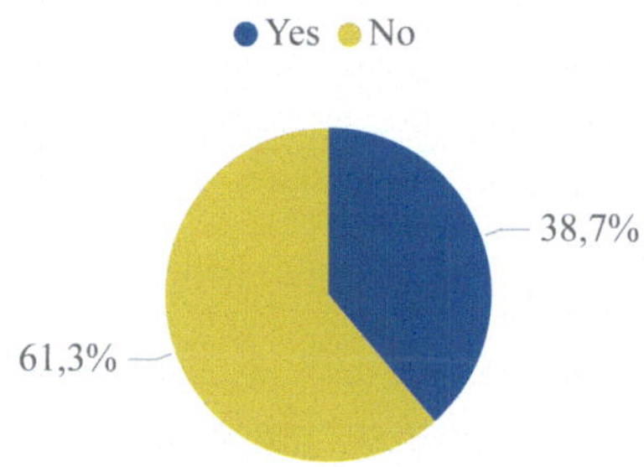


**Fig. 13** Question 11: do you think that it is possible to organize activities with a uniform workload throughout the entire semester?

In the last question, 40% of students stated that the PM course would be helpful to them in the earlier years of study, 54% of students were not sure, and only 6% of them stated that the PM course would be of no help to them at all.

## 4 Analysis

Microsoft Power BI engine was used for analytics and data visualization in real-time, with lots of options and filters, like evaluation of workload per mentor, per grade, and similar, which can be used as a basis for the improvement of the curriculum.

The reference value for the curriculum analysis is the average executed hours per student per ECTS. That value can be used to compare the total ECTS credit load of selected studios. We did assume that instructors in any particular studio would not have the same criteria or the same method. However, the differences in workload among courses and instructors turned out to be higher than expected (see Fig. 15).

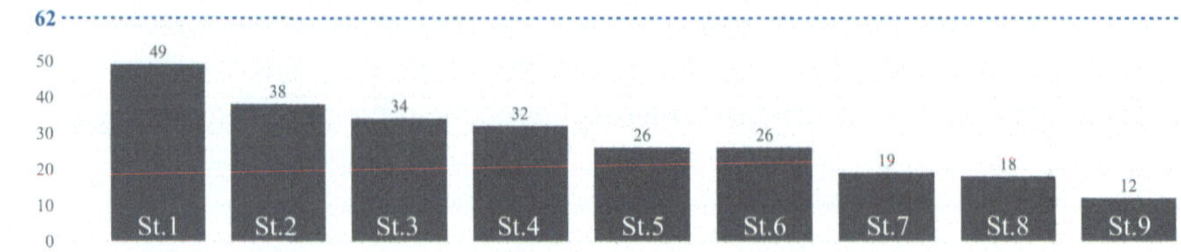


**Fig. 14** Question 12: what do you think could help you achieve a more evenly distributed workload throughout the semester (select multiple answers)?—presented in a diagram is the number of students who selected a specific statement/**St.1**—instructors' awareness of the content and obligations of students in other courses/**St.2**—uniform criteria for studio instructors/**St.3**—students' responsible approach to the fulfillment of obligations/**St.4**—higher structured courses with weekly set milestones/**St.5**—students avoiding procrastination/**St.6**—better and timely communication of students with the instructor and the key persons for the execution of the task / **St.7**—faster decision-making/**St.8**—teamworking in practical courses/**St.9**—investing more time in preparation and research as a basis for elaboration and the successful presentation of the project

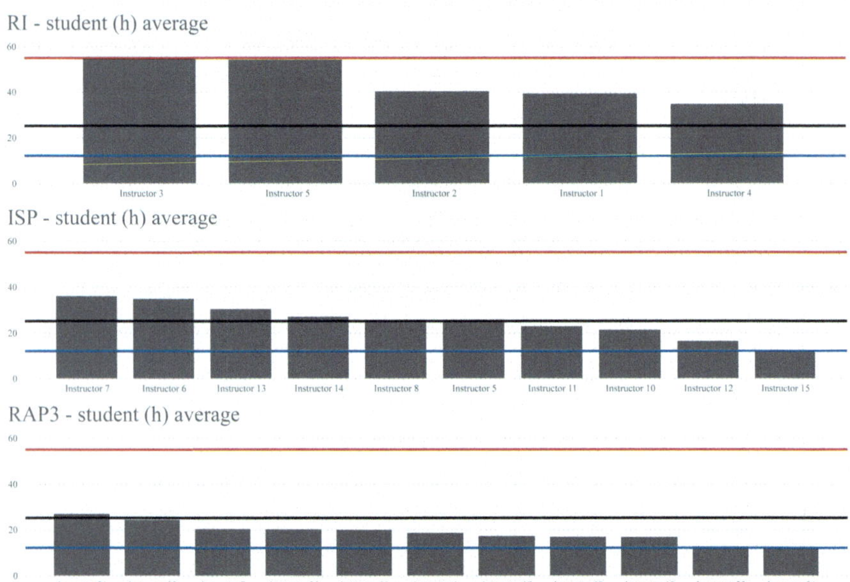


**Fig. 15** Average studio (RAP3—architectural/urban design; ISP—research; RI—interior design) executed workload per Instructor per ECTS. We compared the values with the minimum and maximum values of executed workload with the reference line representing the typically required workload of twenty-five (25) working hours per ECTS [8]—black. The red line represents the maximum measured workload of fifty-five (55) working hours per ECTS, and the blue line represents minimum measured workload of twelve (12) working hours per ECTS

In average values for executed workload per ECTS per studio, RAP3 studio has the lowest value of 19.1 h per ECTS, IPS studio has the value of 25.4 h per ECTS, and RI studio has the highest value of 42.8 h per ECTS.

The lowest average executed workload of twelve (12) hours per ECTS per instructor we see at ISP and RAP3 studio (Instructor 15 and Instructor 24) and the highest fifty-five (55) hours per ECTS we see at RI studio (Instructor 3). That proves the assumption that instructors have different expectations in terms of assignment complexity, typically out of balance with the ECTS workload.

In the students' work, the executed workload is larger in the second part of the semester, so students tend to think that they devote most of their time to their academic obligations throughout the entire semester. However, the average workload distributed throughout the fifteen (15) working weeks during the semester equals 6.97 work hours/day, as presented in Eq. (3), which is significantly less than the ECTS system assumption [8].

$$Avg\ 523\ work\ hours/(15 \times 5\ work\ days) = 6.97\ work\ hours/day \quad (3)$$

## 5 Conclusion

The leading role of architects is crucial in most project design tasks since other professions are not educated as broadly as architects are [9]. Consequently, the curricula should lead to the development of project management skills of architects in the integrated design environment. How we used to collaborate, make decisions, exchange information, and use and organize our time and resources in the past has been full of inefficiencies [10].

Students' reactions to the assignment are not always positive. An occurring myth often appears in students' conclusions that rigid planning affects creative freedom because it limits the time needed to process and reflect on an idea/concept. Despite that, based on survey data we conclude that:

- the introduction of PM courses in the earlier years of undergraduate study could positively affect the quality of the study.
- studio instructors are crucial for the development and adoption of productive work habits. The quality of the execution and the more engaging interest depends primarily on the enthusiasm and leading capabilities of the instructor.
- meticulous planning and guidance throughout the course are necessary in order to achieve a balanced distribution of workload.
- with careful planning, execution, and monitoring of the plan, it is possible to improve efficiency and speed up the decision-making process without affecting the quality of the coursework or design.

The knowledge of PM processes is a positive contribution to the students' efficiency in professional practice, instilling them with a sense of the need for a more evenly distributed workload. Based on the workload data analysis we conclude that:

- the distribution of total workload is uneven throughout the semester. If the current ECTS distribution should be used as a basis for future planning, then it is necessary to adjust the curriculum at the Master in Architecture graduate program at the University of Zagreb Faculty of Architecture to distribute the workload more evenly between the studios or, perhaps, change the methodology of conducting the studios altogether.

We look at the experience gained through this course as the source of knowledge since the integration of PM processes leads to the development of best practice qualities for future work environments. Since AIAS 2002 report [1] many new technologies and processes were introduced through digital transformation of building industry, such as Building Information Modelling (BIM), BIM Management, Generative Design, and Artificial Intelligence, which additionally could help students more easily overcome the myths stated in the report.

To avoid conflicting issues in communication and time planning, both students AND teachers need to learn project management skills in order to achieve satisfactory results in the future. That also implies the necessity to (re)educate educators in project management.

This case study carried at the University of Zagreb Faculty of Architecture, Master in Architecture graduate program, provides insight into the role of project management discipline in architectural education. The authors plan to undertake the analysis and survey conducted in this research, i.e., to repeat it in the earlier years of (undergraduate) study and to compare the findings.

## References

1. Koch A, Schwennsen K, Dutton T, Smith D (2002) The redesign of studio culture. A report of the AIAS studio culture task force. The American Institute of Architecture Students. Washington, DC
2. Leroy S (2009) Why is it so hard to do my work? The challenge of attention residue when switching between work tasks. Organ Behav Hum Decis Process 109(2009):168–181. https://doi.org/10.1016/j.obhdp.2009.04.002
3. Lueth OPL (2008) The architectural design studio as a learning environment: a qualitative exploration of architecture design student learning experiences in design studios from first- through fourth-year. Retrospective Thesis and Dissertations, Iowa State University, https://lib.dr.iastate.edu/rtd/15788, last accessed 2023/04/20
4. al Khalifa F (2017) Autonomy in architectural education: a Bahraini perspective. Int J Architect Res Archnet-IJAR 11(2):24–33
5. Grow GO (1991/1996) Teaching learners to be self-directed. Adult Educ Quart 41(3):125–149
6. Khan AZ, Vandevyvere H, Allacker K (2013) Design for the ecological age: rethinking the role of sustainability in architectural education. J Architect Educ 67(2):175–185. https://doi.org/10.1080/10464883.2013.817155
7. Project Management Institute (2017) A guide to the project management body of knowledge (PMBOK guide), 6th edn. Project Management Institute Inc., Newtown Square
8. European Commission, Directorate-General for Education, Youth, Sport and Culture (2015) ECTS Users' Guide 2015. Publications Office of the European Union, Luxembourg. https://data.europa.eu/doi/10.2766/87192
9. Bachman LR (2003) Integrated buildings: the systems basis of architecture. Wiley, Hoboken
10. Kocaturk T, Kiviniemi A (2013) Challenges of integrating BIM in architectural education. In: eCAADe 2013 conference. eCAADe and Faculty of Architecture, Delft University of Technology, Delft, Netherlands, pp 465–473

# Proximity Between Radical Pedagogies and Utopian Imagination: A Paradigm Shift in Architectural Education

Inês Nascimento(✉) and Alexandra Paio

Iscte—University Institute of Lisbon, DINÂMIA'CET, Lisbon, Portugal
{ines_nascimento,alexandra.paio}@iscte-iul.pt

**Abstract.** The paradigms of the past are being questioned, and the time has come for the architect to equip himself with his most valuable and unique qualities—the ability to provide a vision of a new world and to convey hope. Today, the proximity between the architect and society is remembered, and the Academia has been called to participate in an innovative process. However, Academia is facing a crisis, and the reflection on its repositioning points to the urgency of a paradigm shift. Curiously, the potential of radical and utopian imagination has been recently defended as a response to the current global challenges in the European context. In the 1960s and 1970s, the approximation of these two concepts with Architectural education gave rise to pedagogical experiences that marked the last real innovation in this field. This type of approach and the importance of its return have found resonance in recent theories, sowing the seeds for future pedagogical ecologies in a time when the future of pedagogical radicalism faces new challenges. In this context, and because it is fundamental to recognize the power of this mechanisms, a provocative dialogue between them emerges, warning us to the need for their rooting in the training and practice of Portuguese architects. Therefore, this paper will present a critical reflection through a literature review and original interviews, in order to describe a paradigm shift in Architectural education in Portugal, offering a new perspective on radical and utopian concepts. This reflection underscores the urgency of their urgent (re)approximation by understanding and validating their potential.

**Keywords:** Architectural education · Utopian imagination · Utopia · Radical pedagogies · Utopian pedagogies

## 1 Introduction: The Role of the Radical and Utopian Architect

The paradigms of the past are now being questioned. At a time when society lives a moment of distrust and disillusionment, that has taken away the dimension of dream and hope, utopia reveals itself as the fundamental element for us to overcome it [1]. We now have a desire for rebirth, we aspire to a new vision of the world, and the time has come for the architect to equip himself with his most valuable and unique qualities—the ability to provide a vision of a new world and to convey hope [2]. In this


M. Blanco Lage et al. (eds.), *Towards a New European Bauhaus—Challenges in Design Education*,
https://doi.org/10.1007/978-3-031-40188-6_11

framework, for architects, visualizing the physical space is no longer enough. Architects are now required to produce new narratives about new ways of operating, within new social landscapes [3], by (re)formulating questions, provoking actions or reactions, and sculpting new panoramas of Architecture [4].

Today, the proximity between the architect and society is remembered by Ursula Von Der Leyen when presenting the creation of the New European Bauhaus (NEB) as a response to the need to recover from the effects of the COVID-19 pandemic in Europe [5], and the Academia has been called to participate in this innovative process. Seen as the utopia that Europe needed, this holistic concept reveals itself as a "project of hope" and a "creative and interdisciplinary initiative to design future ways of life" [5].

Utopia should be the engine of these radical processes/developments, and pragmatic utopias need radical experimentation... but is Architectural education up to the challenge? The search for answers around such matters creates a dialogue that opens the way to trace a path towards the objectives pointed out by the NEB, demonstrating and emphasizing the potential of these concepts in various contexts that fall under the subject of the research. Therefore, the concerns that surround this article[1] come in line with the guidelines presented by the High-Level Roundtable in the *New European Bauhaus Concept Paper* (2021), in particular the point 2.4. Innovation and learning-by-doing as a key part of new models of design education at all levels within Europe and beyond.

## 2 Problem and Opportunity: Teaching Architecture Today

### 2.1 Dystopian Reality

The Academia, which has been called to participate in this innovative process (NEB), is currently facing a crisis [6]. Alarmingly, Universities are becoming hyperspecialized, and students build their own curriculum through a "menu of possibilities" [7], even when the problems we face today are systemic and require multidisciplinary responses, rather than an alliance of disciplines [8]. We want our students to be innovative and to transform the future of the architect's profession [9, 10], but we continue to offer ingredients that prioritize market-driven goals and do not encourage the freedom to imagine alternative possibilities [7].

In another point of view, although "design students remain some of the most forward-thinking individuals", there are still a long way to go when it comes to radicalizing Academia, and "a radical approach is not in materials nor the content of the work produced, but by challenging one's teaching practice" [11].

Today, Architectural pedagogies are described as "static, boring and dull" [7]. In Portugal, Architectural education is considered outdated [12]. Both institutions and the profession—that have a retrograde mentality, based on models and processes that do not address current challenges [13]—are in crisis [14], and reflecting on their repositioning highlights the urgency of a paradigm shift. But have utopians lost the ability to imagine radically different futures [15], even for education? Or are we no longer capable of dreaming, let alone making those dreams come true?

[1] This article is part of an ongoing Ph.D. research project.

### 2.2 Utopian Horizon

"There is no utopia without Architecture" [16], and although its potential in teaching the discipline has already been recognized [17–22], this concept remains dormant in the teaching of Architecture [18]. In Portugal, despite some occasional theoretical references, this absence is noticeable. However, the unexpected (re)emergence of counterculture experimental teaching practices in Portugal—by the hand of Academia professors themselves, outside the mandatory curricula of Architecture courses—, critical of the limitations of current Portuguese Architectural education, raises doubt about whether it is fulfilling its mission [13].

In this context, and because it is fundamental to recognize the power of radical and utopian mechanisms, a provocative dialogue between them emerges, warning us to the need for their rooting into the training and practice of Portuguese architects. Therefore, this paper will present a critical reflection through a literature review and original interviews, aiming to describe a paradigm shift in Architectural education in Portugal through a new perspective on radical and utopian concepts—which have become distant from both each other and from Universities. This reflection underscores the urgency of their (re)approximation by understanding and validating their potential.

## 3 Methodology: Connections, Dialogues and Visions

This qualitative methodology aims to shed some light on the undeniable connection between utopian imagination and radical pedagogies through a literature review on the subject and by inviting four experts to share their perspectives on utopian imagination and Architectural education. In line, it is expected that the framework of the literature review introduces theoretically the concepts under observation which are proponed to be clarified, while the interviews reveal points of agreement and disagreement that might demonstrate the presence/absence of utopian imagination in Architectural education.

Thus, for this research, an individual sample as well as semi-structured and not in-person interviews, with open and mixed answers, were conducted. The interviewees were chosen based on their practical involvement, knowledge of the subject in question, and mastery of the concept, being relevant for providing both national and international perspectives.

Four utopians were invited, all of whom are professors in the Academia, to look and reflect upon the concept of utopia throughout their route. They are: the Portuguese teacher and philosopher with a transdisciplinary background, Jacinto Rodrigues; the architect, professor, and main international reference on the topic, Nathaniel Coleman; the Portuguese architect, editor, and professor Pedro Leão Neto; the Portuguese architect, curator, author and award-winning professor, Pedro Campos Costa.

# 4 Results: Proximity Between Radical Pedagogies and Utopian Imagination, a Contribute for a New Paradigm in Architectural Education

## 4.1 A Theoretical Overview

> (…) radical utopianism offers a form of resistance to dominant constructions of reality and our complicity with them. Radical utopianism confronts 'realism' with possibility. (...) The role of the teaching of desire is to make change conceivable, while the role of educated hope is to believe that it is possible. [23]

In the 1960s and 1970s, the approximation of the concepts of radical and utopia with Architectural education resulted in pedagogical experiments that marked the last significant and real innovation in this field [7]. These pedagogical experiments, which played "a crucial role in the formation of the Architecture's speech and practice during the second half of the twentieth century" [7], are known as Radical Architecture practices, and are radical in the purest sense of root—*radix*. These pedagogies questioned the principles and foundations of Architecture, with a speech—now asleep—that appeared through a contested Architecture and the questioning of the teaching matrix. Thus, education assumed itself as a vehicle for subversive actions and for fostering new alternative visions generated through progressive pedagogical initiatives [7].

This type of approach and the importance of its return has found resonance in recent theories, sowing the seeds for future pedagogical ecologies, at a time when the future of pedagogical radicalism faces new challenges. The potential of radical and utopian imagination has been reaffirmed as relevant and recently defended as a response to the current global challenges, such as the Sustainable Development Goals (SDGs) in the European context [24], to the barriers imposed by COVID 19 [25], being part of the debate on Architecture's ability to respond to rapid changes on a global scale [*What is Radical Today*, 2019], gaining dimension both at teaching [26, 27] and research [28, 29] levels.

However, a lingering question remains unanswered: Can these concepts truly be apart from each other when discussing pedagogies?

## 4.2 A Pedagogical Perspective

> One of the primary tasks of radical liberating critical pedagogy is to work on the legitimacy of the ethical political dream of overcoming an unjust reality. It is to work on the genuineness of this struggle and the possibility of change, it's worth saying, it is to work against the force of the dominant fatalist ideology, which encourages the immobility of the oppressed and their accommodation to the unjust reality, necessary for the movement of the dominators. [30]

The concept of *Utopian Pedagogies* was coined by Paulo Freire in 1970, being described as "a process of denunciation and annunciation; a critical interrogation of the present situation coupled with 'a utopian vision of both the man and the world'" [31]. Utopian pedagogy is meant "to guide and direct a collective and collaborative process of memory through a process of convocation and extrapolation in order to produce something new; a substantive utopian vision" [31]. As for its' method, applying utopian imagination to transform the present into a better place "is the only way to ensure that our future world provides adequate pleasure to support a good life for all" [32].

Regarding *Radical Pedagogies,* this concept can be understood according to two definitions: "the notion that secures the changing views about cognitive capacities, conditions, and factors forming human exposure that indicates the system of pedagogical measures and solutions", as well as "an uncompromising commitment to achieve real transformation in accordance with the views on the development of socio-political, ideological and economic spheres" [33]. Furthermore, this type of pedagogies is presented as a way to address contemporary challenges by "bringing to light a panorama of past attempts to subvert the status quo and reveals work to build upon and ideas wanting to be taken again", thus highlighting today's need for a reform in education [34].

In short, radical pedagogies with utopian imagination prioritize social change. With this approach, educators encourage students to challenge the status quo and work toward creating a more just and egalitarian society, by helping them to imagine a better future and work to make it a reality. These pedagogies offer an inspiring and motivating vision of the future, stimulating creativity and innovation, although they might be perceived as unrealistic and neglectful of the current reality. Utopian pedagogies, in turn, give importance to the collective dream and are seen as a more idealistic approach, believing in the proposal of an idealized and transforming vision of reality, allowing experimentation and the creation of concrete alternatives. On the other hand, they can be considered impractical or not applicable to current realities, as they may overlook the concrete difficulties and challenges for their implementation.

However, both practices share a student-centered approach, with the aim of developing critical, creative, and emancipatory skills, as well as fostering engagement with social, political, and cultural issues to raise awareness of social responsibility. They emphasize interdisciplinarity, collaboration, experimentation, reflection, and practice, as well as the development of technical, technological, communication and critical thinking skills. While radical pedagogies can be more pragmatic and solution-oriented—with a focus on activism, political engagement, practice-based learning, and studio/laboratory approaches—, utopian pedagogies may be more technical/conventional and oriented towards the labor market and professional practice, emphasizing the importance of utopian imagination and speculative thinking. In summary, utopian pedagogies are future-oriented and highlight utopian imagination as a tool for social change, while radical pedagogies focus on a critical analysis of existing social and political issues. Despite receiving different denominations, both pedagogical practices embrace the concepts of radical and utopia at their core (eventually only with different relative proportions).

### 4.3 The (*Non*) Place of Utopia in Architectural Education

**Through the Interviewees' Lens.** Utopia, or should I say, hope [35–37]? Often interpreted as a manifestation "of the principle of hope in which future elements allow a dynamic over the stagnant past", utopia possesses two characteristics: a "constructive and critical capacity" and the "ability to bring hope and draw a path capable of solving problems and improving the world, whether on a global or local scale, either individually or as a collective" [37]. This concept is seen as "essential for human life, to carry desires, to elevate the dream, to build life" [38]. However, "anything claimed as the result of, or harboring, utopian imagination will not constitute utopia, which although necessarily desired, is impossible to achieve, other than in fragments" [36].

From a methodological perspective, utopia can serve as a facilitator for the "cultivation of utopian imaginaries that are grounded—pragmatic, concrete, bounded to use and bodily experience and that the emphasis is on resilience, survival, reaching beyond the limits of the given, and to cultivating a utopian mindset that desires utopia while accepting the impossibility of its manifestation (in any total way)" [36]. In this context, it aims to "free ourselves and then confront each other", being compared to "a process to free students from the excessive weight of standards" [38].

Architectural education requires a prospective notion, supported by "transdisciplinary support, which is not in line with the usual futurology (theory of scenarios) almost always limited by past causes, but which invents inspired poetics, involving past and future in the same reality" [35]. It needs an approach that "concentrates on works, use, the concrete, and the everyday (actual bodily experiences), rather than on products, exchanges, abstractions and idealizations (aesthetic fantasies)" [36]. Architects strive "to explore and idealize new ways of using the space to give an unconventional response to the program, trying to improve the lives and work of the people who make use of these spaces" [37], and their proposals can embody "a bigger or smaller degree of utopian imagination", depending on the process behind "an innovative program" [37]. However, can we truly say that utopia is present in today's Architectural education?

Although being "essential that the utopian element should be present throughout the Academia, so that the desire for transformation and the need for social improvement allow, in a participatory way" [35], it is agreed that "there is an absence of utopia in conservative mentalities, both in educators and in students" [35], which raises the concern that "Architectural education is decisively impoverished by the near total prohibition against utopian imaginaries" [36].

In essence, "if utopia, aligned with hope, is always a risk, bound up with acknowledgement of something missing and desires to respond, professional Architecture education, disciplined according to the demands of professional practice and the marketplace, must limit considerations of utopia (or much else) beyond basic problem solving, aesthetics, exchange, novelty, and technological concerns" [36].

## 5 Conclusion: A Framework for a Paradigm Shift in Architectural Education Towards a New European Bauhaus

Radical and utopian approaches to Architectural education have a core essence and have yielded promising responses to the challenges we face today. The proximity between these concepts and pedagogical practices has sparked change through a bottom-up system that emerged when both students and teachers shared the genuine desire to alter reality. Therefore, these realistic utopias or radical experiments—which are idealistic in their dreams (radical), but pragmatic in their actions (in experimentation)—are here evoked, for being exactly what contemporary utopians need: to "experience the future in the present" [15].

Apart from labels or distinctions, the outcomes of these "radical experimentations" have provided a "clear blueprint of the ethical imperatives (…) that drove the impulse for change" [34], with utopia being the common thread among them. "The role of story-making in radical pedagogy has been stressed many times before" [31], and it's our duty to construct "visions of alternative ways of being, recognizing that substantive programmatic visions of the future (blueprints) are needed in order to inspire and guide transformative hope and action" [31]. Radical pedagogies challenge students to go beyond the boundaries of standard/conventional education while envision a better society—by fostering critical thinking, experimentation, cooperation, and questioning of the status quo—, qualities especially relevant in the context of the NEB, since it emphasizes the need for a paradigm shift towards a more ecological, inclusive, and human-centered design.

In summary, it can be said that utopian imagination has long been a vital component of radical pedagogies in Architectural education. These pedagogies go beyond typical classroom instruction, engaging students in participatory, interactive, and reflective learning experiences, as well as empowering them to explore alternative ways of thinking and designing. Utopian imagination plays a pivotal role in shaping the future of Architectural education, enabling students to envison and build new paradigms for the practice and the profession. By challenging the status quo, pushing future architects to think/act beyond the boundaries of conventional knowledge/practice, and addressing contemporary socio-environmental concerns, these approaches have the potential to lead us toward a more just, equitable, and sustainable future.

However, in the conducted interviews, we are left with a warning: "the absence of precise considerations of utopia as a method in relation to Architecture and the city represents a significant absence, or gap, in teaching of Architecture" [36]. In Portugal, after analyzing these interviews—even though opinions about the presence/absence of utopian imagination in Architectural education may differ—, it is widely agreed that there are deficiencies in architects education. Also, "it is through the design of the project that is essentially the basis for a new teaching of Architecture and urbanism and thus, those can overcome the non-places of this dystopian society, building a new reality, a new place that has become an achievable utopia—eutopia—i.e., achievable aspirations (achievable inspiration, imagination and intuition)" [35]. Nevertheless, more important than the absence or (meager) presence of these concepts in the Academia is the consensus on the need of its existence in the education of both the citizen and the architect.

Above all, it is clear that "teaching is a labyrinth" and the obligation of both students and educators is to "be together in the discovery" [38]. In doing so, we can "imagine and build together a sustainable and inclusive future that is beautiful for our eyes, minds, and souls" [5]. In conclusion, I believe that the practice of radical pedagogies carries within the seed of utopia, strengthening the possibility of cultivating a new paradigm in Architectural education and nurturing the growth of a new generation of architects capable of transforming the world.

## References

1. Santiago L, Melânio P (2014) Pedro Bandeira: Proposta de Relocalização da Ponte D. Maria Pia no Porto. Arqa: Arquitetura e Arte 112:104–107
2. Stead N, Oldfield P, Knapp C (2020) Designing hope: the role of architecture education in the climate crisis. Architecture Australia. https://architectureau.com/articles. Last accessed 25 Feb 2023
3. Cutieru A (2020) Arquitetura especulativa: quais são as versões contemporâneas do pensamento radical dos anos 60? ArchDaily. https://www.archdaily.com. Last accessed 27 Feb 2023
4. Fagundes D (2016) Micro-utopias: Notas sobre radical e critical design. ArchDaily. https://www.archdaily.com. Last accessed 27 Feb 2023
5. New European Bauhaus Homepage. https://new-european-bauhaus.europa.eu/index_en. Last accessed 21 Feb 2023
6. Pirondi C (2017) O ensino da arquitetura ou a crise silenciosa. ArchDaily. https://www.archdaily.com. Last accessed 25 Feb 2023
7. Colomina B (2014) Towards a radical pedagogy: lecture. Strelka Institute for Media, Architecture and Design. https://www.youtube.com. Last accessed 23 Feb 2023
8. Vieira F (2020) Os Engenheiros São Os Utópicos Ideais. Há Engenharia Fora Da Caixa. Ordem Dos Engenheiros Região Norte. https://haengenharia.pt/noticias. Last accessed 20 Feb 2023
9. Salama A (2021) Tranformative pedagogy in architecture and urbanism, 1st edn. Routledge, London. https://doi.org/10.4324/9781003140047
10. Jedenov K, Afonso F (2017) Crisis as the new normal: preparing architecture students for uncertainty in social, economic and ecological dynamics. Charrette 4(1):40–57
11. Abdulla D (2019) Radicalise me. In: Modes of criticism 4—radical pedagogy. Onomatopee, pp 25–32. ISBN: 978-9493148130
12. Beirão J N (2017) Sobre o ensino da Arquitectura e o futuro profissional do Arquitecto: O papel da Arquitectura nas sociedades criativas. J-A Jornal Arquitectos: Crónicas. http://www.jornalarquitectos.pt. Last accessed 23 Feb 2023
13. Correia L (2018) Pedagogias de Verão: outros modos de ensinar arquitectura. J-A Jornal Arquitectos 256. http://www.jornalarquitectos.pt. Last accessed 20 Feb 2023
14. Baía P, Labastida M (2014) Onde Pára o Ensino. J-A Jornal Arquitectos 251. http://www.jornalarquitectos.pt. Last accessed 22 Feb 2023
15. Vieira F (2021) Utopias Realistas para a Construção Social Europeia. In: Vasconcelos A (ed) Utopias Europeias : O poder da imaginação e os imperativos do futuro. Fundação de Serralves, Porto, pp 41–53
16. Coleman N (2005) Utopias and architecture, 1st edn. Routledge, London. https://doi.org/10.4324/9780203536872
17. Araújo J, Araújo A F (2006) Utopia e educação. Revista Portuguesa de Pedagogia 40–1:95–117. https://doi.org/10.14195/1647-8614_40-1_4

18. Coleman N (2012) Utopic pedagogies : alternatives to degenerate architecture. Utopian Stud 23(2):314–351. https://doi.org/10.5325/utopianstudies.23.2.0314
19. Coleman N (2014) Architecture and dissidence: Utopia as method. Architect Culture 2(1):44–58. https://doi.org/10.2752/175145214X13796096691481
20. Ganjavie A (2012) Role of utopia for design of future cities: utopia in urban planning literature. Stud Lit Lang 5(3):10–19. https://doi.org/10.3968/j.sll.1923156320120503.1922
21. Hammond C (2017) Hope, utopia and creativity in higher education: pedagogical tactics for alternative futures, 1st edn. Bloomsbury, London. https://doi.org/10.1080/07294360.2018.1467589
22. Yurtkuran S, Kırlı G, Taneli Y (2013) An Innovative approach in architectural education: designing a utopia. Procedia Soc Behav Sci 89:821–829. https://doi.org/10.1016/J.SBSPRO.2013.08.939
23. Storey J (2019) Radical utopianism and cultural studies: on refusing to be realistic, 1st edn. Routledge, London. https://doi.org/10.4324/9781315201580
24. Vasconcelos A, Vieira F, Ribeiro R J, Hatoum M, Ribeiro F M, Relvas J, Rodrigues A, Marcelo G, Moreira J, Jalali C, Valente I M F, Leitão F M A, Azevedo A, Granja I, Lourenço P (2021) Utopias para Europeus. In: Utopias Europeias: O poder da imaginação e os imperativos do futuro, 1st edn. Fundação de Serralves, Porto. ISBN: 9789727393909
25. Lima P V (2021) Pedro Leão Neto: "Ter uma visão da utopia, não como um conceito cristalizado na perfeição, mas mais como uma visão futura." In: U.Porto Press. https://www.up.pt/press/pedro-leao-neto-entrevista. Last accessed 20 Feb 2023
26. Harriss H, Froud D (2015) Radical pedagogies: architectural education and the British tradition, 1st edn. RIBA, London. ISBN: 9781859465837
27. Mitrović I, Auger J, Hanna J, Helgason I (2021) Beyond speculative design: past-present-future. University of Split, Croatia. ISBN: 9789536617562
28. Colomina B, Kotsioris E, Galán I G, Maria-Meister A (2015) The radical pedagogies project. In: Volume #45: Learning. https://archis.org/volume/the-radical-pedagogies-project/. Last accessed 28 Feb 2023. ISBN: 9789077966457
29. Mitrović I, Šuran O (2016) Speculative : post-design practice or new utopia? Ministry of Culture of the Republic of Croatia Designers Association, Croatia. ISBN: 9789536778157
30. Freire P, Freire N, Ferreira de Oliveira W (2014) Pedagogia da Solidariedade, 2nd edn. Paz e Terra, Rio de Janeiro. ISBN: 9786555480313
31. Webb D (2017) Educational archaeology and the practice of utopian pedagogy. Pedagogy Culture Soc 25(2):551–566. https://doi.org/10.1080/14681366.2017.1291534
32. Waterman T (2018) Making Meaning: utopian method for minds, bodies, and media in architectural design. Open Libr Human 4(1):4. https://doi.org/10.16995/OLH.109
33. Fedotova O, Nikolaeva E (2015) Radical Pedagogy: theoretical concept and/or alternative practice? Procedia Soc Behav Sci 186:785–789. https://doi.org/10.1016/J.SBSPRO.2015.04.010
34. Colomina B, Gálan IG, Kotsioris E, Meister AM (2022) Introduction. In: Colomina B, Gálan IG, Kotsioris E, Meister AM (eds) Radical pedagogies, 1st edn. The MIT Press, Cambridge, pp 11–20. ISBN: 9780262543385
35. Rodrigues J (2022) Interview and translation by the author, May 2022
36. Coleman N (2022) Interview by the author, May 2022
37. Leão Neto P (2022) Interview and translation by the author, May 2022
38. Campos Costa P (2022) Interview and translation by the author, May 2022

# About Education in Architecture: Towards an Integrative Pedagogy in the Teaching of Communication Strategies for Architectural Design and Photography

Pedro Leão Neto(✉)

Faculty of Architecture of University of Porto (FAUP), Porto, Portugal
pneto@arq.up.pt

**Abstract.** This paper is based on the didactic experience and refinement of three curricular units (CUs) in FAUP, which are *Photography and Design Communication*, articulated with *Photography of Architecture, City and Territory*. In addition, we consolidate research work on the uses of diverse visual strategies and representation methods for communicating architecture, city and territory, an integrative pedagogical approach with a special focus on photography as a transversal media within these areas of study. The text describes a collaborative and dynamic environment allowing for an informed appraisal of architecture and city space, understanding the diverse forms of its appropriation and giving students the possibility to participate in the discussion of the transformation of these territories. The teaching process adopted encourages the interaction between teachers, researchers and students, leading to a dynamic of discovery built collectively. All this favours a dialogical learning process, exploring the multifaceted richness of the territories under study and the way in which public spaces are transformed and architecture is designed. It will be shown the significant connection that is established between the CUs curriculum and the research developed in the Architecture, Art and Image research group. It is believed that this articulation between CUs and research coming from R&D project teams reinforces and provides conditions to stimulate the beginning of activities related with investigation, as well as the development of students' critical reflexive competences, creativity and autonomy. Moreover, we also ensure that the articulated work of the CUs with research is open to various national and international institutions.

**Keywords:** Integrative · Pedagogical · Architecture · Visual representation strategies · Photography


M. Blanco Lage et al. (eds.), *Towards a New European Bauhaus—Challenges in Design Education*,
https://doi.org/10.1007/978-3-031-40188-6_12

## 1 The Theoretical Universe of the Curricular Units (CUs) in FAUP Speaking of the Authors, Works and Theories that Constitute the Foundation of Their Curriculum

The uses of photography as a design tool and research instrument in the fields of architecture, city and territory is based, in this paper, on the CUs didactic experience and on the investigation undertaken in AAI research group, namely Visual Space of Change, an interdisciplinary research project combining contemporary photography and visual documentation with georeferencing [1].

Visual Spaces of Change (VSC) attempts to improve the understanding and uses of photography as a research tool in the fields of art, architecture and image, by implementing contemporary photographic projects in a direct relation with their locations and general public, which can contribute to a new understanding and the activation and sharing of individual and collective memories regarding uses of existing architectures.

From this perspective, the use of photography can contribute to the creation of a knowledge-enabling environment that allows a specific study of architectural forms and spatial realities, its transformations and appropriations, rendering visible aspects of urban spaces where people socialize and interact which would be difficult to perceive without the use of images and photography.

### 1.1 Communication and Representation Process in Architectural Design: Photography and Drawing

Regarding the CUs of the first cycle FCAAD I, II, the students explore at the very beginning of the Communication and Representation Process in Architecture Design the potential of the visual photographic narrative in a book/booklet support. This means, creating through photography visual narratives that explore and communicate, via an architectural and fictional point of view, the real space and their experiences, telling us a story about the spaces through a journey composed of successive moments, ones that are quite close to a real experience thinking in Le Corbusier's architectural promenade [2]. The weight and intentionality contained in the act of photographing proposes that photographic images can be instruments of representation and conception in the practice and discipline of architecture.

The production of the book is a generator for several collaborative work meetings and a mental instrument and artifact capable of materializing certain knowledge, namely to support ideas of photography in architecture and art [3] (Fig. 1).

In a later phase, which is the second semester, it is possible to deepen the development of the visual narrative previously created and to enrich the Process of Communication and Representation of Architecture Project. The photography component is thus articulated with the photomontage component, with the aim of ideation and exploration of architectural ideas and construction of an atlas/image laboratory [4]. Students must be aware of the relationship between perception and vision, exploring these two universes, trying to show how they interconnect and influence our relationship and understanding of the real world [5].

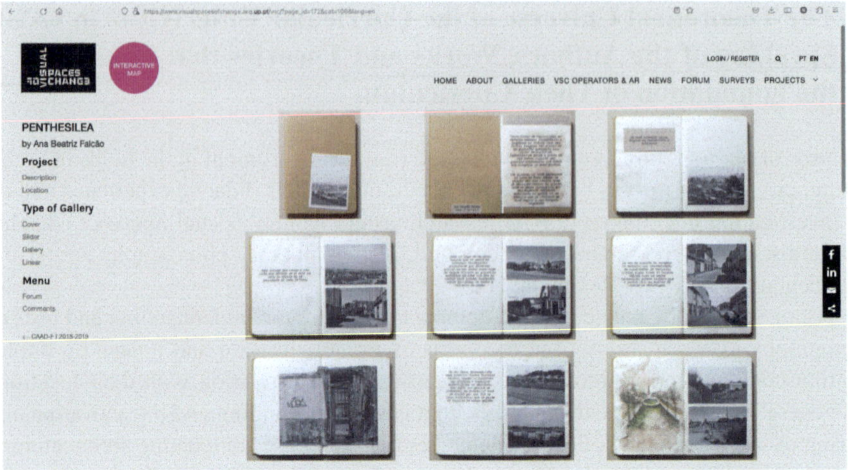


**Fig. 1** Project of student Ana Beatriz Falcão in book format in the teaching gallery of visual spaces of change platform

Through these approaches, the articulation of photography with other representation methods and analogue or digital supports of image are explored (Fig. 2). The combined use of these resources addresses particularly useful aspects in the study of architectural objects and urban landscapes, providing the students with essential research tools and communication strategies as seen in the work of, e.g., Martino Stierli; Luca Galofaro; Klaus Bollinger, Florian Medicus and Kiesler Privatstiftung Wien; Eric Margolis and Luc Pauwels; Theo Van Leeuwen and Carey Jewitt and Gillian Rose [6].

**Fig. 2** Production of synthesis images that combine photography with 3D modelling, using photomontage and collage techniques, allowing the exploration of various design ideas

## 1.2 Photography of Architecture, City and Territory

The aim of the second cycle CU is to deepen the study and practice acquired in FCAAD I, II on the use of photography and its potential to question and problematize the universe of Architecture, City and Territory.

Thus, students are led to think and use photography as a free investigation tool with the purpose of discovering new perspectives about the public space and its architecture. This potential of photography to rediscover new spatial and architectural realities and thus influence the way we perceive and understand our spaces and culture is something that is present in the work of many authors and in very different contexts. This interests us, above all, because a didactic of this kind leads students to explore and problematize the potential of photography and image as a way to represent and question the urban reality and allows them to achieve, during the CU time, a good artistic and technical domain about image and photography (Fig. 3).

**Fig. 3** Students study of composition, colour and light dimension for his photography project

Above all we advocate using photography as a critical territory [7], recognizing the importance of the work coming from several visual artists who use photography not as a mode of documenting reality but rather a mode of engaging with its multiple dimensions (Fig. 3).

Photography has the ability to move between reality and fiction, rendering visible innovative visions and introducing new links between realistic representations, fictional worlds and symbolic meanings. These features are identifiable in the work of several photographers, for example: Filip Dujardin´s work addressing the potential of digital manipulation of images to represent new spatial forms and architectures and creating new imaginary spaces [8]; Bas Princen´s work rendering visible transformations of urban landscapes that have not yet occurred and altering notions of time [9]; Paulo Catrica´s work on landscape and architecture that constitutes a synthesis and insightful perspective of a recent past, and present realities of hybrid city spaces [10]; Paolo Rosselli´s and Philip Schaerer´s work creating exploratory and fictional images using specialized digital montage techniques [11]. All this reveals in many ways how an intentional use of photography can provide us with a vocabulary specific to a place that can germinate spatial or formal ideas and that an intentional photograph can have embedded languages which can actually generate ideas for an architectural project [12].

It is believed that we offer students a pedagogical space that leads them to build their own social consciousness, always linked to space, through the knowledge of the multifaceted wealth of perspectives that exist on these themes.

## 2 Methodology Adopted

Besides the open classes that allowed the critical debate around diverse visual strategies and works from different authors and students, we have also promoted a series of open talks, public presentations and exhibitions of contemporary photography projects. These initiatives intended to broaden the discussion about how architecture transforms and is transformed by trends and ways of living, using as its subject of study Porto's Metropolitan Area (Fig. 4).

**Fig. 4** Author explaining his photography project to students and public in metro station

The methodology developed in this pedagogical experiment builds up on previous research combining blended learning and e-learning with visual research methods and photographic techniques that enable students to manage the whole process of conception, development and implementation of photography projects in a collaborative learning environment. Furthermore, besides this former research this paper also reinforces the results of the first case study implemented within the research project Visual Spaces of Change [13], which means besides other things that we developed further the former open public and pedagogical experiment by expanding the communication and interaction capacities between students and diverse academic and non-academic authors, all related with the universe of design communication and photography through open classes and other initiatives, opening academia to society and allowing the creation of synergies between them.

## 3 Conclusive and Prospective Synthesis: Teaching, Research, Internationalization and Opening of the School of Architecture and Porto University to Society

It is believed that we offer students a pedagogical space that leads them to build their own social consciousness, always linked to space, through the knowledge of the multifaceted wealth of perspectives that exist on these themes. In fact, this is the result of the integrative pedagogy of our program teaching programme where a significant connection of the CUs to the research focused on Architecture, Art and Image (AAI), which is the nuclear universe of all the research developed at AAI. Thus, this pedagogical space is where the various projects are integrated as well as several pedagogical actions related to them as open classes and workshops, as well as the active use of the online VSC collaborative and communication platform.

We believe that this close articulation between CUs, open classes and research reinforces and provides conditions to stimulate the beginning of scientific activity and the development of critical sense, creativity and autonomy of students. In fact, all these actions allow a closer approach between research and teaching and the integration of students from FAUP's curricular units and from other partner institutions linked to the universe of Architecture, Art and Image, through academic internships, reinforcing and giving conditions to stimulate the beginning of the scientific activity and the development of critical sense, creativity and autonomy of the students through their integration in R&D project teams [14]. Furthermore, these various activities are based on a strategic understanding of the image and the use of photography as a means of representing ideas, exploring digital art and various communication strategies that enable new ways of interacting with different spaces and audiences.

It is important to call attention to the pedagogical innovation work that has already started, through the VSC research project and its collaborative platform used by the CUs at FAUP and other University courses as the Master in Image Design (MDI) at FBAUP and the Photography II CU of the School of Media Arts and Design (ESMAD). This pedagogical innovation work aims to harness the potential of architecture and arts students to use and explore a collaborative work environment and specific operators for the CUs. These operators and the collaborative platform motivate students to explore the use of image and photography to observe, analyse and theorise different dimensions of space based on visual research and visual evidence and through analysis and representation exercises of architecture and urban space (Fig. 5).

**Fig. 5** VSC platform collaborative environment: Diachronic Landscape (2020)

We believe that the evolution of this pedagogical innovation project may contribute to broaden the horizons for both academic teaching and learning and research. A project that can be the basis for developing a learning program specifically designed for the production, organization and communication of visual information related to the transformation and appropriation of architecture and public spaces.

Photography is therefore apprehended by the students as an essential element of study, analysis and representation of territory and its spaces and experiences. Thus, it can be seen, how the developed works explore photography as a fundamental instrument for: (1) the documental and fictional record of places, making it possible to understand and communicate the identity of places and to build a critical and analytical visual narrative of the space, fundamental for the architectural design; (2) the production of synthesis images that combine photography with 3D modelling, using photomontage techniques, allowing to explore fictional forms or foresee the impact of the interventions in places; (3) the free exploration of spaces and forms, using image editing techniques, combining methods of representation and drawing, from the sketch, through the physical model and up to the final synthesis communication project that may have as final outputs physical supports—such as posters and publications in book format both analogue and digital.

We argue that this pedagogical methodology addresses the aspects essential to the critical teaching of photography on architecture and public space by providing students with essential tools and strategies vital to the design and communication of architecture. The pedagogical approach attempts to improve the understanding and uses of photography as a research tool in the fields of architecture, which can contribute to the construction of new forums for political discussion and social intervention for the general public and academic communities when engaging with those visual narratives, as happened with the VSC Exhibitions. This paper suggests the value of implementing photographic projects in a direct relation with their locations and general public, which can contribute to the activation and sharing of individual and collective memories regarding uses of existing architectures in the Porto Metropolitan Area. Furthermore, we recognize the importance of the work coming from several visual artists who use of photography not as a mode of documenting reality, but rather a mode of engaging with its multiple dimensions, using photography as a critical territory [7].

Finally, the VSC exhibitions both in FAUP and in the Porto Metro Stations as shown by passed research and in this paper proved to be important initiatives to open university to society and in promoting the debate of students and the public on the current dynamics of use and appropriation of certain spaces in Porto's Metropolitan Area. While it is still difficult to clearly define a pattern with full certitude regarding the opinions of the students and public about the impact of the VSC project in the debate on architecture, public space and urban transformation, since the samples of surveys are not enough representative, there is enough evidence and theoretical research to deduct pertinent issues and evolve with the present research and explore further paths for this investigation.

## References


1. See Visual Spaces of Change research project supported through national funds by the FCT—Fundação para a Ciência e a Tecnologia
2. Samuel F (2010) Le Corbusier and the Architectural Promenade. Birkhäuser, Basel
3. Cityzines (2015). Pedro Leão Neto (ed.), Ana Aragão, Elias Redstone, Ivo Poças Martins, Joaquim Moreno, Nuno Grande, Pedro Leão Neto, José Luís Neves. Scopio Editions. ISBN 978-989-976-99-3-9
4. Galofaro L (2015) An Atlas of Imagination. Damdi. ISBN13 9788968010378; Eduardo Souto de Moura: atlas de parede, imagens de método/ ed. por André Tavares e Pedro Bandeira, Dafne Editora, Porto, 2011, sp
5. Rémy Zaugg: The Question of Perception Cat. Museum für Gegenwartskunst Siegen Exhibition catalogue, edited by Eva Schmidt and Javier Hontoria texts (English) by Mathilde de Croix, Javier Hontoria, Eva Schmidt and texts by Rémy Zaugg: https://www.snoeck.de/book/336/R%C3%A9my-Zaugg%3A-The-Question-of-Perception
6. Stierli M (2018) Montage and the metropolis. Yale University Press. Yale; Galofaro L (2015) An atlas of imagination. IdeaBooks, Amesterdam; Bollinger K, Medicus F, Kiesler PW (eds) (2015) Endless Kiesler. Birkhäuser
7. It is significant to refer the outstanding exhibition (2014) held at the Barbican Art Gallery in London—Constructing Worlds—, which assembled more than 250 works by 18 important photographers, constituting a significant curatorial and expository undertaking able to demonstrate the extent to which the practice and discipline of photography has the capacity to contemplate architecture by going beyond the merely documentary perspective, in other words, integrating the documentary and the artistic. See Pardo A, Redstone E (eds) (2014) Constructing worlds: photography and architecture in the modern age. Prestel, p. 43
8. Gadanho P (2014) Filip Dujardin: fictions. Hatje Cantz, Berlin
9. Princen B (2019) Ringroad (Houston), 2005: the construction of an image. Sophia J 4(1):37–46
10. Teixeira S (2020) Interview Paulo Catrica. Scopio Network, 2018. https://www.scopionetwork.com/blog/2018/11/5/interview-paulo-catrica
11. Rebelo C (2020) On the surface: photography and architecture, crossing borders shifting boundaries—interview Philipp Schaerer. Scopio Network, 2016. https://www.scopionetwork.com/blog/2018/11/5/entrevista-philipp-schaerer
12. Vassallo J (2019) Epics in the everyday: photography, architecture and the problems of realism. Park Books & Rice Architecture. ISBN 978-3-03860-162-3; (…) In the second half of the

book, however, the nature of the collaborations between artists and architects becomes much more active and involved, hinting towards a gradual dissolution of roles and the emergence of new common grounds." p. 65 "In that perspective, we appreciate and really admire that a picture by Bas Prlncen is also a work on Itself," he continues on a long-term collaborator, photographer Bas Prlncen. "It's never a documentary of architecture. He finds a space and takes a picture of It In his way with certain framing, and that's how he looks at the world. You could say the same about all the other works that are there. This kind of looking at the world, in a certain way, is also what architecture has in common. That certain way of looking is fundamental to us." See in https://tlmagazine.com/offlce-kgdvs-way-of-seelng-thlngs

13. Neto P (2021) Visual spaces of change: the use of Image for rendering visible dynamics of urban change in contemporary cities. In: EAAE-ARCC internacional conference & 2nd VIBRArch. ETSA-UPV, Valência, Espanha
14. In this regard we can indicate as projects, works and initiatives linked to research and open to curricular units, among others, the CONTRAST project, with works from 9 Schools of Photography Teaching in Higher Education, Cityzines, the collection of alternative publications that is housed in the library of FAUP, the peer-reviewed international periodical Sophia Journal on Architecture, Art and Image, the international periodical scopio International Photography Magazine and the book "Álvaro Siza: Foto-grafia Documental e Artística - Um Olhar Contemporâneo sobre a Arquitetura Portuguesa", as well as the research project funded by the FCT VSC (http://visualspacesofchange.com)

# Acting Out Supervison Scenarios Training Doctoral Supervisors in Artistic Research

Claus Peder Pedersen(✉)

Aarhus School of Architecture, Aarhus, Denmark
cpp@aarch.dk

**Abstract.** The paper presents and reflects on a scenario-based training resource for supervising artistic doctorates developed for the ERASMUS + Strategic Partnership Project 'Advancing Supervision of Artistic Doctorates'. It will describe how the project partners developed an inclusive understanding of Artistic doctorate supervision and how the outcomes of the project's other work packages helped identify topics to explore in the scenarios. The paper will present how narrative structures and supervisor and candidate archetypes were developed and included in a scenario-based learning structure inspired by Augusto Boal's Theatre of the Oppressed format. It will unfold how testing led to a redevelopment of the format to a video-based online resource. Finally, it will discuss the feedback from users and identify further research needed to continue the development of the format.

**Keywords:** Artistic research · Artistic doctorate · Doctoral supervision · Doctoral supervision training · Scenario-based learning

## 1 Introduction

The paper presents and reflects on a scenario-based training resource for supervising artistic doctorates. The resource was developed as part of the ERASMUS + Strategic Partnership Project Advancing Supervision of Artistic Doctorates (ASAD) initiated by ELIA's (European League of Institutes of the Arts) working group on artistic research. The project brought together institutional partners from the visual and performative arts, music, design and architecture in a collaboration coordinated by the Academy of Fine Arts in Vienna [1]. Each partner was responsible for developing a partial contribution to the project. The contributions were organised under the headings' Setting the Framework', 'Distinguishing the Actors' and 'Improving Practices' and mapped current conditions of doctoral supervision, identified the partner's institutional and supervision challenges relating to artistic research doctorates and reflections on how to improve supervision practices in the field.


M. Blanco Lage et al. (eds.), *Towards a New European Bauhaus—Challenges in Design Education*,
https://doi.org/10.1007/978-3-031-40188-6_13

### 1.1 An Inclusive Understanding of Supervision

Before ASAD, ELIA initiated the SHARE project to map artistic research across geographical and disciplinary boundaries [2]. SHARE identified substantial differences within the young and still-maturing field of artistic research. In continuation of this analysis, the partners agreed to apply an inclusive understanding of the supervision of artistic research doctorates to embrace the wide range of still-developing research practices and traditions of the field. Rather than searching for specific characteristics of artistic research supervision, the project focused on general principles for good supervision that would be valid across various academic disciplines. This focus also aligned the project better with the extensive contemporary literature on doctoral supervision [3]. The literature studies led to understanding supervision as an activity that reaches beyond the direction and feedback on research activities and outcomes. It touches on a wide range of human interactions, relations and emotions and requires understanding and awareness of complex power relations. Supervision goes beyond interpersonal relationships to involve wider academic and institutional frameworks. ASAD applied a triangular model of candidates, supervisors and administrators collaborating to support the development of the doctoral research project for artistic and academic quality, future career trajectories, well-being and work-life balance.

### 1.2 The Training Resource Team

The Aarhus School of Architecture was the lead partner for developing the scenario-based training resource under the inclusive understanding of artistic doctoral supervision agreed upon by the project's partners [4]. The project team consisted of the Head of the PhD School Claus Peder Pedersen, and PhD candidate Joel Letkemann with invaluable input and feedback from PhD Coordinator Mia Mimi Flodager and the Head of the Art, Research, Support Unit at Akademie der Bildenden Künste Wien Michaela Glanz. The final online interactive resource was developed and directed by Gisa Fellerer and Lorenz Tröbinger [5]. The project team represented substantial experience in organising, supervising and participating in (artistic) doctoral education. The training resource was developed from this hands-on perspective. The project team sought to translate their academic experiences with the outcomes of other ASAD work packages and PhD supervision literature into an easily accessible resource its members would feel encouraged to use.

## 2 Methods

### 2.1 Scenario-Based Learning

The ASAD partners agreed to base the resource on scenario-based learning. This type of learning enables participants to interact with the intricate and ambiguous situations that are typical of the artistic, academic, administrative, and social complexities involved in doctoral supervision [6]. Scenarios are based on storylines that allow participants to empathise with unfamiliar roles and behaviours and engage in non-linear trajectories with numerous feedback opportunities. The learning format encourages critical thinking and problem-solving skills in a safe, real-world context that allows participants to explore different ways of engaging with challenges without consequences. The project team articulated three ambitions for the training resource:

- It should support the user's playful exploration of doctoral supervision's complex and entangled character.
- It should be non-normative and open-ended. The resource should not provide solutions, guidelines or exemplify best practices but encourage users to test and reflect on different approaches to supervision based on their context and experiences.
- It should be scalable to support moderated group sessions, supervisor and candidate dialogues and individuals reflecting on supervision.

## 2.2 Literature Studies

Literature studies were conducted on doctoral supervision as part of the training resource development. Gatfield's paradigm of supervisory styles informed the development by approaching supervision from a managerial perspective rather than an academic one [7]. The paradigm describes supervision according to two dimensions: the degree of structure that the doctoral research has to follow and the level of support offered. By distinguishing between high and low levels of structure and support, the paradigm identifies four supervision styles: pastoral, laissez-faire, contractual, and directorial. A style may reflect disciplinary and institutional traditions, personal preferences, as well as agreements between supervisors and candidates. According to Gatfield's paradigm, supervisory styles are contextual and dynamic, and are likely to shift depending on the developmental stage of the doctoral research. The paradigm highlights the significance of the managerial perspective, which involves implicit power dynamics and dynamic relationships that may evolve over time, depending on the specific context and stage of the doctoral study.

## 2.3 Incorporating Material from Other Work Packages

The result of other ASAD work packages played an essential role in determining the focus of the scenario-based learning resource. The Prague Academy of Fine Arts produced a mind map of a wide range of interpersonal relations and actions that could be part of doctoral supervision [8]. The study and the accompanying interactive graphical representation specified and expanded the project team's understanding of the scope of supervision. This understanding was elaborated to gain more specificity to artistic doctoral supervision with the help of a survey produced by the Zurich University of the Arts [9]. In the survey, ASAD partners were asked to identify the challenges they faced in supervision procedures and practices. The project team examined the responses from partner institutions and categorized them into four categories: Academic Development, Personal Support, Interpersonal Negotiation, and Logistical Support. These categories were chosen to reflect different aspects of the supervision process (Fig. 1).

**Academic Development** Developing Supervisor cohorde and supervision skills / Artistic research competencies / Relation between academic and artistic research / Language skills / Cross-disciplinary understanding / Internationalization / Supporting Progression / Structuring PhD training

**Personal Support** Accounting for fellow's personal needs / General human competencies /

**Logistical Support** Commiting to time frames / Fulfilling requirements

**Interpersonal responsibilities** Time consumption for supervision / Frequency of supervision and availability / Matching of expectations / Individuality and responsibilities /

**Fig. 1** The Zurich University of the Arts conducted a survey among ASAD project partners to identify supervision practices and challenges. Based on the responses, the project team identified recurring challenges in the supervision process

# 3 Methods

## 3.1 Constructing Scenarios—Narrative Structure

The project team examined the initial studies and identified pertinent topics, which they translated into narrative structures that could serve as the foundation for scenarios. These narrative structures are built around interactions between a candidate and a supervisor who encounter various supervision challenges throughout the PhD process, starting from recruitment and culminating in the defence. The encounter between Moe, a doctoral candidate with an artistic practice as a painter and an interest in science fiction pulp art and their supervisor Dr Poe, a scholar of art theories, is an example of this:

> About one year into the project, Moe becomes enthusiastic about the transdisciplinary field of science fiction studies as a way to frame artistic research. While Dr Poe has a good grounding in contemporary critical theory, their knowledge of SF studies is non-existent. The American Studies department at their university has a specialist in SF television, Dr Toe, and Moe wonders about the possibility of external supervision from Dr Toe. [10]

This scenario deals with the challenge of managing interdisciplinary knowledge and explores the possibility of co-supervision. It is designed to address various supervisory challenges across different scales. It can imply institutional policies on the involvement of external supervisors. It could also raise an academic dilemma about whether an interdisciplinary perspective will strengthen the research project or undermine an already established research trajectory. The encounter might even question the supervisor's motivations: are they willing to include other research fields in the supervision that might challenge their academic authority? To encourage reader engagement and reflection, the project team created four different responses for each scenario, proposing different solutions to resolving the situation. One of the responses addresses the academic dilemma while also questioning the supervisor's motivations:

> Dr Poe discourages Moe from seeking additional supervision, as they have a particular method and research agenda planned out. After all, Moe came to Dr Poe to learn their specific technique and area of knowledge. Any new input, especially from a different discipline, will only disrupt the ongoing project. [11]

### 3.2 Populating the Scenarios—Archetypes

A set of supervisor and candidate archetypes were developed to populate the supervision training scenarios [12]. Supervisor and doctoral fellow archetypes support the enactment of supervision training scenarios. The archetypes are conceptual personae. They are deliberately exaggerated or caricatured characters that incorporate different supervisory styles. 'The Doktor Vater/Mutter', for example, introduces an authoritarian academic-led approach to supervision:

> The Doktor-Vater/Mutter is the experienced master of their specific research area. They have a clear research method and do not allow discussion or dissent. They have all the answers already, and when they deign to accept a PhD candidate, the student must become a disciple of the Doktor-Vater/Mutter's specific methods and theories. The successful defence is in the candidate's ability to fit the mould defined by the Doktor-Vater/Mutter. [13]

In opposition to 'the Psychologist' who focuses on the candidate's personality traits and well-being:

> The candidate's psychological well-being is the primary concern for the Psychologist supervisor. The Psychologist emphasises personal emotional support at the expense of academic development. While the candidate may feel safe and nurtured, the Psychologist may not adequately challenge the candidate by being unwilling to provoke the kind of self-doubt and self-reflection that leads to learning. [14]

The project team sought to develop archetypes with features and behaviours that supervisors and candidates would recognise from their supervision, even if they were unlikely to identify fully with the caricatured archetype. The hope was that they would introduce a playful tongue-in-cheek mood to the scenarios that would encourage participants to engage and explore roles they would not take on in real life.

### 3.3 Acting Out the Scenarios—From Theatre of the Oppressed to a Video-Based Resource

The project team loosely adapted the Theatre of the Oppressed format to support the explorative enactment of the narrative structures and archetypes [15]. Brazilian director Augusto Boal developed this format in the 1950s and 1960s to initiate social change by engaging the theatre audience through interaction, discussion, critical thinking and action. The audience steps out of the role as passive spectators and becomes active participants in the drama as 'spect-actors'. They intervene, change the direction and propose solutions to the scenes acted out on stage. A facilitator, called the Joker, manages this process and organises a concluding discussion where spec-actors and actors identify shared insights from the scene.

The project team adapted and simplified Boal's model for the scenario resource. Two participants would be selected to act out a scenario structure where the scenario narrative, selected archetypes, and one of the four responses gradually would be introduced to help structure and unfold the scenario (Fig. 2).

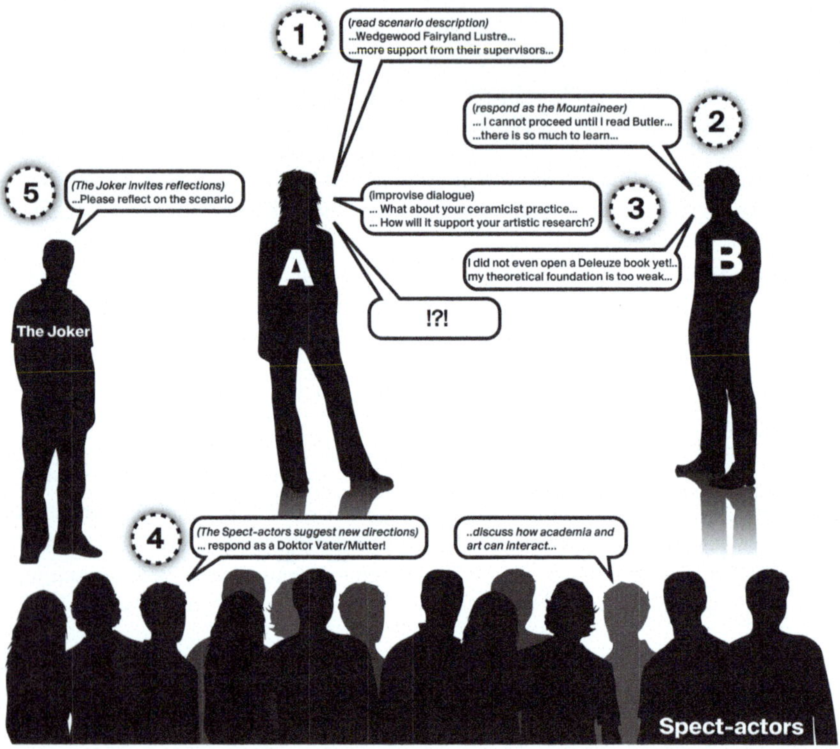


**Fig. 2** Scenario-based training adapted from Boal. Two participants, A and B, act out a scenario and explore one or more responses as supervisor and candidate according to randomly assigned archetypes. A moderator (The Joker) directs the scenario enactment, and the remaining participants intervene as an active audience (Spect-actors)

This setup was tested with supervisors and PhDs at the Aarhus School of Architecture. The test demonstrated that the narratives and particularly the archetypes, supported the scenario enactment as intended. They engaged the participants immediately and led them to reflect on their experiences as supervisors. It did, however, prove difficult to carry this engagement into the enactment and continuous exploration of the scenarios. The scenarios never developed into vivid, dynamic dialogues and ceased to progress after the initial enactment of the manuscript. The project team reflected that they most likely overestimated the participant's capacity to perform their roles without adequate preparation and guidance. The team also underestimated the importance of engaging with a trained facilitator with skills and tools to direct and stimulate a continuous dialogue.

Consequently, the project team changed the scenario format into an online video-based resource where professional actors perform the scenarios and the four responses. This change narrowed down the free exploration of the scenarios intended by the initial Theatre of the Oppressed format. In turn, the online format increased the accessibility and usability of the training resource for larger or smaller groups of users who were no longer dependent on a physical workshop setup. Gisa Fellerer and Lorenz Tröbinger developed the final online interactive format. They wrote dialogues for the scenarios, instructed and filmed actors performing them, and developed the web-based format enabling participants to navigate the scenarios. Participants adopt the role of "spect-actors," utilizing questions interspersed between the video sequences to contemplate, comment on, and discuss the scenarios while contextualizing them with their own experiences (Fig. 3).

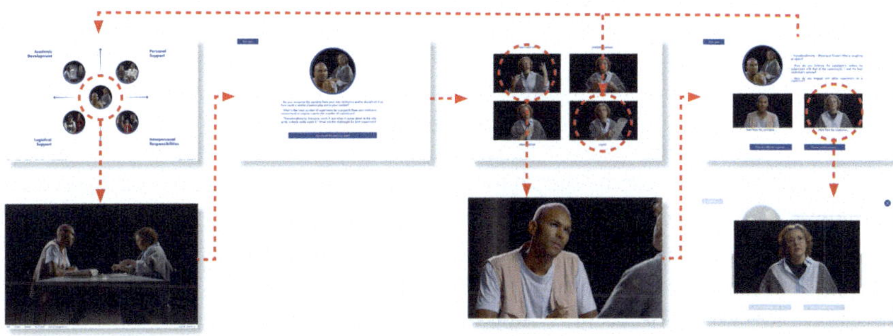

**Fig. 3** Flow diagram showing the navigation of the online interactive training resource from selection of scenario to guiding questions and responses

# 4 Discussion

## 4.1 Testing and Reception of the Video-Based Training Resource

The scenario-based training resource was tested at ELIA's online transnational training workshop on 15th September 2021. The test led to minor adjustments before the final version of the resource was presented at the in-person multiplier event 'Advancing Supervision for Artistic Research Doctorates & Beyond' at the Academy of Fine Arts, Vienna, 21st–22nd October 2021 [16]. The participants, including international doctoral candidates, supervisors and administrators, explored the resource in workshops. The feedback was largely positive. The online resource successfully introduced a framework that let the participants share their supervision experiences and explain the institutional frameworks for doctoral studies at their home institutions while discussing how to respond to the presented scenario.

### 4.2 Inclusivity and Caricature

During the development of the scenarios, the project teams carefully considered using personae, humour, and caricature. Studies have indicated that incorporating humor in an educational setting can enhance awareness, strengthen social cohesion, and reduce stress. However, it can also lead to negative consequences such as exclusion, particularly for individuals who are targeted or foreign to the culture of the humor [17]. The team used exaggeration and humour to create easily recognisable situations, hoping that participants would feel free to challenge and explore supervisor roles playfully. However, the team also considered that some participants might not appreciate or perceive the intended humour and interpret the personae as mocking or ridiculing particular positions. This concern is particularly relevant for the candidate personae who represents the weak part of the supervision relationship and incorporates specific artistic supervision challenges:

> The Mountaineer is awestruck by the prospect of entering the field of academia and research ruled by notions of objectivity and 'truth'. The climbing of this academic mountain is a monumental task that the Mountaineer feels utterly unprepared for. Consequently, they engage in every available book or course on the history of research, research methodology and study technique while postponing the engagement with their research topic and artistic practice. [12]

The Mountaineer represents the challenge of balancing artistic practice and (sometimes imagined) academic requirements for artistic research. This persona reflects the project team's experiences with a challenge that artistic and design-driven research doctoral candidates frequently face. Although this approach enhances the relevance of the persona in exploring artistic doctoral supervision, candidates encountering this challenge may perceive themselves as exposed or ridiculed.

None of the participants who participated in the workshops responded negatively to the characterisation of the candidates. However, a few participants criticised stereotypical representations of characters and relations in the filmed scenarios. According to one participant, a scenario in which a white, middle-aged male supervises a young white female candidate portrays a stereotypical gendered power dynamic. Another participant criticised the same white male's outdated and cliched representation of a painter as someone wearing an artist's beret and a smock with paint brushes sticking out of the chest pocket. Yet another scenario was criticised for its ethnic representation of an Asian woman supervised by a young white female. This criticism happened despite the earnest efforts of the team to secure a balanced gender, age and ethnicity representation across the scenarios. They point to the exposed and vulnerable situation of supervision and the sensitivity of the power relations.

### 4.3 Conclusion and Further Studies

The development and testing of the scenario-based training resource have shown that it is an effective tool for engaging participants in sharing their thoughts and experiences about artistic doctoral supervision. The use of humour and caricatures to promote a light-hearted, explorative and playful entry into the scenarios has generally been successful.

The few negative reactions show this approach's risks, as some participants might feel excluded or misrepresented.

The evaluation shows that the resource might support a group of supervisors to develop a shared understanding of supervising in their disciplinary and institutional settings. Alternatively, help a newly started candidate and their supervisor to share and align their expectations about the supervision. Nevertheless, further testing in institutional settings is needed to explore whether the tool is effective and how further developments and tweaks could improve it. It is also relevant to consider how the training resource can incorporate training sessions or supportive tools that will allow supervisors to develop their supervision skills.

## References

1. Advancing supervision for artistic research doctorates Homepage. https://advancingsupervision.eu. Last accessed 25 Nov 2022.
2. Elkins J (2013) Six cultures of the PhD. In: Ruiten S, Wilson M, Borgdorff H (eds) Share: handbook for artistic research education. ELIA, Amsterdam, Holland
3. Taylor S, Kiley M, Humphrey R (2017) A handbook for doctoral supervisors, 2nd edn. Routledge, London, UK
4. Pedersen C, Letkemann J (2021) Learning from scenarios: exploring artistic doctoral supervision. In: Glanz M, Schober S (eds) Undoing supervision: a compendium of key issues in supervising artistic research doctorates. Academy of Fine Arts Vienna, Vienna, Austria, pp. 38–47
5. Advancing supervision for artistic research doctorates Homepage. https://advancingsupervision.eu/supervision-scenarios/#Home. Last accessed 28 Nov 2022
6. Errington E (ed) (2003) Developing scenario-based learning: practical insights for tertiary educators. Dunmore Press, Palmerston North
7. Gatfield T (2005) An investigation into PhD supervisory management styles: development of a dynamic conceptual model and its managerial implications. J High Educ Policy Manag 27(3):311–325
8. Advancing supervision for artistic research doctorates Homepage. https://advancingsupervision.eu/mindmap/. Last accessed 29 Nov 2022
9. Schiesser G (2021) Doctoral supervisors: some reflections on the process, the research results and the delivery of the work package. In: Glanz M, Schober S (eds) Undoing supervision: a compendium of key issues in supervising artistic research doctorates. Academy of Fine Arts Vienna, Vienna, Austria, pp. 32–37
10. Advancing supervision for artistic research doctorates Homepage. https://advancingsupervision.eu/wp/wp-content/uploads/Scenario-descriptions.pdf. Last accessed 28 Nov 2022
11. Ibid.
12. Advancing supervision for artistic research doctorates Homepage. https://advancingsupervision.eu/wp/wp-content/uploads/Architypes.pdf. Last accessed 28 Nov 2022
13. Ibid.
14. Ibid.
15. Boal A (1979) Theater of the oppressed. Urizen Books, New York
16. Advancing supervision for artistic research doctorates Homepage. https://advancingsupervision.eu/outputs/multiplier-conference-vienna/. Last accessed 29 Nov 2022
17. Banas J, Dunbar N, Rodriguez D, Liu S (2011) A review of humor in educational settings: four decades of research. Commun Educ 60(1):115–144

# Learning from the New European Bauhaus: Co-creating the Curriculum with Wellbeing at Its Core

Ana Souto(✉)

Nottingham Trent University, Nottingham, UK
ana.souto@ntu.ac.uk

**Abstract.** This paper presents a work-in-progress project which started in 2021 when students in the final year of Architecture, inspired by the New European Bauhaus, designed their final projects with sustainability at the core. One group focused on building sustainably; the second, which I led, embraced sustainability as wellbeing, as an opportunity to build and support resilient communities, promoting a sense of community and belonging, creativity, communities, and connectivity. This work was further explored thanks to a Student-Staff collaboration funded project that aimed to co-design the curriculum of Architecture through the lens of wellbeing. Two academics and a final year student of Architecture, working as partners, reviewed module guides, the submissions to the degree show (2021 and 2022), our current the course documentation, and gathered feedback from current students in the degree. The findings and recommendations highlighted a gap in the course (lack of mentioning of wellbeing) and suggested the implementation of wellbeing in the course documentation, physical changes to learning spaces and the creation of wellbeing spaces across the University. Some of these changes are currently being implemented, to enhance students' awareness of wellbeing during their studies (to study well), and applied in their projects (to build for resilient communities). The aim of this project is to further promote an awareness of well-being within the framework of sustainability, and support the revision, reflection, and co-creation of architectural education with our students as partners. This approach could be applied to other architectural degrees, enhancing students' experience and commitment to their education.

**Keywords:** NEB · Wellbeing · Architecture · Education · Co-design with students

## 1 Introduction. Learning from the NEB: Co-designing the Curriculum of Architecture with Wellbeing at Its Core

This paper reflects on the delivery of the final year course of Architecture in 2021–2022, whereby the modules on Design Studio, Technology and Architecture in Con-text explored Architecture centered around sustainability which, following the spirit of The New Bauhaus, aimed to:


M. Blanco Lage et al. (eds.), *Towards a New European Bauhaus—Challenges in Design Education*,
https://doi.org/10.1007/978-3-031-40188-6_14

> […] imagine and build together a sustainable and inclusive future that is beautiful for our eyes, minds, and souls. […]. Enriching, inspired by art and culture, respond-ing to needs beyond functionality. Sustainable, in harmony with nature, the environment, and our planet. Inclusive, encouraging a dialogue across cultures, disciplines, genders, and ages. [1]

This paper reflects on the context of the New European Bauhaus, its opportunities and challenges within academia; it highlights concerns around the wellbeing of our students; and analyses the work produced by students who embraced wellbeing as manifesto. This analysis was done through the visual analysis of their final projects, and through a report that two academics and a student, awarded with a Student-Staff co-creation fund (2022), suggesting improvements to our learning and teaching contexts using a wellbeing lens.

Several methods were used between 2021 and 2022 to support the gathering and analysis of the primary data. First, in summer 2021, the author analysed the use of green infrastructures and other design elements to support/ enhance the wellbeing of their users across all the projects submitted to the Architecture Degree Show website (2021). This was followed by a selection of interviews with students who identified these strategies at the core of their studio design projects. The analysis of this data had an impact on the design and delivery of the same modules the following academic year, 2021–22.

Internal funding was awarded to continue this analysis (2022), using the Students as Partners approach to co-create the curriculum: a student, supported by two members of staff, reviewed the degree's documentation, module guides and assessment briefs; and organised a focus group with students which, together with her own auto-ethnographic analysis, reflected on the significance of wellbeing in the course. This is an on-going project, currently on its second phase, based on the implementation of the recommendations raised by the students (2023). This phase includes the implementation and evaluation of these findings, as well as the application for new internal funding to keep reflecting, evaluating, and co-creating architectural education with our students.

## 2 The Context of the New European Bauhaus (NEB)

The announcement of the New European Bauhaus (NEB) by the President Ursula von de Leyen in September 2020 renewed the interest of architects, designers, educators, governments, and users in our cities. The interruption of Covid 19 served as a turning point to rethink our cities, how we use them, and how we encourage an international initiative which recognizes the sustainability, aesthetics, and inclusion dimensions of urban design [2]. The fact that inclusion is one of the main goals is a very significant aspect of the NEB, which, to 'regain people's confidence, it can and must include a social, artistic and architectural dimension, including everyone' [3]. Inclusivity becomes the frame of reference, something we must apply as well on architectural education.

Even though the original Bauhaus remains as a main object of inspiration, the NEB substitutes the 'avant-garde, non-conformist and revolutionary ideas' with a project that 'advances its Green Deal and Renovation Wave agendas': there is no room for the *tabula rasa* proposed by the original Bauhaus, but for an inclusive approach [4]. This allows for a wider representation and group of stakeholders sharing the responsibility of the

NEB, which ultimately embraces a new framework of philosophy, ethics, as well as a belief system, for example, thinking about resources, reuse and repurpose towards a more sustainable landscape [4]. Moreover, as highlighted by the European Union, '[…] the act of building in cities and rural areas must be transformed from a source of environmental predation into a restorative force for terrestrial health and well-being' [5].

To achieve this, education and society must embrace these changes. As an architectural educator, I feel the responsibility to re-educate my practice and knowledge and share this with the students in Architecture. This has been highlighted by other academics [6, 7], to ensure a swift transition towards a holistic innovative approach to education, 'a renewed profession that is enriched by all kinds of cultural, scientific and environmental values' [7], that aligns with the messages from the New European Bauhaus and the UN SDGs, between others.

## 3 Architecture and Wellbeing: The Future of the Profession

Better health and wellbeing are at the core of the NEB, and it should also be the focus of architecture and urban planning, especially by promoting more inclusive approaches, addressing 'complex societal problems together through co-creation' [1]. The World Health Organization defines mental health as 'a state of well-being in which an individual realizes his or her own abilities, can cope with the normal stresses of life, can work productively and is able to make a contribution to his or her community' [8]; moreover, 'a society's well-being can be determined by the extent to which they are resilient, build capacity for action, and are prepared to transcend challenges' [9]. This should apply to large organisations, like universities, embracing and educating our students on these challenges. The World Health Organisation (WHO) is instigating the 'health in all policies' which should result in a 'form of spatial medicine, whereby the design of built environments positively contributes and facilitates human and planetary health and wellbeing' [10].

Architects, and students in architecture should be well versed in mental health and wellbeing to design built environments that align with these ideas. However, the profession is still not facing this taboo: mental health problems remain a stigma in the architecture [11]. Similarly, as explored by Kirkpatrick in her dissertation (2018), architecture students seem to accept stress and bad mental health as part of their studies, due to the studio culture, expectations of working through the night, as well as financial debt due to the length of the course [12]. Along these lines, a 2020 study demonstrated that the mental health of students of architecture was significantly worse that students in other degrees, highlighting the same reasons [13]. Even though the situation is still quite worrying, several institutions are trying to make a change in architectural education and practice culture: this change has already started [13].

We believe that there is a gap between students' experience in architectural education (stressful) and their engagement with architectural design that enhances wellbeing. We must ensure that, as educators, we recognise the value of wellbeing as part of our curriculum: to provide the right learning environment for our students, and to expect that our students enhance and promote a sense of belonging, community, and wellbeing, in their projects. This is what Rice and others have coined as 'nurturing architecture'

[14]. As the RIBA highlights, architects must 'consider how your work might impact on lifestyle diseases and mental wellbeing' [15]. In what follows I will present examples of some of the students in my institution who have used greening infrastructures and other design approaches in their projects to promote better health and wellbeing.

## 4 Learning from the Bauhaus and Embracing Wellbeing

As part of my personal interest in wellbeing, resilient communities and the connection between heritage, representation and sense of belonging, I have encouraged my students to consider these ideas since October 2019. However, it was only during the pandemic that we talked more openly about wellbeing, and access to green infrastructures during the lockdowns of 2020 and 2021. In September 2020, one of the Design tutors, Rebecca Pallett, proposed the topic of healthy architecture to promote healthy communities, to which students had to respond with designs that 'facilitate much needed social engagement, inclusivity and wellbeing, provide flexibility of usage and adopt principles of sustainability all without losing sight of the experience of the place'. I supported the development of the Design studio module through Architecture in Context 3, sharing with the students my own research on sacred spaces and wellbeing, facilitating a reading list on wellbeing, emotional geographies, green and blue landscapes, etc. We noticed a real appetite from students to engage with these topics, and their projects at the Degree Show demonstrated this.

'A place to grow', by Maddie Rutherford-Browne (Fig. 1) and Greening the City for Well-being, by Zana Kentish Brade (Fig. 2) are examples of design proposals that introduced green and blue landscapes to support health and wellbeing. As part of this research project, I interviewed these students, who reflected that the decision to use this architectural approach was based on their personal interest, rather than knowledge acquired during their degree. These responses triggered a more targeted approach to wellbeing in the delivery of Architecture in Context 3 and Design studio in the following academic year 2020–21.

In Summer 2021 I applied for internal funding in the Department of Architecture and was successful in creating 5 summer scholarships to explore 'Access to Green Spaces in Sneinton', a clear response to the lockdowns. This involved visiting green spaces, doing face-to-face interviews, and many conversations and reflections on wellbeing and access to green spaces, since all of us, sadly, had become experts forced by the circumstances.

Wellbeing as sustainability was embraced again by the Design tutors in September 2021, offering a unit on The New Bauhaus, with a special 'focus on the social aspect of sustainability from the perspective of the user, the idea of resilient cities and the importance of wellbeing through design'. As the previous year, we explored ideas of wellbeing and education in architecture. By this time, final year students had experienced a very disjointed degree, and were very enthusiastic about being at the university, and exploring the connections between wellbeing, sense of community and belonging and architectural education. There was a clear increase in the number of projects in the degree show that, not only identified wellbeing as a main design focus, but also used different ways to enhance this through architectural features. For example, Alex Evans (Fig. 3) regenerated an old 1960s building to build sustainably, and foment a sense of

**Fig. 1** A Place to Grow, by Maddie Rutherford-Browne: "to resurrect our nature-society relationship in a rewilded landscape, as a haptic experience, imitating pond reeds and bulrushes". https://2021.architecture-ntu.com/work/a-place-to-grow, accessed on 24th November 2022

**Fig. 2** Greening the City for Well-being, by Zana Kentish Brade. Vertical Farming methods, Traditional Farming methods, Botany, Beekeeping and making products from beekeeping". https://2021.architecture-ntu.com/work/sustainable-farming-education-greening-the-city-for-well-being, accessed on 24th November 2022

continuity through the reuse of heritage, reinforcing the sense of community, belonging, and consequently, wellbeing. Louis Huband (Fig. 4a, b) used biophilic design principles to enhance the learning experience of his School of Architecture.

In June 2022 I obtained the TILT Student-Staff Collaboration Fund to reflect on our curriculum and learning spaces. This project was led by Alex Evans, a student who had already participated in the summer placement 2021, Design & AIC3 unit on wellbeing, and was about to graduate in July 2022. Evans, with Tom Hughes and the author, did a thorough analysis of the course and its spaces, as well as a focus group to gather a broader view from other students. Evans analysed the Degree Shows of 2021 and 2022

**Fig. 3** Regenerate, by Alexandra Evans. A 1960s, non-listed lab building is regenerated into an innovative school of Architecture, where sustainability, heritage and creativity are key principles. The new structure is of Passivhaus standard, with timber columns growing into complex timber grid shells. The design demonstrates the key methods of adaptive reuse; preservation, insertion and integration. https://2022.architecture-ntu.com/work/regenerate, accessed on 24th November 2022

and saw an increase on projects connected to wellbeing, from 29% in 2021, to 38% in 2022. This visual and textual analysis was based on the information published on the Degree Show website, which includes a succinct version of the text and images of their final projects. There is a clear correlation between the design and research units, and students adopting these ideas. However, after reviewing the course documentation, it became clear that wellbeing was only explored by some of the tutors, but was never acknowledged within the course guides. After running a focus group with students in the BArch and MArch, Alex recommended the following changes:

1. To increase the students' knowledge of the connections between Wellbeing and Sustainability.
2. To support the wellbeing of students and members of staff; and
3. To highlight wellbeing as teaching material and applied to the learning spaces.

Reflecting, reviewing and co-creating the curriculum with Students as Partners has helped me and my colleagues to keep working on this initiative. And we are in the process of doing it. It is a work in progress.

(a) (b)

**Fig. 4** **a** and **b** The Biophilic Micro-City, by Louis Huband. The school becomes an urban organism that embodies the complex and diverse thresholds of the city, of past, present and future, where nature permeates the building morphology. https://2022.architecture-ntu.com/work/the-biophilic-micro-city, accessed on 24th November 2022

## 5 Conclusion: Still a Work in Progress

As explored in this paper, in collaboration with colleagues and students, I have been exploring issues of belonging, wellbeing, and better approaches to education in Architecture. Inspired by the NEB and the UN SDGs, we can start imagining and co-creating our future School of Architecture:

1. The Bauhaus remains a starting point: we cannot start from a tabula rasa, but we embrace creativity and co-creation as design principle and pedagogic approach.
2. We are inspired and will follow the guidance of the NEB, the New Urban Agenda, Quito (2016) and the UN Sustainable Development Goals (2015); in particular goal 3: Good health and well-being, and goal 11: Resilient cities and communities.
3. Wellbeing should become an essential principle, at the core of the curriculum, expanding to the learning spaces, students, and members of staff (as explored and supported by the RIBA, aae, AJ, ABS, ARB, etc.).[1]
4. Finally, we must collaborate with our students, as partners in developing and evaluating the curriculum, as we are currently doing.

[1] RIBA: Royal Institute of British Architects; aae: association of architectural educators; AJ: Architects' Journal; ABS: Architects Benevolent Society; ARB: Architects Registration Board.

The main conclusion aligns with that of Kirkpatrick, in 2018: our graduates in architecture should leave the university with:

> healthy working habits, knowledge of mental health and how to take care of their own wellbeing, eventually the profession will benefit from this as a whole and the culture could start to shift towards a more positive outcome. [11]

Our next step is to reflect and evaluate on the initiatives put in place in our School and analyse whether they have had an impact on our student and staff population, whether working with students as partners in this project also brings a stronger endorsement from all stakeholders.

## References

1. New European Bauhaus homepage. https://new-european-bauhaus.europa.eu/index_en. Last accessed 20 Feb 2023
2. Widuto A (2021) Cohesion policy contribution to the New European Bauhaus. Eur Parliam Res Serv PE 679:105
3. Figueroa J (2021) A conference for the future: from Bauhaus to the New House—post COVID landscapes. Docomomo 65:120–121
4. Ness D (2021) The shift from the new build to regeneration. Can the New Bauhaus transform architecture and design to meet global challenges? AGATHON 9:22–31
5. Commission E (2022) Horizon Europe—New European Bauhaus nexus report. European Commission, Brussels
6. Sadowski K (2021) Implementation of the New European Bauhaus principles as a context for teaching sustainable architecture. Sustainability 13:10715
7. Rosado-Garcia MJ, Kubus R, Arguelles-Bustillo R, Garcia-Garcia MJ (2021) A New European Bauhaus for a culture of transversality and sustainability. Sustainability 13:11844
8. World Health Organisation (nd) Health and wellbeing, https://www.who.int/data/gho/data/major-themes/health-and-well-being. Last accessed 21 Feb 2023
9. World Health Organisation (nd) Promoting well-being. https://www.who.int/activities/promoting-well-being, Last accessed 21 Feb 2023
10. Rice L (2020) After Covid-19: urban design as spatial medicine. Urban Des Int
11. Newman V (2017) Coming out of the darkness. https://www.ribaj.com/culture/mental-health-for-architects-virginia-newman-culture. Last accessed 21 Feb 2023
12. Kirkpatrick, M (2018) Mental wellbeing and the architecture student—dissertation by Melissa Kirkpatrick, Architects Benevolent Society (absnet.org.uk). Last accessed 20 Feb 2023
13. McClean D et al (2020) Mental health in UK architecture education. An analysis of contemporary student wellbeing, an initial study. Robert Gordon University, UK RIBA Research Grant. https://researchportal.northumbria.ac.uk/en/publications/mental-health-in-uk-architecture-education-an-analysis-of-contemp. Last accessed 20 Feb 2023
14. Rice L, Meraz F, Marco E, Drożyński C (2021) Nurturing architecture: education, research and practice for health and wellbeing. aae Charrette 7:2
15. Russell J, Thompson M, Jones A (2021) RIBA: study architecture well. RIBA, London

# The New European Bauhaus in Architecture Education? An Anthology for the Invisible

Hanne Van Reusel[1,2](✉), Dag Boutsen[1], and Michela Barosio[3]

[1] Department of Architecture, KU Leuven, Leuven, Belgium
hanne.vanreusel@kuleuven.be
[2] Faculty of Design Science, University of Antwerp, Antwerpen, Belgium
[3] Department of Architecture and Design, Politecnico di Torino, Turin, Italy

**Abstract.** Looking at the ambitions of the New European Bauhaus (NEB) initiative this paper looks into what the skills and professional behaviours—ways of working—are that are sought for. We identify an informal and incomplete list of skills and behaviours required to contribute to a sustainable innovation that weaves together the core values of aesthetics, sustainability and inclusion. Focusing on architecture education (AE) the paper builds on the research findings of the Erasmus + "Architecture's Afterlife" (AA) research and discusses how AE delivers graduates that find their place in diverse professional sectors (multi-sector), for which the obtained skills and behaviours remain close despite the different professional trajectories graduates take. Looking into the matches and mismatches between skills and behaviours—ways of working—trained in AE and the ones that could support the NEB ambitions the study reveals a significant match is to be found. In this, the paper likes to—among others—address the role personal competences play in this relation. As a conclusion, the argument is made that a key role is to be played recognising the beyond architecture and dive into the ways of working, the personal competences, the skills and behaviours that are lying beyond.

**Keywords:** Architecture education · New European Bauhaus · Innovative curricula · Architecture graduates · Invisible skills

## 1 The New European Bauhaus and Architecture Education

Beautiful, sustainable and together; the three core values of the New European Bauhaus (NEB) aim to weave together the search for innovative approaches to tackle the complex environmental, economic and cultural needs of the European Union. With the European Green Deal as its backbone, the NEB initiative looks to support and foster a sustainable transition as "innovation in action" [1, 07:20]. The session on the "New European Bauhaus: Collaboration, Community and Culture for Innovation" as part of the EU Science & Innovation Research Innovation Days organised in 2022 [1], presents some of the underlying ambitions and projects in progress. Key in this endeavour will be to find sustainable innovation with the support of academia and with the cultural and creative


M. Blanco Lage et al. (eds.), *Towards a New European Bauhaus—Challenges in Design Education*,
https://doi.org/10.1007/978-3-031-40188-6_15

sector on board as an indispensable contributor. Broad and ambitious as the NEB is, it is clear that many have an active role to play in what is promising to be a shared and impacting movement. Yet, it also raises the question by whom and how a contribution can be delivered. Member of the European Parliament Marcos Ros Sempere draws the link to the importance of education and more particularly the need for innovation in curricula:

> We need to also incorporate education. Education has always played a key role. The education is very present in the report that the European Parliament adopted a couple of weeks ago [September 2022]. And we have called for innovative curricula to be developed in line with the principles and objectives of the New European Bauhaus for cultural education and the development of spatial and creative skills. [1, 25:25]

Subsequent Having the importance of spatial and creative skills addressed while referring to cultural education and alluding to the 20th Century Bauhaus movement, the role of architecture education (AE) cannot be ignored in the framework of the NEB initiative. How can AE contribute to the NEB movement Europe is investing in? In order to contribute to the debate around this question this paper looks into:

1. What are the skills and professional behaviours—the ways of working—the NEB is looking for?
2. What are the skills and professional behaviours architecture graduates display?
3. A discussion on the (mis)matches between both with suggestions on how to take things further from the current state of affairs.

## 2 Looking for the Bauhaus Ways of Working

In order to explore the skills and professional behaviours—ways of working—the NEB is aiming at, our study has looked into (1) the official website of the New European Bauhaus initiative [2], (2) the recording of the NEB session during the Research and Innovation Days of 28–29 September 2022 [1], traces of the on-going discussion on education and the NEB as documented in (3) a prepary note for parliamentary work [3] and (4) the recording of the "Education and the New European Bauhaus... let's imagine!" session hosted during the NEB Festival in Brussels on the 10th of June 2022 [4]. In addition, the study also looked into the awarded NEB prizes for 2022 [5] and the more general 2016 "New skills agenda for Europe" [6].

Looking to learn more about the key skills and professional behaviours—ways of working—discussed in the framework of the NEB, it shows that these questions—especially in relation to education—are still rather open. While the 2016 "New skills agenda for Europe" clearly articulates ambitions to advance digital skills and entrepreneurship to strengthen human capital, employability and competitiveness [6], the NEB discourse integrates these, yet brings up many more aspects. These are significantly less strongly formulated and directed. In the consulted sources on the NEB we found references to the need for an interdisciplinary approach, to combine the local with the global, to act at the intersection of the old and the new, to have a multi-layered approach, to work place-based, to be innovative and creative, to develop green skills and to allow for life-long learning.

When diving into the projects awarded by the NEB Prize in 2022 [5], the NEB platform provides quite some insight on how the selected projects provide inspiration in successfully obtaining the objective for a sustainable approach, aesthetics and inclusion. Yet it remains undiscussed which type of experts—let be which type of skills and competences—are behind. There is no explicit information on the organisations or institutions that are behind those initiatives, nor how they have worked together (interdisciplinarity), nor how they have worked at intersections between the old and new, etc. As inspiring as the awarded cases can be, the given descriptions fail to provide insights on how—through which skills and professional behaviours—the initiators behind the projects have managed to innovate for a sustainable, beautiful and inclusive future.

When asking Artificial Intelligence to screen the internet to reply to the question: "What are the skills and professional behaviours the New European Bauhaus is looking for?" ChatGPT generates the following description [7]:

> The New European Bauhaus initiative seeks to promote sustainable, inclusive, and beautiful design in Europe. The initiative is seeking a range of skills and professional behaviours from designers, architects, and other professionals involved in shaping the built environment. Here are some of the key skills and behaviours that the New European Bauhaus is looking for:
>
> 1. Sustainability: The New European Bauhaus values sustainability, so professionals with knowledge of sustainable design practices are particularly sought after.
> 2. Interdisciplinarity: The initiative is looking for professionals who can work across disciplines, such as design, architecture, engineering, and the arts, to create truly innovative solutions.
> 3. User-centricity: The initiative seeks to prioritize the needs and experiences of users, so designers and architects who focus on user-centric design are valued.
> 4. Collaborative: Collaboration is a key aspect of the New European Bauhaus, so professionals who can work effectively in teams and collaborate with others from different backgrounds and disciplines are particularly sought after.
> 5. Innovative: The initiative values creativity and innovation, so professionals who can bring new and original ideas to the table are highly valued.
> 6. Inclusivity: The New European Bauhaus values inclusivity and diversity, so professionals who can create designs that are accessible and welcoming to all are particularly sought after.
> 7. Communication skills: Professionals who can communicate their ideas and designs effectively, both visually and verbally, are highly valued.
>
> Overall, the New European Bauhaus is seeking professionals who are passionate about sustainable, user-centric, and innovative design, and who can work collaboratively across disciplines to create beautiful and inclusive spaces.

When it comes to defining the skills and professional behaviours—ways of working—requested in the framework of the NEB there is no clear-cut list yet our research brings the following attributes to the front:

- Interdisciplinarity (and collaboration),
- innovation (entailing creativity),
- sustainability (including green skills),
- a multilayered approach (combining the local and global, the old and the new),
- working place-based,
- life-long learning,
- user-centricity (in an inclusive approach) and communication skills.

## 3 Looking into the Architectural Ways of Working

To see how AE is and could be contributing to the skills and professional behaviours—ways of working—that are needed to support the NEB movement this paper will source from the results of the "Architecture's Afterlife" (AA) study. Diving into the question on what can be the multi-sector impact of an architecture degree within the European context, the AA research has studied the obtained and missing skills and competences as identified by architecture graduates. The Erasmus + Strategic Partnership between the Royal College of Art, the University of Antwerp, the University of Zagreb, the Politecnico di Torino, the Universitat Politècnico de València and the KU Leuven investigated diverse trajectories of architecture graduates. Finalised in December 2022, the research involved a quantitative study—a survey that elicited over 2.500 responses—and a qualitative study based on 49 in-depth interviews with participation of architecture graduates in Europe. Revealing many thought-provoking insights on AE and the understanding of acting, thinking and being as an architect, a selection of the findings of the study will be here discussed in relation to the NEB ambitions and its relation to AE [8].

As a first finding, the AA survey identified that around 40% of the architecture graduates work beyond the architectural practice in its strict—related to building—sense (see Fig. 1). Studying more qualitatively those graduates that made it to related and unrelated sectors, it shows that the story is more complex. Many graduates combine professional work in building architecture with many different professional orientations and defining what can be considered "working in relation to architecture" seems tricky. Furthermore, the study showed that also among the "conventional" architects, a huge variety of professional trajectories could be found. Depending on the type and size of architecture office, architects find themselves in charge of the more conceptual part of the design, or being focused solely on the work site and the financial management, or to be in charge of the relationship with clients. Working as an architect in practice can mean many things. Likewise, as the interviews revealed, very few of the in the study participating architecture graduates feel they "left" architecture, even when their current profession is far from a practice in building architecture. The architectural way of acting, thinking and being remains highly present, even when working in the event sector or when being a mayor. Among other insights, the study revealed that the architecture degree opens a pathway to agilely work at the intersection of and beyond disciplines.

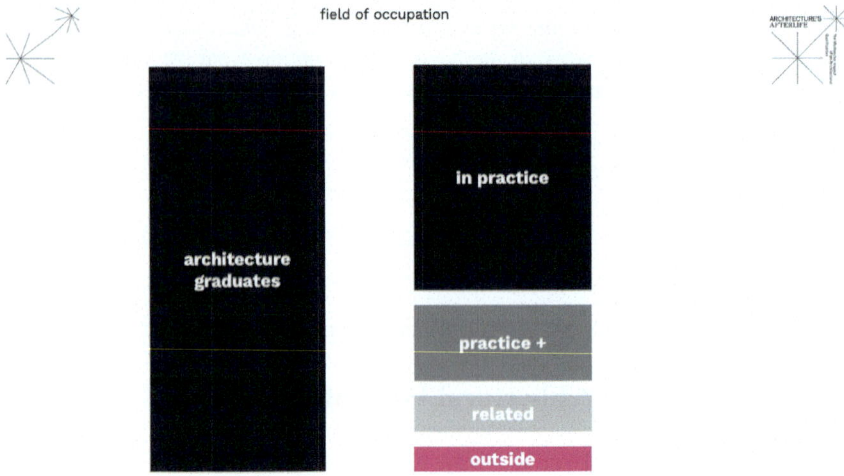


**Fig. 1** The Architecture's Afterlife questionnaire shows that of the participating architecture graduates 62% are working in practice (building architects), 21% in practice plus other, 10% in sectors related to architecture, and finally 7% in non-related sectors, by the research team defined as "outside" of architecture

Second, the Afterlife study helped to identify the (mis)matches in the skills acquired through AE and the used skills in diverse professional trajectories. Despite having various different sectoral occupations, the levels of (mis)matches remain rather coherent. Whether architecture graduates end up in the "strict" architectural discipline or beyond, it seems like the experience of acquired and used skills remains rather close.

The AA questionnaire has, not so surprising, shown that AE delivers rather high skills in design thinking and spatial skills which live up to the current and diverse professional needs. Maybe less evidently, the questionnaire also revealed a rather high obtaining of skills concerning: constant learning and self-improvement, flexibility, determination, endurance, being critical, handling criticism, being passionate about architecture, and theory and history (see also Fig. 2). We furthermore want to note that the biggest mismatches in skills are: digital skills, project management skills, working with clients, empathy, and business management.

As a third and last point of findings from the study that we would like to discuss here, has been the importance of the so called "soft skills". The AA research involved a preliminary "test" survey, which highlighted the not anticipated importance of skills related to personal competences, involving endurance, determination, work ethic, handling criticism, constant learning and self-improvement, flexibility, and dealing with uncertainty. As one of the conclusions on the final survey the AA's report states [9]:

> Generally, respondents reported that personal competences were the competences that they acquired the best during education. They also needed them the most in their current jobs.

**Fig. 2** The Architecture's Afterlife questionnaire shows the level in which skills were acquired (in grey) in AE in relation to how much they are used (in magenta) in the current professional occupation

## 4 A (Mis)match?

The data from the AA survey does not explicitly look into the skills and behaviours brought up in the context of the NEB. Yet—as shown in Table 1—comparing the results of this quantitative research, some parallels can be found rather directly in relation to the skills and behaviours brought up in the framework of the NEB; an interdisciplinary (multi-sector) way of working, the need for innovative and creative skills (design thinking, flexibility, being critical, dealing with uncertainty), sustainability (determination, endurance, work ethic), a multi-layered approach (theory and history), place-based working (spatial skills), the importance of life-long learning (constant learning and self-improvement), and an element of inclusivity (work ethic). Furthermore, the comparison shows that where AE provides mismatches in relation to the graduates' current job occupations—such as digital skills, working with clients and empathy—there is also a relation to be found to the skills and behaviours sought by the NEB involving sustainability and user-centricity.

There is clearly still space to reinforce the connection of skills and behaviours as trained in AE in relation to the NEB, yet a significant match can be identified. Without making a claim about the relevance of other degrees and professional ways of working, this study hints at understanding AE and the skills and behaviours it trains as a relevant potential lever to make up and strengthen the NEB movement.

**Table 1** Comparing NEB skills and behaviours with obtained skills in AE

| NEB skills and behaviours | Obtained skills in AE | Mismatches AE |
|---|---|---|
| Interdisciplinarity (and collaboration) | Work at the intersection of and beyond disciplines | |
| Innovation (entailing creativity) | Design thinking, flexibility, being critical, dealing with uncertainty | |
| Sustainability (including green and digital skills) | Determination, endurance, work ethic | Digital skills |
| Multilayered approach (combining local and global, old and new) | Theory and history | |
| Working place-based | Spatial skills | |
| Life-long learning | Constant learning and self-improvement | |
| User-centricity (inclusive approach) and communication skills | Work ethic | Working with clients, empathy |
| | Handling criticism, being passionate about architecture | Project management skills, business management |

In order to further illustrate the parallels, we like to bring in the anecdotic evidence from the qualitative part of the AA study. In addition to the qualitative survey, 49 in-depth interviews were conducted with architecture graduates working as practitioners as well as active in other sectors. The results of the interview samples with architecture graduates situated in Europe had been processed using methods of Constructivist Grounded Theory [10]. The results emphasise architecture as collaborative, as interdisciplinary, and as having blurred boundaries [11]. The interviews did not explicitly research any link to the NEB skills and behaviours, yet the results show insightful parallels.

Multiple participants expressed their passion and desire to improve the world around them, to create nice and beautiful buildings and projects, and more crucially to work "for people, for whom you want to improve their life and give them something to cherish" as stated by one of the interviewed architecture graduates who combines architectural practice with event organisation.

An architecture graduate working as vice-director of an environmental association addresses how imagination shows to be a main skill from his architectural way of working:

> Every time that I deal with something, I try to imagine how that place, that the project could change in a better way. And I always try to imagine something different. I think this is the main skill that I bring to every kind of situation. And I think it comes from [the fact] that I am an architect.

An interior architect working in urbanism and scenography gives a testimony on how she works across scale:

> On one hand, in architecture there are a lot of scales and levels of scales. Sometimes you work for some project in millimetres and then on a bigger scale you work in metres. And sometimes, if you work on the city level, you even work in kilometres, and so you have a really big range.

The architectural lightning desires describes how she works in an innovative—versatile and creative—way which she relates back to her architectural upbringing:

> Everything that I do, I feel that I'm an architect and that is the reason why I'm doing things that way. Apart from all the other skills, experience, blah, blah, blah, I feel myself an architect. When I do something, I can recognize my architecture background in what I'm doing. I mean, only an architect or somebody with an architecture background could have done that specific project in lighting design in such a way.

She continues:

> For instance, somebody asked me to do a report to be taken to the court in order to clarify a situation. It is something that really never crossed my mind to do such a thing. And then well, I put all my skills onto this report, and then we end up winning the cause. That made me think… Even things that you don't really appreciate, you can end up to be very good, just because you do things differently. I think because you bring into the subject a fresh perspective or at least one or two small details that make a huge difference.

An architect having her own office combining this with a position as teaching assistant and doctoral research describes her understanding of being an architect as carrying ethical responsibilities striving for inclusivity:

> An architect is a very well educated intellectual, a public intellectual who always tries to project for public good. It is a person who is very sensible towards the environment and towards people, and who has ethically high standards and a high moral ground. An architect is someone who fights for the weaker, who can really help out with the knowledge and improve given situations. Someone who strives for beauty because he believes that beauty is healing. This can be applied in designing everything: fashion, program briefs... It's someone who lives his profession with its entire body and mind.

AE delivers skilled professionals that are able to develop creative and spatial skills, yet also show to work beyond disciplines, across scales and with the capacity to dive into complexity, while looking to contribute to improve the living environment, striving for inclusivity. If anything, we hope this collage of quotes—in addition to the results from the questionnaire—can illustrate how architecture graduates have trained a range of skills and behaviours that in multi-sector forms show a parallel with the ways of working that could be in support of the NEB movement. These 'architectural ways of working' not only show to question the very disciplinary boundary of the profession of the architect, yet entail a promising potential in Europe's NEB search to bring about a transition that is beautiful, sustainable and developed together.

## 5 Beyond Architecture

In this paper we looked into the skills and behaviours—ways of working—that could be at support of the NEB movement, resulting in an informal list that brings up: Interdisciplinarity (and collaboration), innovation (entailing creativity), sustainability (including green skills), a multilayered approach (combining the local and global, the old and the new), working place-based, life-long learning, and user-centricity (in an inclusive approach) and communication skills. Followingly, looking at the skills and behaviours trained in AE and practised by architectural graduates—as studied through the AA research—a set of (mis)matches could be identified. Overall, the study of this paper suggests there is a significant match to be found between the 'architectural ways of working' and the orientation directed by the NEB.

This paper thus provides a promising starting ground to further explore the question on how AE can play a role in the future Europe is aiming for. From an optimistic interpretation of this paper's results, we would suggest to further invest and advance AE—and the broad architectural practice—as an important—maybe even crucial—lever for a more sustainable, beautiful and inclusive Europe. A suggestion, which is not out of tune with the direction the NEB has taken, bringing (among others) design and architecture to the front.

Yet, we would also like to address the limitations of this study. First and foremost, the identification of the skills and behaviours sought for in the NEB movement are still rather open and undefined and would need further clarification and clear directions from the European institutions behind the initiative. As a second note, it is important to address that the AA study from which the paper draws, has not been directed with the NEB skills and behaviours in mind. Due to these two elements the relation between skills and behaviours brought up by the NEB and those trained in AE, are not straightforward. As a third, and last limitation we would like to address this paper has only looked into the AE, thus no comparison has been made to other (design) degrees and the relevance whether or not of the skills and behaviours trained there.

To conclude, we would like to provide three suggestions to further research and advance the potential that can be found in the parallels between the NEB ways of working and those trained through AE.

As a first suggestion we would address the need for more clarity on and/or recognition of the skills and behaviours that are in support of the NEB's ambitions for sustainability, inclusion and beauty. More precisely we would like to address how projects that are put in the spotlight as more exemplar—such as those awarded by the NEB prize—could be discussed with more attention to the skills and professional behaviours underlying them. Looking into the ways of working behind, this study left us wondering on the how of the realisation of the project, the types of experts and expertises that were employed. How did collaboration and interdisciplinarity take place? What drove the place-based approach? We thus would call for the NEB project descriptions to be more inclusive of the often left invisible underlying process of realisations, the ways of working. We would specifically like to draw attention to personal competences (such as being critical, dealing with uncertainty, determination, endurance, constant learning and self-improvement, and work ethic) that tend to have a driving force—at least as is the case for architecture graduates—yet currently remain out of the spotlight. We thus would like to invite the NEB movement to question our western notion focused on the technological, the supposed straightforward objective-result process and the emphasis on the finished project.

Secondly, we would like to suggest building further on the potential of what is already there in the AE. It is this paper's argument that AE already has—yet often unconsciously—fostered a range of important skills and behaviours that allow architecture graduates to work in ways that show to be in line with the NEB ambitions. More precisely, the broad way of multi-sector directions architecture graduates take, show the study and the architectural ways of working have a potential that can reach far beyond what is traditionally understood to be the 'capacity' of an architectural degree. Hence, investing in AE might train—often unnoticed—ways of working that can help to set up the needed innovations that embrace a creative way of addressing the need for sustainable, inclusive and aesthetic solutions. We would suggest further research on this connection between AE and the NEB, among others by identifying the participation of architecture graduates—and those of other degrees—at diverse levels of the implementation of the NEB. Furthermore, we would like to suggest to focus the discourse more on the (architectural) ways of working, rather than the profession of architecture (yet still crucial to involve).

As a third and final suggestion, we would like to call for spotlighting and strengthening the soft skills / personal competences in EA, and beyond. These showed to be both strongly developed in AE, yet also leaving architecture graduates with still insufficient capacities to work with clients or have sufficient empathy as requested by their current occupations. Endurance, life-long learning, working on different scales, an ethical drive... These types of skills are not explicitly addressed by curricula in AE, yet show to be most impacting in the experience of architecture graduates. These skills tend to remain implicit and invisible in our western society, yet have shown to—although remaining under the radar—be of significant importance to architecture graduates. They tend to appear more clearly in the qualitative than quantitative part of the Architecture's Afterlife study and are not explicitly represented in the curricula of AE. Policy documents at EU level express objectives to redevelop curricula focused on competitiveness and employability in order to achieve the Green Deal [3], we would like to add a more fundamental recognition of the potential of training personal competences in EA—and maybe in higher education in general. These skills and behaviours—ways of working—show to relate to what the NEB discourse is hinting at.

To conclude, we would like to call for looking beyond the architecture and dive into the ways of working, the personal competences, the skills and behaviours that are lying beyond.

## 6 An Anthology for the Invisible Beyond Architecture

In support of the last two points of suggestions made, we conclude with a representation of five—out of many more—beyond architects and how they describe their ways of acting, thinking and being.

SIGRIED KELLENS

| | |
|---|---|
| sector | architecture + creative sector |
| based in | Brussels, BE |
| nationality | Belgian |
| age group | 20–40 |
| identifies as | female |

#03.01
– INTERIOR ARCHITECT
– URBAN DESIGNER
– prof. IN SCENOGRAPHY

"But because it's all going around spaces, so what is special to my profile, is this working on a lot of different levels. ... In all these levels, actually you work in different ways. And I remember once that somebody told me that actually, what I have as profile is a T-profile and it means that you are not a globalist, and not a specialist. It's somebody who is in-between, somebody who overviews."

"On one hand, in architecture there are a lot of scales and levels of scales. Sometimes you work for some projects in millimeters and then you turn to bigger scale projects in meter. And sometimes you even, if you work on the city level, you work in in kilometers, and so you have a really big range.

[...] And, I don't know if I can say really, I find it hard to say... Because they say I have a profile who is working in this a lot, there is kind of baseline that is the same."

"I would also add a way to think and to have a critical opinion. Because actually as an architect you learn to ... what are the other people doing, looking to references, try to put a position where you put yourself in it in comparison with what's already happened and so you, I think that is even even more important in urbanism you really need to motivate, to find a good argument, to have a vision, and all these things I think you already learn, on a more little scale, already in the field."

"And I think the type of studios that you still have also in our schools who are part of a university it is really something that is coming from our schools. Because you, you don't have that in engineering or other fields that are quite close to us. They work in a really classical way as a university, but not with projects that you need to develop and that you need to come to a teacher who helps you to develop the project. And I think, that's another thing that is good about architecture schools is that you learn to work on projects. You actually, now more and more in a lot of fields you need to be able to work project-based and that needs also, a really other, type of knowledge or soft skills."

## JOAO COELHO

sector creative sector
based in Lisbon, PT
nationality Portuguese
age group 40–60
identifies as female

## #03.07 – ARCHITECTURAL LIGHTENING DESIGNER

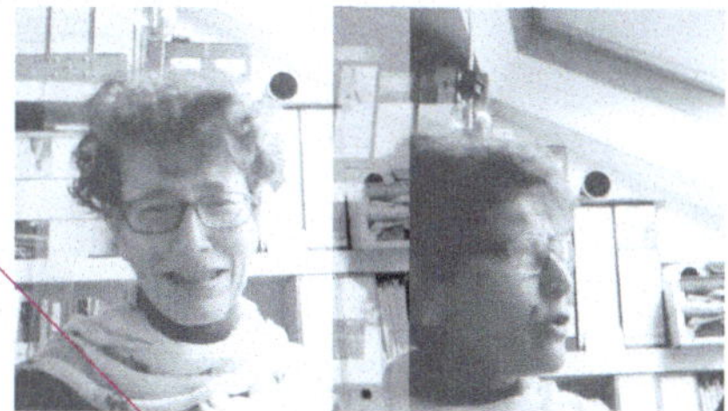

"So in the end we somehow have to combine or introduce our worries. And that, that is interesting. It's difficult sometimes, but it's all about communication. And that is one of the skills that I think that in the university should be promoted.

And going a bit backward; communication. That is something that I've not learned at university at all, because nobody asked me about anything, but I've had to learn with experience. And I think it's the key for almost everything; is to be able to communicate the way you see things, the way you want to sort it out the way you identify the problem: communication."

Considering the conceptual idea that lies behind 'soft skills' [...] I carried on thinking about possible more revealing words. [...]
– Skillsbow
– Synapse Skills
– Connectome Skills
– Wired skills

"For sure. So everything that I do, I feel that I'm an architect and that is the reason why I'm doing things that way, apart from all the other skills, experience, blah, blah, blah, blah.

I feel myself as an architect, when I do something and I can recognize my architecture background in what I'm doing. I mean, only an architect could have done something or somebody with an architecture background could have done that specific specific project in lighting design in such a way. So yes, it is there, it's a way of thinking of looking at things of behaving, of communicating. It's a way of being really."

## NILS GAETHALS

| | |
|---|---|
| sector | architecture + creative sector |
| based in | Antwerp, BE |
| nationality | Belgian |
| age group | 20–40 |
| identifies as | male |

## #03.08 – ARCHITECT + EVENT ORGANISATION AND DESIGN

"I am a 'ondernemend persoon' [entrepreneurial / venturous person], like a person that that knows how to deal with stuff. And that's also why people ask me for these kind of positions or jobs. It's also because you know how it works to be self-employed and how that goes.

Of course architecture brings also other skills that are very handy, not only organization -and this like being a company and working a lot- but it's also the creative thinking, designing and also like the graphical part."

"You put yourself on the line and it's also – how to say it –, it's also a dangerous path to go on sometimes. Because when you're self-employed, you're doing stuff for yourself and it can always be more and better and faster.

And in the end, I do know that yeah, from other architecture offices or people that I talk with, that there is a very thin line between doing just the right amount and doing too much because this also exists "

"And in that sense, as an architecture, you're able to deal with these kind of things [complex projects such as event organization and design].

[...]

So in that sense, when you come into these other fields [other than architecture], it's something for me that is very, very clear. Okay, now I will deal with it like that. And I do feel that some people that I work with, they really don't have that kind of dealing with complex problems. And they just go in head first and it will be fixed in the end, but they lose themselves. It is really a skill. One that's very, very valuable for sure."

## ANDREEA FELCIUC

| | |
|---|---|
| sector | other sector |
| based in | London, UK |
| nationality | Romanian |
| age group | 20–40 |
| identifies as | female |

## #03.12 – VENUE DESIGNER

"Well, my first experience [working in a team with no other architects] was I think, quite a blunt one and it was quite different. When I came in and I had my first meeting about a project, I was the only person in the meeting room who brought some paper and a pen. I wanted to sketch what we were... I mean, I will not take notes. I will try to sketch what I'm discussing, because it's much more easier for me to remember and to relate to that. And that was something completely new for them."

"Well, this is the interesting thing [working as part of an interdisciplinary team]. Everyone is doing something a little different. Everyone specializes in something completely different. I'm the only one in the team [that is an architect]. So for example the director of this team, he's a professional violinist, a conductor. And at the same time he has 20 something years of experience doing this [building and designing venues]. And I tend to find that he knows more about architecture than most of the people that I know.

[...] it's a bunch of people doing different things. Most of them are part of acoustics or part of technical equipment, or they do tend to have something related to management, arts and culture, management... But no one is actually doing the thing that they specialized in. So we somehow do lots of things and we learn a lot from collaborating with other people."

I think critical thinking is the most important. I do tend to draw a lot to sketch a lot, but most of the time is about analyzing and responding to things fast and as precise and insightful and helpful and constructive as it can be. And that is maybe the most challenging thing.

## DANIELE RATTI

| | |
|---|---|
| sector | creative sector |
| based in | Milan, IT |
| nationality | Italian |
| age group | 40–65 |
| identifies as | male |

## #03.07 – PHOTOGRAPHER

"When you draw, you draw what you have in mind, but there is a gap from the drawing and the reality, right? [...] It's an imagination. It could be also very accurate, but it's a drawing, it's something that exists on paper.

Right now you pretend to show your clients exactly what [the project] will look like, and you don't let your brain work. [...] In Italian we say "Pappa pronto" - when the meat is ready, you have to go and sit at the table and eat. There is no... you are not cooking. The drawing is the cooking."

"I have to do a lot of things for two reasons, because today it's hard to. I would like to do one work well, being well paid. But today it's, I have to do 10 works to be well paid. This is a contradiction today, or maybe it's also, it's also me. I have to do things, to think things... And the fact is that I like what I do, for me it is not work.

This is the real problem; for me it is not work."

"I think that I'm not a person that can do everything by himself. So one of my characteristic is that I always find people who are, who I need. Who -I don't think this is correct in English- are 'usable to one another'. And I try to find the best people, the best ones."

## References


1. EU Science & Innovation (Director) (29 Sep 2022b) New European Bauhaus: collaboration, community and culture for innovation [Video recording]. https://www.youtube.com/watch?v=5je_zj2lAYY&ab_channel=EUScience%26Innovation
2. European Union (nd-a) New European Bauhaus: About the initiative. Retrieved 16 Oct 2022, from https://new-european-bauhaus.europa.eu/about/about-initiative_en
3. European Parliament (2021) Education and the New European Bauhaus (p. 2) [Background report]. https://www.europarl.europa.eu/RegData/etudes/ATAG/2021/689363/EPRS_ATA(2021)689363_EN.pdf
4. EU Science & Innovation (Director) (10 Jun 2022a) Education and the New European Bauhaus… Let's imagine! https://www.youtube.com/watch?v=qnJnBuDU3d0
5. European Union (2022b) Winners of the New European Bauhaus prizes 2022. https://2022.prizes.new-european-bauhaus.eu/

6. European Commission (2016) A new skills agenda for Europe: working together to strengthen human capital, employability and competitiveness (Communication from the Commission to the European Parliament, the Council, the European Economic and Social Committee and the Committee of the Regions, p. 18). https://eur-lex.europa.eu/legal-content/EN/TXT/?uri=CELEX%3A52016DC0381R%2801%29&qid=1669629599020
7. Open IA (2 Apr 2023) What are the skills and professional behaviours the New European Bauhaus is looking for? ChatGPT. https://chat.openai.com/chat. ChatGPT lists the following five sources as "reference for where to find information on the skills and professional behaviours the New European Bauhaus is looking for": (1) European Commission (2021) New European Bauhaus. https://ec.europa.eu/info/law/better-regulation/have-your-say/initiatives/12919-New-European-Bauhaus. (2) New European Bauhaus (2022) Manifesto. https://neweuropeanbauhaus.eu/manifesto/. (3) New European Bauhaus (2021) FAQ. https://neweuropeanbauhaus.eu/faq/. (4) European Cultural Foundation (2021) The New European Bauhaus: a cultural project for a new European Green Deal. https://www.culturalfoundation.eu/library/new-european-bauhaus. (5) European Parliament (2021) The New European Bauhaus initiative. https://www.europarl.europa.eu/news/en/headlines/society/20210311STO99113/the-new-european-bauhaus-initiative
8. Architecture's Afterlife (In press) Architecture's Afterlife: professional identities beyond the degree. Routledge
9. Architecture's Afterlife (2021) Intermediate report on the second Afterlife survey. http://architectures-afterlife.com/en/outputs?q=63
10. Charmaz K (2017) Constructivist grounded theory. J Posit Psychol
11. Architecture's Afterlife (11 Nov 2022) Architecture's Afterlife: the multisector impact of an architectural qualification. Valencia International Biennial of Research in Architecture. http://architectures-afterlife.com/files/file/dissemintation/MEO6_2022_11_UPV.pdf

# NEB and Contemporary Design

# Mutualist Associations' Contribution to a New European Bauhaus

Paula Cristina Barros[1(✉)], Ana Patrícia Duarte[2], and Margarida Perestrelo[1]

[1] University Institute of Lisbon, Dinâmia'cet-iul, Avenida das Forças Armadas, 1649-026 Lisboa, Portugal
pcsbs@iscte-iul.pt

[2] University Institute of Lisbon, BRU-Iscte, Avenida das Forças Armadas, 1649-026 Lisboa, Portugal

**Abstract.** This chapter discusses how mutualist associations (MAs) in Portugal can contribute to communities and regions' sustainability in terms of health, well-being, welfare, and resilience, as stipulated in the United Nations' Sustainable Development Goals 3, 11, and 17. The research was based on a conceptual model of human beings' role on this planet and its consequences for diverse territories. This study integrated multi- and transdisciplinary perspectives and concentrated on MAs' contributions in Portugal to determine whether and how these associations can provide solutions that can be used to update the New European Bauhaus. The findings are based on a literature review, national secondary data, and case study of two MAs. The results reveal that these organizations are key players in efforts to promote territories' sustainability and communities' health, well-being, and welfare, acting in partnership with other regional stakeholders. Acknowledging these associations' significant contribution to communities and regions' well-being and development could encourage other groups to create other similar organizations. Their activities would contribute to varied communities and territories' sustainable development, thereby generating multiple gain spirals.

**Keywords:** Communities · Regional resilience · Mutualist associations · Portugal · Sustainable development

## 1 Introduction: The Relationship Between Mutualist Associations (MAs) and Regional Resilience

A fuller understanding can be formed of the relationship between MAs and regional resilience in Portugal by examining the generic process by which the European Commission (EC) identifies these organizations. The EC defines MAs as non-profit organizations that seek to satisfy shared human needs by providing life and non-life insurance services and the benefits of complementary social security schemes or social services.


M. Blanco Lage et al. (eds.), *Towards a New European Bauhaus—Challenges in Design Education*,
https://doi.org/10.1007/978-3-031-40188-6_16

These organizations can be formed via two types of mutualization. The first covers risks such as illness and death, which are subject to specific legislation. The second is more closely related to insurance that covers all types of risks. The latter is normally subject to the general laws applicable to insurance [11].

This chapter focuses on how MAs are defined in Portugal, namely, as associations[1] with an added mutualist component.[2] These organizations also provide social services, for example, nursing homes, daycare for children and senior citizens, pharmacies, and health clinics (i.e., complementary services that increase well-being). One MA also owns a bank, so its role is more complex than the EC's interpretation of mutual societies would usually allow.

In Portugal, these organizations are called "*associações mutualistas*," which can be translated directly into English as "mutualist associations." Regional resilience in this country is intrinsically linked to MAs' history, so their operations permeate these territories and communities. However, MAs' contribution to regions and communities' strength and sustainability has not always been fully understood or acknowledged. Given this tendency, the remaining sections explore how MAs can contribute to meeting the challenge of updating the New European Bauhaus.

## 2 Conceptual Model: Human Beings' Role on This Planet

Barros [5] describes the model presented in "Developing the Bridges Theory between Well-being and Well-feeling" as follows:

> Th[is …] is a proposal on which a more in-depth academic study should focus, exploring a wider perception of the human being in communion with the planet [. …] The model is based on an understanding that links the past, present, and future correlated with generations who must grow up in [… communal] contexts based on solidarity. (p. 47)

Responses to this call for further research have included, from a generalized human perspective, studies of the complex problem of generational gaps[6] and the identification of social and solidarity economy interactions [1] based on reciprocity. Another related investigation focused closely on Polanyi's work [27].

The present research concentrated on MAs because of their presence in Portugal's regions for 720 years [35] as dynamic actors. These organizations were started as a reflection of individuals' interest in democratic processes in a collective format. Portugal's MAs contribute to sustainability, that is, the United Nations' (UN) Sustainable Development Goals (SDGs) 3, 11, and 17 (UN, 2015). In addition, these associations help strengthen the three pillars defined by Elkinghton (2004)—environmental, social, and economic welfare.

---

[1] An association is a group of people who, of their own free will, legally organize themselves to achieve a common objective.

[2] The MAs seek to mutualize life risks, savings, and longevity in ways that are complimentary to the available welfare services.

Barros [5] also highlights these three aspects and further includes the management of humanity's common heritage in a third dimension identified in the *International Dictionary* as "the other economy" (as cited in [8]). Figure 1 presents the resulting conceptual model.

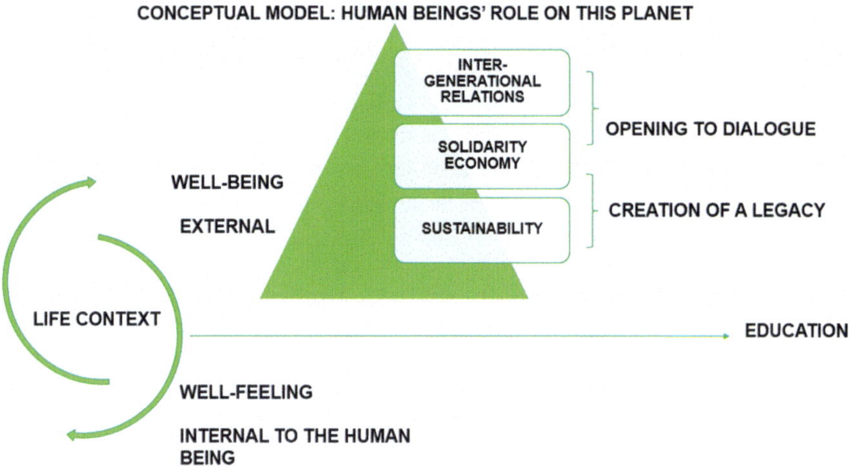


**Fig. 1** Conceptual model of human beings' role on this planet. *Source* Adapted from Barros [5]

## 3 Literature Review and Research Question: MAs' Contribution

### 3.1 MAs in Portugal

This chapter[3] seeks to address a single question to promote a greater consensus on the research topic: How can MAs in Portugal contribute to communities and regions' sustainability in terms of health, well-being, and welfare? Welfare, in this context, is related to living conditions and quality of life [36] due to factors external to humans, which are linked with the conceptual model [5] in Fig. 1. Health and well-being, in turn, are connected to satisfaction and subjective well-being [34] based on humans' internal coping mechanism, which are also related to the conceptual model [5].

According to Portuguese legislation, MAs fall within the cooperative and social sector categories. These organizations are thus regulated by the Social Economy Basic Law n.º 30/2013 of May 8th [9] and the more specific Decree-Law n.º 59/2018, which sets out the Code of Mutualist Associations [31]. The latter legislation specifies that:

> To strengthen the mutualist category, the concept of mutualist association has been redefined to highlight, first, this type of organization's associative nature and mutualist function and […, second,] its integration into the combined spaces of private institutions of social solidarity and the even broader social economy. (p. 3740)

[3] This chapter is part of a doctoral thesis.

The cited code's Article 8 adds to this definition by stating that the use of the MA label is exclusive, which contrasts with the EC's definition. Given multiple understandings of this kind of association [3], the Portuguese Council of Ministers [, p. 3744] clarifies further that "[t]he use of the terms 'mutualist association', 'mutual aid association', or 'mutuality' is reserved exclusively for the institutions covered by this regulation" [31].

The code guides and operationalizes the financial aspects of MAs' governance model, among other elements, describing key principles in Article 9 [31].[4] As a result, this legislation is fundamental to the current investigation. Notably, the code stipulates that the Supervisory Authority for Insurance and Pension Funds is in charge of supervising and regulating some of these organizations' activities for a transitional period of 12 years.

The European Parliament regulates insurance and reinsurance activities (i.e., [29]), but the current guidelines exclude MAs in Portugal and only recognize their mutual insurance services[5][12]. However, in Portugal, these organizations' guardianship (i.e., approval of their Statutes and Benefit Regulations) is the central government's responsibility, namely, the Ministry of Labor, Solidarity, and Social Security (i.e., Article 6, point 5) [31]. This rule contrasts with the EC's interpretation of MAs [11] since this specific point of the Code of Mutualist Associations is exclusively applicable to MAs. More specifically, their human (i.e., number of members) and financial (i.e., surpluses) capital needs to be regulated to safeguard of its members.

Regardless of these complex features, the EC treats all solidarity organizations in Portugal the same even though MAs' particularities unequivocally meet their associates' needs in the field of health—some with pharmacy services and others with social programs for the elderly and/or children. Other needs are met through welfare services providing ways to ensure survival, savings, longevity, and death risk coverage and, finally, enhanced well-being (i.e., education and quality of life).

In addition, the present research confirmed that some MAs carry out their activities in emblematic buildings that help preserve Portugal's architectural heritage. Their installations' conservation add to the historical value of not only the relevant cities but also these institutions. A prime example is the Edifício dos Antigos Armazéns Grandela, in Lisbon, Portugal's capital, which is an historical building built in 1907 and destroyed by the 1988 Lisbon fire and which reopened in 1996 [7]. Located at 42 Rua do Carmo, this structure houses the Montepio Geral Associação Mutualista's services and operations. Other examples are the headquarters of the Associação de Socorros Mútuos Familiar Vimaranense, in Guimaraes, in the north of Portugal, where this institution has had "a home since 1935" [4].

## 3.2 The Study: Two MAs in Portugal

The universe of MAs in Portugal, according to the government's Social Security national database, comprises 106 associations.[6] According to Decree-Law n.º 59/2018, 1,100,000

[4] The six principles of MA are presented in Article 9 as freedom, democracy, equality and non-discrimination, independence and autonomy, solidarity, and responsibility.

[5] In Portugal, only one MA carries out this activity: Mútua dos Pescadores.

[6] More specifically, 105 MAs are registered in mainland Portugal and the Azores [26, 32] and Madeira has one association [16].

associates in Portugal are members of these MAs. Since 2018, the Portuguese population has stabilized around that year's figure of 10,276,617 [22], which means that 10.70% of the Portuguese population is part of an MA.

Two associations were selected to understand more fully these organizations' contributions within Portugal. These two MA were chosen for this study for specific reasons. César Oliveira wrote in 1974:

> The great period of associations [...] refers to [...] the period between 1851 and 1872. Moreover, this surge [...] accompanied urban growth, [...] the growth of the working class, and the emergence of new salaried professions in large urban centres, [...] especially in Lisbon. (pp. 17–18)

These observations provide two important pieces of information. First, MAs are related to workers and, second, these associations are normally restricted to specific regions. Thus, the two MAs were chosen because they provide social services in the areas of health, social needs, and welfare, among other workers' benefits, to city hall employees.[7] The associations are located in São Pedro do Sul, in the Dão-Lafões subregion, and Gondomar, in the Oporto Metropolitan subregion.

These public sector workers had specific motivations for organizing themselves on a legal level. A specific factor in Portugal is that, after MAs are constituted and their registration approved, these organizations acquire the statute of private social solidarity institutions to which the Portuguese government confers benefits. For example, MAs are exempted from social security taxes or pay a reduced rate if they hire citizens who are vulnerable and thus their employment is considered "socially necessary" [16].

Information was gathered on these MAs and their regions for the present study. The Workers of the São Pedro do Sul City Hall (MUT) MA was formed in 2002, and, by 2022, 10.65% of the workers who qualified to do so had joined the association [28]. The surrounding subregion has a resident population of 252,793 citizens [24], which is 2.44% of Portugal's population.[8]

An analysis of the data available confirmed that the São Pedro do Sul municipality is part of the Viseu district, which is located, according to the Nomenclature of Territorial Units (NUT), for statistical purposes in the NUT-II Central Region and the NUT III Dão-Lafões subregion. This territory occupies 3,237.74 square kilometers (km$^2$) [20], with a population density per km$^2$ of 43.6 [23].

[7] According to Portuguese law, the city hall is the municipality's executive body. It implements and defines policies that develop and improve the surrounding region in areas such as social action, the environment, external cooperation, culture, housing, land-use and urban planning, heritage, municipal police, development, health, and leisure activities.

[8] Data for 2021.

The region is known for its thermal spas both as a water resource and tourism attraction, and the area is crossed by the Vouga River and the Ribeira da Água, among others. The landscape's morphology is uneven, with a mountain range over 1,000 m high. The climate is oceanic, and temperatures rise higher in lower altitude areas [15].

The Municipality of Gondomar MA [AMUT] was established in 2011, and, by 2022, 55.35% of the eligible workers were associates [2]. The surrounding subregion has a resident population of 164,277 [24], which is 1.59% of Portugal's population.[9]

Gondomar is a municipality in the Oporto district. According to NUT II, Gondomar is located in the North Region, and NUT III places this city in the Oporto Metropolitan Area. Gondomar comprises an area of 2,041.27 km$^2$ [21], with a population density per km$^2$ of 1250.3 [23]. Urban agglomerations appear across the landscape. These settlements are shaped by Douro River's right bank, a high rugged terrain separating them into valleys, and fertile soils fed by abundant water [14].

These MAs were next examined to understand their main influence on city hall policies and missions. The research also focused on the relationships established between the internal and external factors affecting the human beings involved. After these topics were explored, conclusions could be drawn about the two MAs' contributions to their respective territories.

## 4 Methods: Strategic Analysis of Actors and Wellness Index

Barros's conceptual model [5] was applied to understand how the MAs under study are linked to communities and regions' sustainability in terms of health, well-being, welfare, and resilience, as outlined by SDGs 3, 11, and 17. Based on a multi- and transdisciplinary perspective, the methodology selected assesses how workers who have joined the MAs can individually contribute to regional and community resilience. More specifically, the data are being collected in interviews and processed using the strategic analysis of actors approach and a wellness index.

### 4.1 Strategic Analysis of Actors

According to Godet and Roubelat's [13] "Creating the Future: the Use and Misuse of Scenarios":

> [T]he use of simple formal tools like structural analysis, actors' strategy analysis, morphological methods or probability analysis, illustrated with case studies on steel and iron industry, are useful to avoid distractions and to explore all possible scenarios. (p. 1)

---

[9] Data for 2021.

The present study's methodologies thus facilitated the identification of MAs' strategic objectives, channels of action, constraints, power relations between varied actors, and expectations of future partnerships. The analyses, therefore, focused on these associations' influence, relationships, and cooperation, among other aspects.

Perestrelo writes [30]:

> The "appropriation" by urban and territorial planning of the assumptions of strategic business planning is due, on the one hand, to the growing complexity of forms of territorial occupation, [and] to the diverse interests, often conflicting and divergent, that manifest themselves, as a defense of a [...] more innovative [state, namely], a more "encouraging" state than a managerial state. (p. 15)

With this complexity in mind, the selected methods were applied to understand how these organizations' workers, stakeholders, communities, and other actors can contribute to well-being, subjective wellness, sustainability, and generational legacy. As Barros's research [5] highlights and Sect. 3.1 discusses, these patterns are an integral part of Portuguese territories.

### 4.2 Wellness Index

The well-being index chosen for this research is used primarily to ascertain quality of life, health, and satisfaction. The Commission on the Measurement of Economic Performance and Social Progress's (2009) report provides the indicators for quantifying the Gross National Happiness index used by the UN [33]. In 2004, the Instituto Nacional de Estatística [25] (i.e., the National Institute of Statistics) published a report on Portugal. This institute [18] had previously clarified its index's purpose:

> The objective of the Well-Being Index is to make available, on a regular basis, results that make it possible to monitor the evolution of well-being and social progress in two key aspects – material living conditions of families and quality of life, through [... changes] in [from] three [... to] seven domains of analysis, respectively: a) economic well-being; economic vulnerability; [... and employment]; b) health; work-life balance; education, knowledge and skills; social relationships and subjective well-being; civic participation and governance; personal security; and environment. (p. 9)

The current study applies this index to clarify how these workers perceive themselves in terms of health and well-being, more specifically, with regard to satisfaction and subjective well-being, as previously mentioned in Sect. 3.1.

## 5 Results: A Path to Follow

This chapter reflects undergoing research, so the results presented above are only preliminary. However, they provide a wider perspective on MAs and discuss these organizations' potential contribution to their regions. The literature review results reveal that these organizations are key players in efforts to promote territories' sustainability and communities' health, well-being, and welfare, acting in partnership with other regional stakeholders, thereby responding to the European Association for Architectural Education 2022 conference's call for a clarification of the New European Bauhaus's role in meeting SDGs.

## 6 Conclusions: Towards a New European Bauhaus

Von der Leyen (as cited in European Commission, 2021) argues for the importance of clarifying "how we want to live together" [5, sec. 0:47], "respecting the planet" [5, sec. 0:49], "bringing generations together" [5, sec. 0:55], and "matching sustainability with style" [5, sec. 1:10], as well as ensuring that "[i]t's about all of us" [5, sec. 1:12]. The present research thus sought to develop a new understanding of territorial resilience and, more specifically, of MAs' valuable role as private organizations with a public function and as part of the social and solidarity economy.

MAs are created to address a particular group's communal situation, risk, or vulnerability and to offer assistance that complements government and private sector services. These associations' complementary offers can provide essential help to ensure communities gain access to primary services at fairer prices. In addition, the responsibility is shared by all members through democratic processes.

This chapter also sought to apply Barros's (2019) conceptual model of human beings' role on this planet, thereby ensuring a multi- and transdisciplinary perspective on the research topic.[10] The above findings, therefore, address in part the challenges of updating the New European Bauhaus.

## References

1. Amaro RR (2009) Economia Solidária Apresentação do Conceito. Revista de Economia Solidária, 1
2. AMUT Homepage. https://amut.pt/historia/. Last accessed 6 May 2022
3. Archambault E (2010) Mutual organizations/mutual societies BT. In: Anheier HK, Toepler S (eds) International encyclopedia of civil society. Springer US, pp. 1015–1021. https://doi.org/10.1007/978-0-387-93996-4_50
4. Associação de Socorros Mútuos Familiar Vimaranense Homepage. https://www.afvimaranense.pt/dt_gallery/sede-familiar-vimaranense/. Last accessed 25 Jan 2023
5. Barros PC (2019) Desenvolvendo a teoria das pontes entre o bem-estar e o bem-sentir [ISCTE-IUL]. http://hdl.handle.net/10071/19654. Last accessed 24 Aug 2022
6. Bengtson VL (1970) The generation gap: a review and typology of social-psychological perspectives. Youth Soc 2(1):7–32. https://doi.org/10.1177/0044118X7000200102. Last accessed 22 Aug 2022
7. Câmara Municipal de Lisboa Homepage. https://informacoeseservicos.lisboa.pt/contactos/diretorio-da-cidade/edificio-dos-antigos-armazens-grandela. Last accessed 25 Jan 2023

[10] That is being studied in the doctoral thesis of which this paper is part.

8. Cattani AD, Laville J-L, Inácio Gaiger L, Hespanha P (2009) Dicionário Internacional da Outra Economia (S. Edições ALMEDINA (ed)). Almedina. http://hdl.handle.net/10316/80208. Last accessed 21 Aug 2022
9. Decreto do Presidente da República n.o 58/20130 de 8 de maio - Lei no30/2013 - Lei de Bases da Economia Social. In Diário da República, 1.a série—88(88)
10. EU Debates (2021) Ursula von der Leyen eudebates the New European Bauhaus. eudebates.tv. https://www.youtube.com/watch?v=ZYFojt2zi9w. Last accessed 25 Nov 2022
11. European Commission Homepage. https://single-market-economy.ec.europa.eu/sectors/proximity-and-social-economy/social-economy-eu/mutual-societies_en. Last accessed 24 Aug 2022
12. European Parliament and Council (2009) Actividade de seguros. Jornal Oficial Da União Europeia, 1–155. https://eur-lex.europa.eu/legal-content/pt/TXT/?uri=CELEX:32009L0138. Last accessed 25 Jan 2023
13. Godet M, Roubelat F (1996) Creating the future: the use and misuse of scenarios. Long Range Plann 29:164–171
14. Infopédia (2022a) Gondomar. Infopédia, Dicionários Porto Editora. https://www.infopedia.pt/apoio/artigos/$gondomar. Last accessed 07 Aug 2022
15. Infopédia (2022b) São Pedro do Sul. Infopédia, Dicionários Porto Editora. https://www.infopedia.pt/apoio/artigos/$sao-pedro-do-sul. Last accessed 07 Aug 2022
16. Instituto de Segurança Social (nd) Instituições Particulares de Solidariedade Social - Associações Mutualistas. Retrieved Dec 1, 2022, from https://www.seg-social.pt/associacoes-mutualistas. Last accessed 1 Dec 2022
17. Instituto de Segurança Social dos Açores (2022) Instituições Particulares de Solidariedade Social Registadas. https://portal.azores.gov.pt/documents/36925/768021/ListadeIPSSnaRAA.pdf/69aeb49e-b760-e942-b61a-677499387ea4. Last accessed 5 May 2022
18. INE (2019a) Homepage. Documento Metodológico - índice de bem-estar (versão 2.0) - Instituto Nacional de Estatística
19. Instituto Nacional de Estatística (2019b) População residente (N.º) por Local de residência. https://www.ine.pt/xportal/xmain?xpid=INE&xpgid=ine_indicadores&contecto=pi&indOcorrCod=0011166&selTab=tab0. Last accessed 20 Nov 2022
20. Instituto Nacional de Estatística (2020a) Portal do INE. https://www.ine.pt/xportal/xmain?xpid=INE&xpgid=ine_indicadores&contecto=pi&indOcorrCod=0008350&selTab=tab0. Last accessed 6 May 2022
21. INE (2020b) Homepage. Portal do Instituto Nacional de Estatística. https://www.ine.pt/xportal/xmain?xpid=INE&xpgid=ine_indicadores&contecto=pi&indOcorrCod=0008337&selTab=tab0. Last accessed 24 Aug 2022
22. INE (2019) Estimativas de população residente em Portugal - Instituto Nacional de Estatística. https://www.ine.pt/xportal/xmain?xpid=INE&xpgid=ine_destaques&DESTAQUESdest_boui=354227526&DESTAQUESmodo=2. Last accessed 20 Aug 2022
23. INE (2020c) Homepage. Portal do Instituto Nacional de Estatística. https://www.ine.pt/xportal/xmain?xpid=INE&xpgid=ine_indicadores&contecto=pi&indOcorrCod=0008337&selTab=tab0. Last accessed 30 Sep 2023
24. INE (2021) Homepage. População residente (N.o) por Local de residência - Instituto Nacional de Estatística. https://www.ine.pt/xportal/xmain?xpid=INE&xpgid=ine_indicadores&contecto=pi&indOcorrCod=0011166&selTab=tab0. Last accessed 20 Aug 2022
25. Instituto Nacional de Estatística (2021) Índice de Bem-Estar. https://www.ine.pt/xportal/xmain?xpid=INE&xpgid=ine_indbemestar. Last accessed 16 Nov 2021
26. Instituto Segurança Social da Madeira (2022) Instituto de Segurança Social da Madeira, IP-RAM. Last accessed 25 Nov 2022

27. Machado NMC, Cardoso NM (2010) Karl polanyi e a nova sociologia económica: notas sobre o conceito de (dis)embeddedness. Revista Crítica de Ciências Sociais 90:71–94. https://doi.org/10.4000/rccs.1771
28. MUT (2022) Estatutos|Associação Mutualista dos Trabalhadores da Câmara Municipal de São Pedro do Sul
29. Solvency II Directive (2009) Parlamento Europeu e do Conselho. Actividade de seguros [Solvency II Directive]. Jornal Oficial Da União Europeia pp 1–155. https://eur-lex.europa.eu/legal-content/pt/TXT/?uri=CELEX:32009L0138
30. Perestrelo M (2005) Prospectiva: planeamento estratégico num contexto de desenvolvimento regional. ISCTE-IUL—Instututo Superior de Lisboa
31. Presidência do Conselho de Ministros (2018) Decreto-Lei n.º 59/2018. In Diário da República n.º 148/2018, Série I de 2018-08-02
32. Social DGdS (2022) Associações Mutualistas Registadas. https://www.seg-social.pt/documents/10152/864429/Lista+AM/053592b8-06be-4067-8546-b56353701b6f
33. Stiglitz JE, Sen A, Fitoussi J-P (2009) Report by the commission on the measurement of economic performance and social progress
34. Strack F, Argyle M, Schwarz N (1992) Subjective well-being an interdisciplinary perspective (Issue January)
35. União das Mutualidades Portuguesas (2017) Notícias do Mutualismo. Portuguesas, Mutualidades, Revista n., pp 1–64. http://mutualismo.pt/portal/images/website/publicacoes/mut/2017/mut10.pdf. Last accessed 21 Aug 2022
36. Veenhoven R (1996) Developments in satisfaction-research. Soc Indic Res 37(1):1–46. https://doi.org/10.1007/BF00300268

# The Question of Housing in the Scope of NEB Goes South

Teresa Calix[1,2](✉), José Pedro Sousa[1,2], and João Pedro Xavier[1,2]

[1] Faculty of Architecture, University of Porto, Porto, Portugal
teresa.calix@arq.up.pt

[2] Center for Studies in Architecture and Urbanism, Faculty of Architecture, University of Porto, Porto, Portugal

**Abstract.** Southern European regions face common climatic and social threats. If they strengthen their ability to cope, they can develop together social and nature-based solutions, sustainable while still inclusive and aesthetically qualified. This challenged six southern architecture schools to create NEBgS. Under the umbrella of education, a series of six co-design thematic events took place in 2021. Starting with an introduction to the NEB goes South initiative, this paper describes the co-design event organized by the Porto School—FAUP, which was dedicated to discussing the "Question of Housing" (QoH). Considering that cities and urban territories are a priority issue, the event started with an inspiration talk by Lacol architects who discussed their work around cooperative housing. Then, more than 40 professors, researchers and students from different countries gathered in 5 parallel sessions to discuss their ideas and experience on specific topics, which ranged from affordable housing, reconciliation with the environment, resilient modes of inhabitation, emerging design paradigms, and modes of education and practice. This paper concludes by summarizing the outcomes of the event. By arguing that QoH is a key topic for our common future, it demands, as before, an answer that goes further the individual action and depends on our cooperation in the construction of holistic and contextual solutions for an inclusive and sustainable tomorrow.

**Keywords:** Architectural education · New European Bauhaus · Housing · Sustainability · Inclusion · NEB goes South

## 1 Introduction

In the twentieth century, technology gained world scale and accelerated our time. We believed that we were freeing ourselves from geographical, corporeal, and temporal ties. However, beyond all the climatic changes that we were devaluing, the recent covid-19 pandemic made our human and vulnerable side more evident. It became clear that we should move towards the twenty-first century and recognize the paradigm shift that leads to a new built and natural environment condition as the New European Bauhaus strives to inspire [1].


M. Blanco Lage et al. (eds.), *Towards a New European Bauhaus—Challenges in Design Education*,
https://doi.org/10.1007/978-3-031-40188-6_17

The global pressure imposed on us (e.g., citizens, professionals, politicians …) to introduce urgent and often drastic changes in our life cannot admit easy (and bad) solutions. Existential threats like those we are facing today are complex and prompt us to question whether they should be addressed differently around the world. It seems clear that we cannot ignore the specificities of different territorial realities (e.g., climate, economic, cultural …), even in a small in the world as Europe is.

## 2 The NEB Goes South Initiative

In Southern Europe, countries are dealing with several problems such as increased signs of extreme climate events, rising temperatures, heat islands, desertification, or drought. These countries also share rich histories and a cultural heritage which need careful protection while addressing the mentioned emergencies. Therefore, it seems logic that a common and coordinated effort must be made in the region.

Based on the previous assumptions, six schools of architecture from six universities of six different countries in Southern Europe looked for building a southern perspective on the New European Bauhaus goals and to contribute to a climate neutral Europe by 2050. Through the combined efforts of students, teachers, architects, designers, engineers, geographers, sociologists, and other scientists and artists, with the ideas of everybody concerned.

As a result, the NEB goes South (NEBgS) initiative emerged in 2021 as a pan-European platform created to broaden and enhance the international debate on co-designing sustainable solutions as the New European Bauhaus strives to inspire [2]. It was jointly organised by the Faculty of Architecture of the University of Porto, the Valencia School of Architecture, the Toulouse School of Architecture, the Department of Architecture, Alma Mater Studiorum of the University of Bologna, the Faculty of Architecture of the University of Zagreb, and the School of Architecture of the National Technical University Athens (Fig. 1).

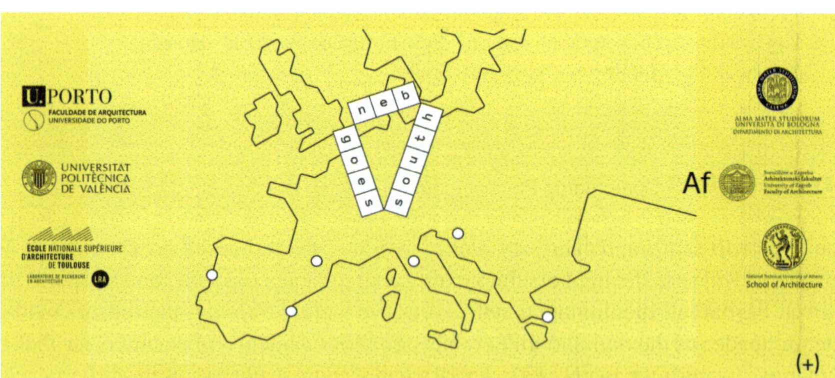


Fig. 1 The NEB goes South consortium

Closer co-operation and co-design between Southern European countries/institutions/actors could strengthen our ability to address our common threats and develop social and nature-based solutions together. This experience is not intended to limit the discussion or reduce the problem to a confined territorial area. It may also influence or even inspire other regions to discuss methods, instruments, and solutions to common planetary challenges. Furthermore, considering the critical role of higher education and research institutions, we believed that efforts need to be taken into promoting inclusive participation, and organizing and disseminating results.

## 3 The Co-design Events

The NEB goes South kick-off event took place in June 2021 during the pandemic period. Hosted online, it counted with the participation of the European Commissioner Elisa Ferreira and Ruth Reichstein from European Commission's Advisory Board I.D.E.A on Green Deal and New European Bauhaus. This session was followed by five other thematic events, each one coordinated by one of the schools and involving individuals (specialists and students) related with the topics that most mattered to the Southern European regions (Fig. 2).

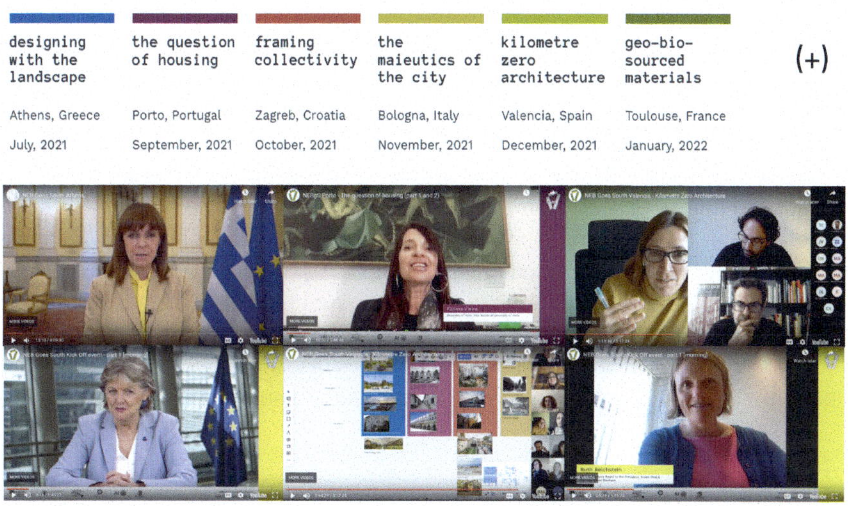


**Fig. 2** The six co-design events organized during the first phase of the NEB goes South

The first event was entitled "Designing with the landscape" and was coordinated and broadcasted from Athens in July 2021. The landscape was the point of departure for this Southern quest, considering that the New European Bauhaus wave is counting on our ability to build places for living, giving the inhabitants a sense of belonging, safety, comfort, harmony with nature, without devaluing the incredible beauty of many natural/human made landscapes.

The second was entitled "The question of housing" and was organized from Porto in September 2021. It aimed at directing our attention to the fact that agenda for climate change consider cities and urban development as priority issues, given that all housing problems converge and are enhanced in the denser areas. The challenges on providing, designing, and re-designing dwellings are particularly complex but the potential for change is also more significant. The outcomes of the discussions held on the scope of this event are the focus of this paper, so this topic will be discussed further ahead.

In October 2021, Zagreb coordinated the third event dedicated to the topic of "Framing collectivity". Participants reflected on how architecture has a lot of resources to contribute to trigger interaction; how green infrastructure is fundamental to compensate energy consumption and promote biodiversity; how public space has to be more inviting to facilitate people's meetings and increase proximity; how the active reuse of public buildings is a powerful resource to diversify; and how architectural education in particular, is fundamental to fight for a better world inspired by New European Bauhaus ideas.

The fourth event was organized in November 2021 in Bologna and focused on "The maieutics of the city". The initiative stressed the idea that we have at our disposal a range of effective solutions that gives us a myriad of instruments to tackle the problems of our cities.

"Kilometre Zero Architecture" was the fifth event that took place in Valencia in December 2021. It brought to the discussion a concept that tries to solve material needs by going first to resources found in the soil itself and, as a second option, to companies as close as possible. It was considered that having an environment duly certified and political friendly reduces the ecological footprint by decreasing the amount of fuel needed to move materials and in addition promotes local development and uses techniques and materials traditionally used locally. The work session also addressed the impact of innovative and sustainable architectures based on the valorisation of cultural and material resources of territories will be addressed along with the idea of sustainable buildings following efficient vernacular principles among many others.

Finally, the last event was coordinated by Toulouse in January 2022 under the topic "Geo-BioSource architecture". It proposed the efficient and sustainable use of materials and design strategies in architecture, discussing topics like, for instance, the use of natural and bio-based materials to reconnect with nature and promote circular economy.

## 4 The NEB Lab

After the conclusion of the six co-design events, the NEB goes South was recognized as one of the first three community-driven NEB Labs. This achievement stimulated a new phase for the initiative, aiming at increasing the international involvement and contemplating distinctive outputs. The current goals are:

- to prepare a set of architectural design workshops—Blended Intensive Programmes—that will happen face-to-face in distinct locations, putting together students, teachers, and local individualities (either from science/experimental fields or from civil society) from our six architecture schools and social or political environments. These intensive programs will considerer practical experiences in real contexts related with the topics discussed in the previous six events.

- to implement changes in the academic curricula and increase projects that will intensify the research in the relevant subjects of present-day challenges.

As a NEB Lab, the NEBgS highlights its mission as a co-creation space at the service of the NEB community. Both working avenues look for achieving tangible changes in local or specific contexts: "the Lab is where the New European Bauhaus is implemented through concrete and tangible projects. It is about what we can achieve when we work together in a spirit of openness and trust. It focuses on connecting people, learning from each other, and exploring experiences".

With this action plan, the NEBgS consortium believes that it is important to formulate a synthesis of what was achieved—a status report—through several publications that summarize the discussions held and the conclusions reached during these months of intensive work and on the one yet to come. Broad and effective dissemination is therefore understood as a relevant dimension for the NEBgS initiative, with a view to promoting and expanding the tangible work in progress.

## 5 The Co-design Event on "The Question of Housing"

Nearly 150 years ago, Engels wrote "The Housing Question", a text we continue to reference today [3]. In 1872, European cities, forged by the industrial revolution, were blighted by overpopulation as families shared one roof in the slums that grew in the back-alleys of city-centres. Denouncing palliative solutions for a systemic problem, Engels sought for the answer in the path for a fairer, more just society. The following century saw these cities undergo a transformation process. Post-war reconstruction and the generalised development of welfare states could find, all over Europe, solutions for the major housing shortages.

Today, we face new challenges. However, the Question of Housing, although fundamentally transformed, still requires today answer that goes further than the individual action. It depends on our co-operation in the construction of holistic solutions for an inclusive and sustainable tomorrow. This understanding was the reason the topic of housing was selected by the Faculty of Architecture of the University of Porto (FAUP) to organize the 2nd NEBgS co-design event in September 2021 [4].

The debate started with the participation of two keynote speakers. The first one was Gonçalo Byrne, who is currently the President of the National Board of the Ordem dos Arquitectos (2020–2022). In his talk, he stressed the menace of human extinction if we do not face the climate crisis. He also defended the urgent paradigm change that leaves no one behind and leads us to a post-carbonic society. In the scope of architecture, the holistic approach is an objective that also means a mandatory need to change to a vision of naturalization of the artefact. The second keynote speaker was Carles Baiges, who represented Lacol Arquitectes, a cooperative of architects established in Barcelona, Spain, that believes the way to transform the city is through the active participation of the people who inhabit it and through proactive action [5]. He presented the built case of La Borda housing cooperative and focused on the importance to create awareness through the involvement of the community in the construction and maintenance of the building spaces. His talk exposed the relation between sustainable solutions that simultaneously assure comfort based on quality and efficiency without depending on energy prices. This condition of low energy consumption is not only fundamental to address the global demands, but also to guarantee inclusive solutions.

These two interventions were followed by five parallel roundtable sessions that focused on the following topics (Fig. 3):

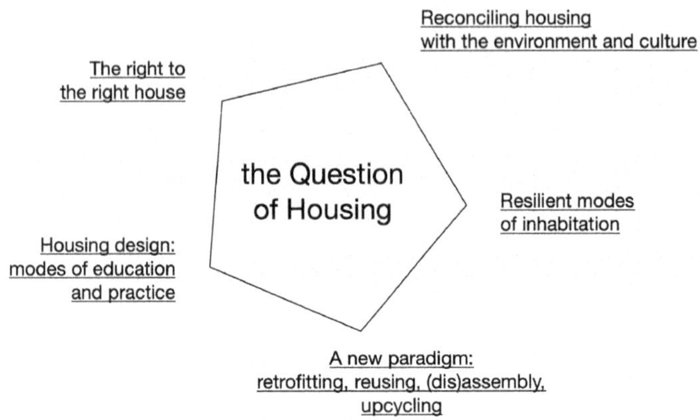


**Fig. 3** The structure of the co-design event on "the question of housing"

1. The right to the right house;
2. Reconciling housing with the environment and culture;
3. Resilient modes of inhabitation;
4. A new paradigm: retrofitting, reusing, (dis)assembly, up-cycling;
5. Housing design: modes of education and practice.

These parallel sessions counted with the participation of more than 40 international professors, researchers, and students, and were followed by a final recap session (Fig. 4).

## 6 The Five Roundtables

The first roundtable "The right to the right house" recognized the prohibitive cost of property and rental markets of an increasing urban population as the main difficulties to access to dwelling. Public policies can and need to play a key role and change land use while architects and promoters need to be aware that the right house is continuously a different house, once what is right today may not be right tomorrow. However, it became consensual that there are emerging social models to answer/accommodate, and the construction actors should not think about the right kind of people.

The second session "Reconciling housing with the environment and culture" highlighted the community as a central topic today, together with environment, heritage, and tradition, which relates to materials, techniques, crafts, skills, and, at the end to reappreciate heritage. The relevance of policies was also mentioned stressing that regulations or laws should not think just in spaces but in conditions.

**Fig. 4** Participants in the roundtables

In the "Resilient modes of inhabitation" roundtable, the idea that inhabiting is an act of appropriation and resilience is the ability to adapt to change was emphasized. So, it was consensual the urgent need to adapt housing to changing conditions, different life patterns, ageing, household structures, among others, bearing in mind that community areas regain importance. To achieve this is fundamental to explore forms of collaborative housing and participatory processes that admit designing with the collective engagement of the communities.

The fourth roundtable was entitled "A new paradigm: retrofitting, reusing, (dis)assembly, up-cycling" and considered that we are facing a multi-scale challenge, from the building to the city. So, smart use of design and (smart) technologies to improve building performance and tackle challenges such as the zero-energy requirement, state funding and mass testing as laboratories for innovation and change were understood as the main challenges to understand the climate change problems in the South of Europe to inform design solutions.

Finally, the fifth session was focused in "Housing design: modes of education and practice" and recognised that housing remains an essential topic for the education of future architects as it is the base and toolbox for many questions involving architectural problems, although it was also accepted the need to build less. During the discussion it was consensual the social responsibility of the architect, as well the role of architecture as a public service fighting to achieve something better and not just an automatic response to market. It was also advocated the importance to close the gap between academic research and public administration (e.g., regulations), which frequently limit the right responses as well as the need to increase international discussions discuss to learn and confront differences and affinities.

## 7 Final Remarks

The co-design event organized in Porto under the NEBgS initiative confirmed Housing as a fundamental question in today's world. This topic was dissected in depth in the different roundtables through the exchange of ideas and experiences from the different participants. The following considerations are just a selection of those who obtained unanimous agreement:

- the right to adequate housing intertwined with the different fabrics as an essential condition for reducing social inequalities and moving forward;
- the design of the solutions is key to ensure resilient and adequate housing;
- the critical dimensions of a new paradigm that imposes the long life of the building stock while ensuring social integration, comfort, and energy efficiency;
- the articulation between intervention measures in the built heritage and the need to revisit, recycle, reuse, retrofit, are key words at the present;
- learning and closing the gap between academic research and the institutional framework is crucial.

Having those in mind it became clear that housing and its social and sustainable environment is today an overly complex, urgent, and demanding topic. The discussion balances between the fierce debate about the need to reform a post-capitalist economic and social model—much more globalized and rhizomatic than in Engels' and Marx's time—and the existential urgency of radically transforming the construction industry around ways of designing and building greener. Being a central question to our common future, it is central to the education of an architect, whose future action, as a professional, will necessarily face this disturbing context. Therefore, it must challenge the current curriculum, which should address new and specific ways of dealing with this multi-scale challenge. Furthermore, technology as a resource—or a tool—to achieve a balanced intervention must be integrated in the teaching programs, while it is also essential that academia gets closer to territorial and public administration to set the right conditions for striving and studying housing.

Like Engels did, we are seeking again for the answer towards a fairer and more just society having in mind that our heritage relies on the recognition of urban history and housing history but on our ability to respect our environment and reinvent our practices.

**Acknowledgements.** The authors would like to thank the more than 40 participants in "The Question of Housing" co-design event of the NEB goes South initiative.

## References

1. New European Bauhaus. https://new-european-bauhaus.europa.eu/index_en. Accessed 2022/12/02
2. NEB goes South. https://www.up.pt/neb-goes-south/. Accessed 2022/12/02
3. Marxists Internet Archive. https://www.marxists.org/archive/marx/works/1872/housing-question/. Accessed 2022/12/02
4. NEB goes South—The Question of Housing. https://www.up.pt/neb-goes-south/the-question-of-housing/. Accessed 2022/12/02
5. Lacol—Arquitectura Cooperativa. https://www.lacol.coop. Accessed 2022/12/02

# The Future of Dwelling: Urban Co-housing in the Time of Climate Change

Camilo Cerro(✉)

College of Architecture, Art and Design, American University of Sharjah, Sharjah, UAE
ccerro@aus.edu

**Abstract.** The UN estimates that 68% of the world population will live in urban areas by 2050, adding 2.5 billion people to already overcrowded cities. And prompting many countries to face challenges in meeting the needs of their growing urban populations. Reconciling urban sprawl, verticality and the lack of housing will force us to adapt to greater density and new housing typologies. With more and more people moving into cities, understanding the key trends in urbanization will be crucial in implementing the Sustainable Development goals proposed by the UN. To address the housing crisis, and to become less car and carbon dependent, countries need to densify its job-rich metro areas so that more people can afford to live there and walk, bike, or take public transportation to get to work and back. A sustainable set of solutions are needed that are designed to address a net positive and self-sufficient approach to dwelling. The idea that technology will fix complex and systemic problems like climate change, poverty, the housing crisis, or healthcare is simplistic if we do not also change our existing models of living. This is why co-housing is well suited to help by promoting social, and environmental sustainability while providing an adaptive new dwelling typology. In its urban form it could be designed to add density to urban areas that are facing a housing crisis while solving other sustainable issues. This paper will present how co-housing can manifest as a tool to help deal with the housing crisis while also addressing climate change and adding to the quality of life of its users. A shift has occurred toward building social bonds and mutual support in the immediate environment after the pandemic, all while reducing resource consumption to lessen our environmental impacts. This frame of mind is making co-housing more appealing than ever before. A housing typology with social justice at its center, that empowers work, commerce and culture should be a human right globally and the future of dwelling.

**Keywords:** New housing typologies · Urban density · Housing crisis

## 1 Introduction

Humanity's ability to cooperate is what has allowed us to live in community. As we moved from a nomadic lifestyle to a sedentary one, early humans began to gather around hearths and inside shelters to eat and socialize. The domestication of plants and animals, and the development of agriculture helped our ancestors grow enough food to feed their


M. Blanco Lage et al. (eds.), *Towards a New European Bauhaus—Challenges in Design Education*,
https://doi.org/10.1007/978-3-031-40188-6_18

communities while freeing some of its members to fulfill other needs. Agriculture had a considerable impact in the development of human society. Rather than moving around, people stayed in one place. And by sharing food, caring for the young, and building social networks, our ancestors were able to meet the daily challenges of their environment, together. One of the earliest examples of structured community living where we can understand these interactions, took place in the neolithic town of Skara Brae (a stone-built Neolithic settlement on the Orkney islands of modern-day Scotland) inhabited between 3180 and 2500 BCE. The village consisted of eighteen open-plan homes with a central hearth built of stone slabs, housing between 50 and 100 inhabitants [1]. It has long been suggested that the standardized house design was a sign that no one person was more important than another [2]. Whether this was true or not, the archeological excavations showcase a self-sufficient community with commercial ties to other townships in their vicinity, that survived for 600 years. People living in a community, bound together by the common things they shared. Humans need a sense of belonging which is what connects us to those relationships that create community [3]. Today, after a pandemic and as climate change becomes evident and humanity finds itself needing to adapt to its consequences, living in community becomes even more important. The fallout of the climate crisis will vary depending on where we live, but how we live together can help us adapt. This was proven during the pandemic where humanity had to rethink the rules that guided almost every aspect of our lives. As we look back, quarantine and living in isolation stand out as themes that required the most adaptability, both subjects that point at our way of dwelling as the first thing that needs transforming [4]. Following, this article will present co-housing as a residential typology that has proven resilient in the pandemic and that for the same reasons would be effective in adapting to climate change.

## 2 What Is Co-housing?

Poet Donne said: "No Man is an island, entire of itself" [5]. This idea seems especially relevant when talking about co-housing. The first attempt to plan a co-housing community took place in Denmark in 1964 but the first communities were not built until 1973. Since then, the ideas behind the movement have evolved and financial institutions have adapted to the peculiarities of the typology [6]. Today's co-housing (or collaborative housing) consists of a horizontal property management model, non-speculative, and sustainable, organized around privately owned homes designed around shared common areas to create a strong sense of community and social engagement. Although there are co-housing developments as small as 2 households, the average co-housing developments will have between 40 and 100 people, consisting in 10–40 households, a number set with the intent to make human interactions as easy as possible. Communal spaces are added to create a balance between privacy and community [7]. These spaces are different from community to community but in its most basic form will consist of a shared kitchen and dining area, laundry, gym, garden, farm, and a multipurpose room designed to share resources and socialize. One key characteristic of co-housing communities revolves around spending time together regularly in the shared kitchen and dining area by eating together a couple of times a week [8]. Residents are also the decision-makers and decisions are often based in consensus. The nucleus of this intentional community starts by establishing the basic principles of community living, cooperation, and shared ownership. This participatory process will co-develop, co-design, and co-organize the project to ensure a high functioning community.

## 3 Quarantine, Isolation, and Loneliness

Environmental sustainability and social sustainability go hand by hand. We are social creatures, wired to come together physically. An idea we take for granted but which is not always the case. Loneliness has become widespread in modern life, and due to the 2019 coronavirus pandemic it has been exacerbated, showing an increase in depression and anxiety in our communities [9]. Loneliness, social isolation and living alone are risk factors for early mortality increasing the likelihood of death by 26–32% [10]. For the elderly and those who live alone, living in isolation can be particularly problematic. Because the quality of housing has proven to be a social determinant of health [11], future housing typologies have a responsibility to address the social aspects of dwelling as they try to tackle environmental sustainability. According to the National Association of Realtors (NAR) in the United States, the coronavirus pandemic led to an increase of 15% in the number of homes purchased for multigenerational residential use between April and June 2020 [12]. This represents the highest number since NAR began tracking multigenerational homes after the Great Recession in 2012. A real state study by Generations United found that the main causes for living in a multigenerational household are: The economy, the need for elderly caregiving, childcare, job loss, a change in the job status or underemployment, healthcare cost, cultural and family expectations, and education expenses [13]. People realized during the pandemic that living in community had benefits. Whether it was multi-generational living, moving back home or co-housing alternatives, people living together kept each other entertained, developing a sense of camaraderie that shielded them from the boredom, loneliness, and isolation of quarantine [14]. While the model of collaborative housing provides social, financial, and environmental benefits, its strength resides in its capacity to combat loneliness.

## 4 Sustainable Co-housing Today

The idea that technology will fix complex and systemic problems like climate change, poverty, health care or the housing crisis is simplistic at best. A radical shift in how we live is needed. One designed for environmental and social sustainability. It's here where co-housing can help. Inherently, this residential typology is better on the environment because it presents a model for self-organized design, development, and management which can contribute to a bottom-up approach to rethinking the idea of smart city and smart neighborhoods [15]. To start, the co-housing model allows for a reduction of $CO_2$ emission based on reduced commuting, due to sharing transportation and providing co-working spaces as part of the communal infrastructure. Cohousing communities are built with environmental factors in mind. Dwellings are smaller than the market average and clustered together creating a smaller footprint on the larger site, leaving unpaved land for gardens, playgrounds, and seasonal produce farms. In some cases, even animals. Clustered buildings, set up vertically or horizontally, also require fewer building materials than normal construction. And careful design consideration for the local climate, natural daylight, and optimizing solar orientation make for highly energy efficient buildings. Because people attracted to co-housing generally have a higher level of environmental awareness, communities often focus on incorporating green design elements and try to lighten their impact on the environment. Co-housing is conservation at its most basic level, each household saves the environmental cost of owning "one of everything" [16] by sharing essential goods. And at the social sustainability level, because behavioral change in humans is not individual, but driven by communities, co-housing supplies a natural social context for its members to learn from each other. On average, residents of co-housing communities have a smaller carbon footprint than normal city dwellers, but can they do more? [17].

## 5 Sustainable Self-sufficiency Co-housing Tomorrow

As we look towards climate change adaptability, cohousing provides a versatile model that can function horizontally or vertically depending on the urban context and the need for higher density habitation. In either case, there are certain fundamental themes where digital and analogue technologies can help the co-housing model go past sustainability and aim for sustainable self-sufficiency. This means having their own sources of energy, water, and food, to gain freedom from market manipulation. While saving money, eliminating utility bills, and limiting the effects of inflation. To reach this goal co-housing would need to: Produce energy through wind or solar and store it for later consumption. Produce, store, clean, and recycle water by using atmospheric water generators, rain catchment systems, and grey water filtering. Have living ponds instead of pools for leisure and to store and clean water. And use some of this water to produce food through hydroponics, fungi-culture, and seasonal gardens. Use permaculture principles in all green areas. Create a waste management closed loop system by composting and reusing the compost in the seasonal garden. All of which creates different forms of passive income. There are also other instances where design can upgrade the co-housing formula by: Providing commercial space for the community to rent, providing revenue back for management costs while having a say on the type of commercial institutions that are welcomed into the neighborhood. And, providing smaller units for rent, bringing constant new blood to the model and space for extended family to join when needed. All these upgrades would not only make the co-housing model more appealing, but it would also provide more self-manage options, while being financially stable and environmentally self-sufficient. The future of dwelling needs houses that give back instead of always taking away.

## 6 Conclusions

Co-housing has existed as a living typology since 1964 when the first versions appear in Denmark. Since then, its ideas and format have spread, and today one percent of the Danish population (about 50,000 people) live in this type of dwelling arrangement. In the United States there are 321 co-housing communities recorded by the US Co-Housing Association [18]. The first one, named Muir Commons (in Davis California) has been active since 1991 [19]. In a similar way, this western concept based in Social Democratic ideals has spread all around the planet, adapting to different cultures, traditions, and socio-economical levels. In 2021, 'What we share: A model of co-housing' was showcased in the Nordic Pavilion of the Venice Biennial designed by Helen and Hard Architects, presenting a framework for designing and building communities based on participation and sharing [20]. Today, co-housing has already proven itself as a good alternative dwelling model. And while the debate around how to fix the global housing crisis mainly hinges on whether more market or more state is required, co-housing can be the middle ground between this two arguments. As it becomes another tool to help deal with the housing crisis, it can do so while also focusing on creating adaptive, resilient, and versatile dwellings to help deal with climate change. Understanding its value the New Urban Agenda by the United Nations [21], highlights the benefits of cooperative solutions, such as co-housing, community land trusts, and other forms of collective tenure. The versatility and adaptability of this type of dwelling format makes it accessible. Co-housing includes all kinds of buildings, existing and new. It is not tied to a particular type of occupancy. And the groups of people it involves can vary considerably in size depending on the format of the community. At its heart it functions around two key principles; Residents do not only live next to each other, but in community with each other, in buildings that comprise shared spaces and facilities. And they participate in, the design and management of their communities. This type of dwelling format is not only enabling people to escape the peculiarities of the housing market. In many cases it also demonstrates the innovative potential that comes with empowering people to take control of their homes and communities. Co-housing presents a housing typology with social justice at its center, that empowers work, commerce and culture, elements that should be a human right globally, and the future of dwelling.

## References


1. Mark J (2012) Skara Brae. World History Encyclopedia. https://www.worldhistory.org/Skara_Brae/. Accessed 07 Mar 2023
2. Childe V (1931) Skara Brae: a 'stone age' village in Orkney. Antiquity 5(17):47–59. https://doi.org/10.1017/S0003598X00005597
3. Baumeister RF, Leary MR (1995) The need to belong: desire for interpersonal attachments as a fundamental human motivation. Psychol Bull 117(3):497–529. https://doi.org/10.1037/0033-2909.117.3.497
4. Evidence Matters (2022) The role of housing in climate change mitigation and adaptation. Office of Policy Development and Research. https://www.huduser.gov/portal/periodicals/em/Summer22/highlight1.html
5. Donne J (1970) No man is an island. Villard. ISBN-13: 978-0394404820
6. Vestbro DU, Horelli L (2012) Design for gender equality: the history of co-housing ideas and realities. Built Environ 38(3):315–335. https://www.jstor.org/stable/23290266
7. Scotthanson C, Scotthanson K (2005) The cohousing handbook: building a place for community. New Society Publishers. ISBN: 0-86571-517-3
8. Shulevitz J (2021) Does co-housing provide a path for happiness for modern parents? The New York Times. https://www.nytimes.com/2021/10/22/opinion/cohousing-mothers-pandemic-community.html

9. Weissbourd R, Batanove M, Lovison V, Torres E (2022) Loneliness in America: how the pandemic has deepened an epidemic of loneliness and what we can do about it, making caring common project. Harvard Graduate School of Education. https://static1.squarespace.com/static/5b7c56e255b02c683659fe43/t/6021776bdd04957c4557c212/1612805995893/Loneliness+in+America+2021_02_08_FINAL.pdf
10. Holt-Lundstad J, Smith T, Baker M, Harris T, Stephenson D (2015) Loneliness and social isolation as risk for mortality: a meta analytic review. Perspect Psychol Sci 10(2):227–237. https://doi.org/10.1177/1745691614568352
11. Krieger J, Higgins DL (2002) Housing and health: time again for public health action. Am Public Health Assoc 758–768
12. Lautz J (2021) Full house: the rise of multigenerational homes during COVID-19. National Association of Realtors. https://www.nar.realtor/blogs/economists-outlook/full-house-the-rise-of-multigenerational-homes-during-covid-19. Accessed 27 Oct 2022
13. Goyer A (2021) Family matters: multigenerational living is on the rise and here to stay, generations united. Washington, DC. https://www.gu.org/app/uploads/2021/04/21-MG-Family-Report-WEB.pdf
14. Sisson N, Willroth EC, Le BM, Ford B (2022) The benefits of living with close others: a longitudinal examination of mental health before and during a global stressor. PsyArXiv preprints. https://doi.org/10.31234/osf.io/v9mc4
15. Szypulski A (2016) Co-housing: abundant potential for sustainable housing and neighbourhood development, designing sustainable urban futures. KIT Scientific Publishing, pp 91–101. ISBN: 9791036538223
16. Durrett C, McCamant K (2011) Creating co-housing: building sustainable communities. New Society Publishers, Canada
17. Szaraz L (2015) Pro-environmental characteristics of urban co-housing communities. Geogr Local Stud 3(1):490–529. ISSN: 2053-3667
18. US Cohousing Association (2023) Communities directory. US Cohousing Association. https://www.cohousing.org/directory/wpbdp_category/comm/. Accessed 09 Mar 2023
19. Sweeney C (2006) The 21st century commune. The New York Times. https://www.nytimes.com/2006/09/10/realestate/the-21stcentury-commune.html
20. Braaten M (2021) What we share: a model for cohousing, its liquid. https://www.itsliquid.com/whatweshare.html. Accessed 09 Mar 2023
21. UN Habitat (2016) The new urban agenda. United Nations. https://habitat3.org/the-new-urban-agenda/. Accessed 09 Mar 2023

# Co-creating Urban Commons Through Community-Engaged Pedagogies

Nadia Charalamboush(✉), Christina Panayi, and Effrosyni Roussou

Department of Architecture, University of Cyprus, Nicosia, Cyprus
charalambous.nadia@ucy.ac.cy

**Abstract.** The importance of empowering and engaging citizens in the shaping of their living environments to ensure a sustainable development and promote a sense of community has been highlighted in recent years. Citizens and professionals are called to adopt new roles within the spatial design and provision process, which often challenges the ability of the latter to respond effectively to the rising need for a community-engaged design approach. The necessity to train future architects in conceptualizing and implementing such approaches, within a transdisciplinary framework, reveals the need to revisit, assess and rework the relevant pedagogical approaches. This paper reflects on a community-engaged housing studio approach and a subsequent design and build co-creation workshop at the Department of Architecture, University of Cyprus (UCY), for the design and development of a public space in a suburban neighbourhood in Nicosia. The housing studio incorporates principles of inclusivity and sustainability aiming to create the conditions for an integrated framework for achieving sustainable urban governance and to initiate a community of practice, fostering a participatory process. Drawing from the tools and methods of live studios and Urban Living Labs, this transdisciplinary process involves all participants at various levels and stages through a circular methodology of assessing and understanding, co-creating, co-designing and implementing and co-evaluating. Reflecting on the outcomes, the paper discusses the opportunities and limitations of a community-engaged design methodology through future graduates' exposure to real world contingencies and interaction with diverse stakeholders, while empowering citizens as decision makers who have a responsibility for their residential environment.

**Keywords:** Co-creation · Community engagement · Teaching and learning practices · Design studio · Transdisciplinarity · Sustainable urban governance · Spatial agency

## 1 Introduction

### 1.1 Overview

This paper reflects on a community-engaged ("live") housing studio approach and a subsequent design and build co-creation workshop at the Department of Architecture, University of Cyprus, in the design and development of a public space in a residential


M. Blanco Lage et al. (eds.), *Towards a New European Bauhaus—Challenges in Design Education*,
https://doi.org/10.1007/978-3-031-40188-6_19

neighbourhood in suburban Nicosia. This studio and workshop aim to create the conditions for an integrated framework of achieving sustainable urban governance and to initiate a transdisciplinary community of practice between students, educators, researchers, enterprises, the community and the local municipality, through co-creation processes. In other words, we endeavour to lay the foundations for and examine the possibility of fostering commoning practices in the Cypriot context, through a sustained and equitable cross-pollination between groups of people with diverse power relations, ideologies, interests, backgrounds and capabilities.

We specifically look into the ways in which such a studio approach affects the participating students' (a) confidence and transversal skills as emerging professionals in the fields of architecture and urban planning/design, within an increasingly volatile socioeconomic, political and environmental reality driven by persisting neoliberal policies, (b) their positionality, i.e. the awareness of their own agency and social role as well as its implications within co-creation processes, and (c) their capacity for reflexivity and critical thinking. In the following sections, we analyse the necessity of such approaches to architectural education generally and in contexts like Cyprus in particular, followed by the methodology adopted, both in terms of pedagogical approaches and assessment of the impact of this approach, as well as the outcomes.

### 1.2 Neoliberal Spatial Politics and the Conceptualization of Architecture and Architectural Education

As the crisis of capitalism, as well as those spawned by capitalism deepen, the neoliberal efforts to keep the system afloat turn increasingly frequently -among others- to spatially manifested authoritarianism [1] and tactics of co-optation of what used to be radical discourses, like environmentalism and sustainability [11, 15]. Architects and other spatial practitioners, bearing the legacy of the modernist conceptualisation of the field, have generally stood aloof, assuming the mantle of the objective, neutral experts, thus severing architecture and spatial design from their political and social implications, and isolating them from their interdependencies [11, 13].

Inevitably, architectural education has been shaped around the aforementioned valorization of neutral expertise. Students are -for the most part- trained in a secluded learning environment [4] and guided to acquire the skills necessary to develop into suitable members of the workforce within the neoliberal market [8]. There are, however, considerable efforts put into reconceptualizing architecture -and consequently architectural education- as "spatial *agency*"; architecture as spatial agency is understood as a political, inclusive, and cooperative process of producing space, a lively terrain of negotiation and mutual knowledge [4].

### 1.3 Revitalizing the Commons Through a Live Studio Approach in Nicosia, Cyprus

The "live studio" has been the pedagogical testbed on which the reconceptualization of architecture as spatial agency is interpreted, put into practice and assessed. The live studio format aims at exposing students to real-world contingencies, through a transdisciplinary framework of collaboration with communities and local stakeholders, public authority representatives, and practicing architects and planners [6]. Whether through participatory/co-creation processes or design and build projects, students are called to leave the safe space of the normative design studio and venture into real-world uncertainty, in an effort that encourages critical knowledge production that would allow students to question their own preconceptions about their role and the context of operations [10].

The co-creation studio at UCY, along with its subsequent design and build workshop, draw heavily from the live studio methodology; structured around a collaboration between Latsia Municipality, Nicosia Development Agency (ANEL) and the University of Cyprus, the aim was to explore the notion of the (urban) commons, through the design and implementation of a neighborhood public space in suburban Nicosia, Cyprus.

Cyprus is a post-colonial state characterised by a tense socio-political history, that strives to balance between its ottoman legacy, a colonial past, and a post-colonial reality [12]. The arrival of the British colonists brought about profound changes in land use and ownership patterns; the modernisation of land rights, befitting a more capitalist worldview, prioritised private land ownership. This approach, paired with a post-colonial developmentalist agenda, led to a highly unsustainable urban growth, carried out in top-down processes that did not allow for public participation [7]. This contrasted with the traditional/ottoman land use patterns that were in use in pre-colonial times, which were often based on a system of communal ownership and shared use of land, that allowed for a more equitable access to and distribution of resources among the members of a community [5]. Therefore, this pattern clearly reflects the recognised value of sharing and safeguarding the commons during the pre-colonial period of Cyprus. In this sense, the community-engaged exploration of the commons that is the goal of the studio, is in essence an effort to revitalise faded aspects of a daily life that is based on sharing and neighbourliness.

## 2 Methodology

### 2.1 Housing Studio's Methodology

To address the aims highlighted in the previous section, the housing studio's methodology builds on the methodology of Urban living Labs (ULL) and incorporates principles of other participatory approaches. ULL are effective in bringing actors together to design, test and learn from innovation in real time and place, giving users a central role to shape the outcomes by contributing with knowledge and experience [9]. According to Menny et al. [9], there are three phases in their development that often seem to overlap: the design, implementation, and evaluation phases, each one involving users at different levels: information, consultation, and co-decision [2].

Building on the ULL methodology and adapting it to an educational framework, the studio's methodology is circular and can be repeated, enhancing reflection and improvement (Fig. 1).

1. Design of the framework; It includes the preliminary site analysis, the stakeholders' identification and engagement, the development of the learning outcomes and toolkit for participation, the design of the process and the action plan

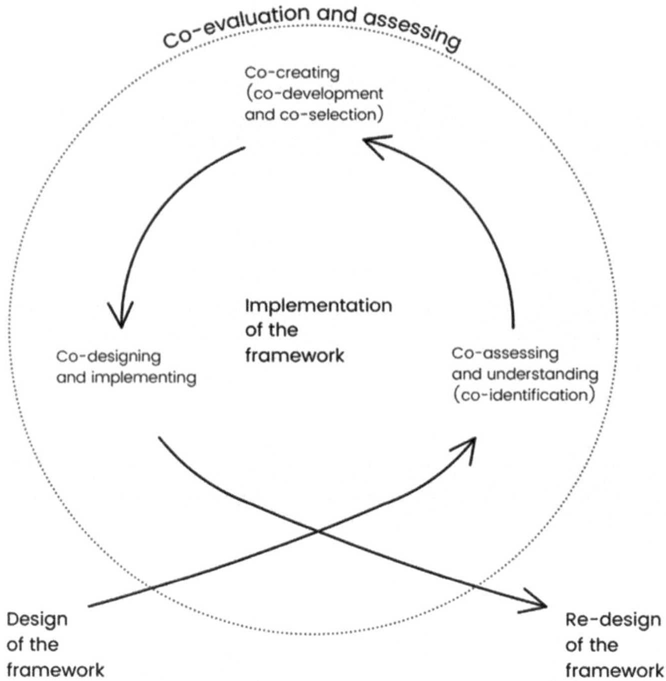


**Fig. 1** Project's circular methodology

2. Implementation.
   i. Co-assessing and understanding; co-identification of all the needs, opportunities and ideas as well as the determination of the available resources, creating a clear view of the existing situation and capacity
   ii. Co-creating; several co-creation sessions for the co-development and co-selection of solutions and ideas, leading to a common and shared vision
   iii. Co-design and implementing; co-design in detail and implementation of the final proposal(s) in terms of pilot models or on site
3. Co-evaluation and assessing; continuous parallel evaluation of the process and the outcomes.

## 2.2 The Role of the Researcher/tutor—Research Methodology

To assess the impact of the process on students in terms of their motivation, attitude and positionality, a participatory action research methodology (PAR) was implemented. Its participatory nature enables all the involved researchers that act also as tutors, to be part of the production of knowledge, connecting the research findings with the teaching practice. The students' spontaneous behavior is observed throughout the project in terms of a. their relationship with the other stakeholders b. their attitude towards the process through observation as a tool for gathering data and the use of a reflection diary. Additionally, focus group discussions highlight feelings, reactions, and informal opinions of the students during different phases of the design studio and during the design and build workshop while semi -structured questionnaires (using Google Forms—Likert scale) at the beginning and at the completion of both the design studio, and the design and build workshop, enrich the outcomes mapping students' attitude through the process.

# 3 Outcomes/Results

Designing the framework included a preliminary site analysis, stakeholders' identification and mobilization, the development of the learning outcomes and participation tools and the creation of a detailed action plan.

Having a balanced and broad representation in terms of age, gender and socioeconomic status was a priority and thus the process of identification was multilevel; communication was hybrid (face to face meetings, Miro platform, Facebook discussion group, Google Drive) and workshops were organized both at the university studio as well as in the area under study, in the Municipal Hall, to ensure as much as possible, an accessible and inclusive process.

During the co-assessing and understanding phase a number of collaborative activities took place as part of the design studio, where the site was analyzed in depth by groups which included students, local residents, Municipality representatives and representatives of the Nicosia Development Agency. The challenges and opportunities of the site as well as the residents' needs and the municipality's plans and vision, were identified and discussed, through questionnaires, SWOT analysis, photographs, diagrams, and mapping. This first phase highlighted the need for a nature-based, open meeting space in the neighbourhood which could be co-created and co-governed by all relevant stakeholders. Through several co-creation sessions, different solutions, scenarios and ideas were co-developed (using Lego, sketches, diagrams and collages), leading to a preliminary concept design. The residents, students, tutors and municipality representatives reflected on, discussed and revised the preliminary proposal during an open co-creation workshop at the Municipal Hall. The final proposal includes a covered meeting space, green and cultivation spaces, encouraging through the design informal playing and exploration.

Following the completion of the housing studio, a three-weeks design and build co-creation workshop took place at the Department of Architecture, focusing on the production of detailed final drawings (lighting and landscape design, construction details and canopy design) and the pilot construction of mobile urban furniture proposed during the housing studio (wooden benches, wooden playset for kids and cinderblocks structures with sitting areas and vegetation pots). Several local experts participated at key stages of the workshop with presentations and hands-on sessions, including carpenters, practicing architects, an urban lighting designer, a landscape architect, the Cyprus Certification Company (for playground regulations) and a local play equipment enterprise. Students, residents and municipality representatives worked on the site on cleaning and leveling, as well as at the university fabrication lab.

## 3.1 Students' Perspectives

Assessing the impact of the community- engaged design and build process on the students, highlighted important information on their motivation, their skills, experience and gains as well as on their overall attitude and positionality towards participatory approaches and design and build before and after the process.

Although students found it challenging to meet the requirements of a design studio which also includes a number of co-creation activities with stakeholders, they were engaged and committed to the process, felt excited, motivated and gradually felt a sense of social responsibility towards the local residents and the Municipality (Fig. 2). At the end of the studio, students felt more confident as regards their communication and collaboration competencies, and pointed out that they gained new knowledge and experience due to their exposure to real-life issues, risks, and constraints. Their exposure to real world contingencies enhanced their awareness of actual social, environmental, and contextual challenges, strengthening their commitment and responsibility towards the process and their own design proposals.

Hands-on activities and 1:1 construction during the design and build workshop, further enhanced students' confidence as regards their practical skills, their creativity and capacity to readily respond to design and construction challenges (Fig. 3). Coming in contact with practitioners, experts as well as students from different years was noted positively, strengthening skills related to teamwork, communication, leadership, and time and resources management through their engagement with professional, practical, and ethical issues.

Regarding the students' attitude and positionality towards participatory activities and community engagement, a shift was noted through the process, from passive listeners to more active and engaged members of the team. The benefits of the involvement of architecture students in their own communities, acknowledging the important role that local contexts, and users play in the design process as well as their future role in the shaping of the built environment, was discussed and reflected on through fresh perspectives and insights (Fig. 4)

## 4 Locating Learning Between Academia and the Everyday

Overall, the need to address the insularity of the design studio and the architecture student from the actual needs and visions of the users and the society at large, as well as from advances in the profession itself, has emerged through the activities of the housing design studio at UCY. The potential of a reorientation of the design studio education towards an engaging approach that also considers the social responsibility of future architects, instilling a sense of involvement in the students, emerges. This involvement refers both to professional activities students will potentially be engaged in, as well as to the community they will eventually serve. Such an approach calls for knowledge that supersedes the disciplinary/academic boundaries and draws on all agents/stakeholders involved [4]. Having said that, as Morrow points out, architectural pedagogy in the form of live studios remains a pedagogical tool with different aims and methods than architectural practice [6].

Exposure to the opportunities and constraints of a real-world project fosters an awareness of the (inter)dependencies of architecture [14]; from understanding the challenges of building materials and regulations to acknowledging user needs, students gain new insights of their role as spatial agents in a constant navigation through negotiations and

a

**Fig. 2** **a** Co-creation activities and design and build workshop. **b** Comparison of students' self-evaluation on relevant skills (before and after the design and build workshop)

dependencies [3]. Such pedagogical alternatives embrace an outward-facing agenda that aims at creating interfaces between academia and the real world through community outreach activities, involving residents in the process exposing students to the real living conditions and providing hands-on, informal learning, collaboration, and interaction with all actors.

Through the transdisciplinary learning environment developed through the housing design studio at UCY, the foundation is laid for a sustained partnership between the municipality, the community, professional practitioners in the field and academia. By nurturing this partnership through a sustained collaboration around co-creation and design & build projects, the aim is to re-introduce the University as an open and vital

b

Fig. 2 (*continued*)

actor in local urban development, that would work to ensure a transparent, inclusive, and equitable process of spatial production. This aim is tightly anchored to a belief in the need for an alternative approach to urban governance and land use, especially in a context such as Cyprus, where the value system around the commons in the pre-colonial era is reminiscing of the contemporary conceptualizations of an anti-capitalist perspective on the right to land and space. This belief will be channeled in building on the UCY housing studio methodology gradually rendering it a key hub of spatial agents, students, and community members alike.

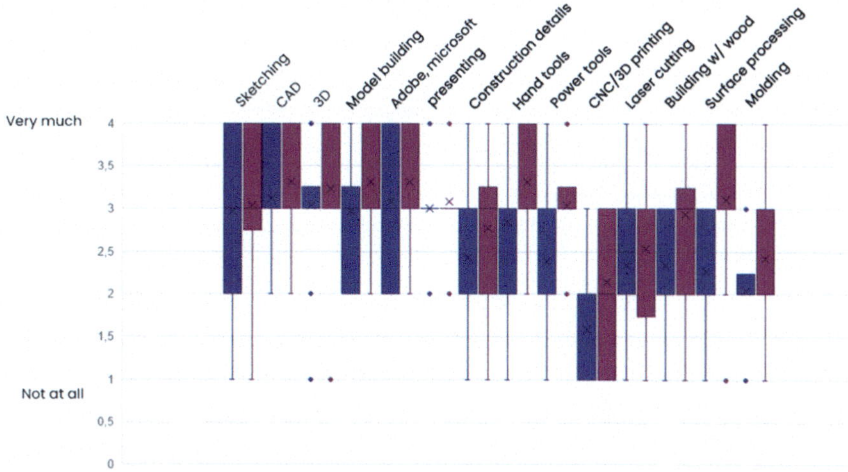


**Fig. 3** Comparison of students' self-evaluation on relevant skills (before and after the design and build workshop)

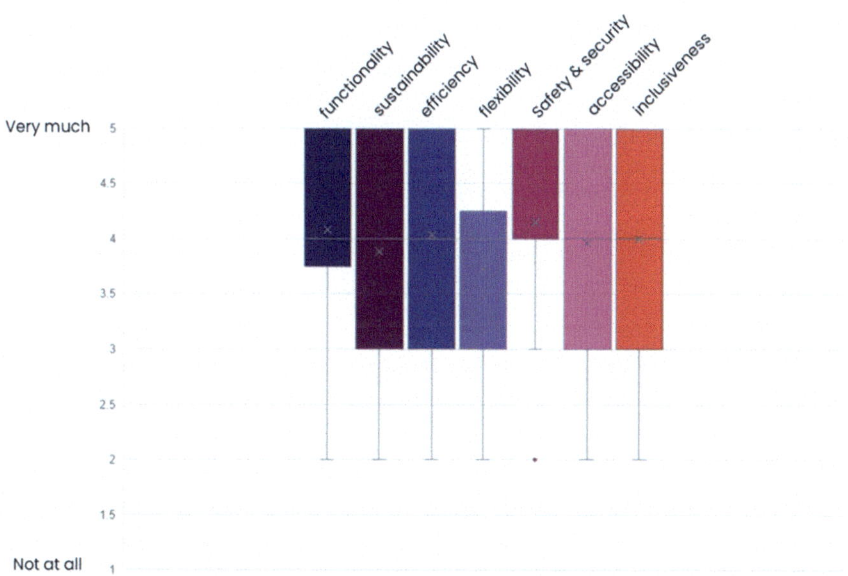


**Fig. 4** Students' rating of their team's performance on different aspects such as efficiency and inclusiveness

## References


1. Apostolopoulou E, Liodaki D (2021) The right to public space during the COVID-19 pandemic: a tale of rising inequality and authoritarianism in Athens. Greece City 25(5–6):764–784. https://doi.org/10.1080/13604813.2021.1989157
2. Arbter K, Handler M, Purker E, Tappelner G, Trattnigg R (2007) The public participation manual: shaping the future together
3. Awan N, Schneider T, Till J (2011) Spatial agency: other ways of doing architecture. Routledge
4. Charalambous N (2018) Responding to uncertainty through the housing design studio pedagogy: implications and opportunities for architectural education. In: Charalambous N, Kyriazis G InterActions. JOVIS Verlag GmbH, pp 20–39
5. Dietzel I (2016) Sharing Traditions of Land Use and Ownership: Considering the "Ground" for Coexistence and Conflict in Pre-modern Cyprus. In: Bryant R (ed) Post-Ottoman Coexistence. Berghahn Books, pp 41–58
6. Harriss H, Widder L (2014) Architecture live projects pedagogy into practice. Routledge
7. Ioannou B (2016) Post-colonial urban development and planning in cyprus: shifting visions and realities of early suburbia. Urban Plann 1(4):79–88. https://doi.org/10.17645/up.v1i4.768
8. Lorne C (2017) Spatial agency and practising architecture beyond buildings. Soc Cult Geogr 18(2):268–287. https://doi.org/10.1080/14649365.2016.1174282
9. Menny M, Voytenko Palgan Y, McCormick K (2018) Urban living labs and the role of users in co-creation. GAIA—Ecol Perspect Sci Soc 27:68–77. https://doi.org/10.14512/gaia.27.s1.14
10. Mezirow J (2006) An overview on transformative learning. In: Illeris K (ed) Contemporary theories of learning, 1st edn. Routledge, pp 90–105
11. Pyla P (2015) Crisis spins. J Architect Educ 69(1):8–12. https://doi.org/10.5334/ah.ay
12. Pyla P, Phokaides P (2020) "Dark and dirty" histories of leisure and architecture: Varosha's past and future. Architect Theory Rev 24(1):27–45. https://doi.org/10.1080/13264826.2020.1753282
13. Standring A (2021) Relational expertise and the spatial (re)production of austerity: challenges and opportunities for progressive politics. Environ Planning C: Politics Space 39(3):555–573. https://doi.org/10.1177/2399654419873674
14. Till J (2009) Architecture depends. The MIT Press
15. Tulloch L, Neilson D (2014) The neoliberalisation of sustainability. Citizensh, Soc Econ Educ 13(1):26–38. https://doi.org/10.2304/csee.2014.13.1.26

# Commoning Practices and Mobility Justice in Data-Driven Societies: Urban Scale Digital Twins and Their Challenges for Architecture and Urban Planning

Marianna Charitonidou(✉)

Faculty of Art Theory and History, Athens School of Fine Arts, 42 Patission Street, 106 82 Athens, Greece
m.charitonidou@icloud.com

**Abstract.** The paper examines the role of commoning practices in data-driven society, placing particular emphasis on their role in the establishment of new agendas in architectural education. Its main objective is to explore the impact of commoning practices on the relations between citizens in data-driven society and how this issue is addressed in architectural education. The paper pays special attention to urban scale digital twins, which are virtual replicas of cities that are used to simulate environments and develop scenarios in response to policy problems. It investigates the shift from technical to socio-technical perspectives within the field of smart cities. Despite the aspirations of urban scale digital twins to enhance the participation of citizens in the decision-making processes concerning urban planning, the fact that they, often, neglect social non-quantifiable aspects concerning how citizens experience urban environments, on the one hand, and are based on a limited set of variables and processes, on the other hand, calls for an exploration of how architecture and urban planning education can respond to these challenges. The paper aims to shed light on the tension between the real and the ideal at stake during this process of abstracting sets of variables and processes in the case of urban scale digital twins. It examines the critiques of 'digital universalism', and reflects upon the role of urban scale digital twins in data-driven decision-making concerning urban policies and urban planning.

**Keywords:** Commoning practices · Data-driven society · Digital universalism · Architecture education · Mobility justice · Urban scale digital twins · Socio-technical perspectives

## 1 Introduction

Urban scale digital twins can help us use big data to enhance social advocacy. By incorporating urban scale digital twins in the decision-making processes concerning urban planning, urban planners can shape new participatory design methods. The paper explores how urban scale digital twins can help us enhance the role of commoning practices, on the one hand, and take into account the opinions of citizens during the


M. Blanco Lage et al. (eds.), *Towards a New European Bauhaus—Challenges in Design Education*,
https://doi.org/10.1007/978-3-031-40188-6_20

decision-making processes. We should not, however, neglect or underestimate the risks of 'digital universalism' [1–3]. Architecture education strategies are confronted, nowadays, with the challenge to help students shape methods in both architecture and urban planning that would offer them the opportunity to combine the use of advanced technologies for decision-making and simulations, such as the urban scale digital twins with approaches such as the 'co-production' and the 'negotiated planning' approaches [4], that aim to enhance social equity and to take into account existing commoning practices of the citizens while shaping design strategies. Patrick Bresnihan underscores, in "The More-than-human Commons: From Commons to Commoning", that "[t]he noun 'commons' has been expanded into the continuous verb 'commoning' to denote the continuous making and remaking of the commons through shared practice" [5–7]. At the core of this paper is Stavros Stavrides's claim that "[c]ommoning practices importantly produce new relations between people" [8, p 2]. Manuel Castells's theory could help us better understand the relationship between big data and urban planning in our data-driven society, on the one hand, and to reshape the agendas of architectural education within the context of the New European Bauhaus, on the other hand [9–13].

## 2 The Role of Urban Scale Digital Twins in Decision-Making in the European Framework: Sustainable Development and Big Data

As Giorgio Caprari, Giordana Castelli, Marco Montuori, Marialucia Camardelli and Roberto Malvezzi remark in "Digital Twin for Urban Planning in the Green Deal Era: A State of the Art and Future Perspectives", the European Union has set the following goals regarding sustainable urban planning strategies: firstly, the empowerment of "urban actors towards common goals; secondly, the development of people-oriented urban planning strategies that aim to contribute to the social equity of communities; thirdly, the development of digital platforms and other digital tools that intend to enhance interactive and proactive approaches in urban planning decision-making, and "the creation of integrated, open, and functional technological infrastructures for the development of programmes and the provision of services (data-driven planning)" [14]. To realize the central role of Europe within the framework of the endeavors to incorporate urban scale digital twins in decision-making concerning urban planning, we should take into account the fact that "Europe is emerging as the main centre of development of urban digital twins, with over 60% of the existing" [15] urban scale digital twins. As Jaume Ferré-Bigorra, Miquel Casals, and Marta Gangolells remark in "The adoption of urban digital twins", among existing urban scale digital twins that are either in operation or under development are the twins of the following cities or districts: Athens in Greece, Plzeň in the Czech Republic, Dublin Docklands in Ireland, Herrenberg in Germany, Vienna in Austria, Zurich in Switzerland, New York in the United States of America, London in the United Kingdom, and Helsinki in Finland [15]. Other cases of urban scale digital twins are those of Cambridge, Gothenburg, Munich, Newcastle, Paris, Rennes, and Rotterdam [14].

Two programs that play a major role in shaping sustainable urban planning methods are the European New Green Deal, the Agenda for Sustainable Development, and its Sustainable Development Goals, which are also known as SDGs. The former—the European Green Deal—is based on the intention to achieve zero net emissions by 2050. Moreover, this program places particular emphasis "on achieving a circular economy by 2050, creating a sustainable food system and protecting biodiversity and pollinators" [16]. As John Hatcher remarks, in "Digital twins can help sustainability", which was published in June 2022 in *Smart Building Magazine*, "60% of organizations across major sectors are leaning on digital twins as a catalyst [...] to fulfil their sustainability agenda" [17] Hatcher also highlights, in the same article, that "digital twin implementations are set to increase by 36% on average over the next five years" [17]. Caprari, Castelli, Montuori, Camardelli and Malvezzi discern the following main characteristics of urban scale digital twins: firstly, their "scalability"; secondly, their "predictability", which becomes possible thanks to the use of simulation algorithms; thirdly, their capacity to integrate new elements thanks to the use of IoT sensors, and data updated concerning in situ real-time data, and, finally, their capacity to enhance cooperation due to the fact that they can be broadly accessible. He also underscores the fact that citizens can download and upload data enhancing in this way social equity, on the one hand, and participatory design methods, on the other hand [14]. Gordon S. Blair, in "Save Share Reprints Request Digital twins of the natural environment", distinguishes three challenges concerning the creation of digital twins: firstly, the challenge of "bringing the environmental assets together in one logical place, including both data assets and modeling assets"; secondly, the challenge of allowing different assets to work together as part of a larger digital twin architecture", and, thirdly, the challenge of ensuring "that the necessary storage and processing capacity is available when it is needed, especially given the sizes of the challenges and the associated potentially very large datasets" [18].

## 3 Smart Cities and the Shift from Technical to Socio-Technical Perspectives: Thinking Locally Means Thinking Critically

To enhance social equity when we introduce Big Data in urban planning decision-making, one should not forget that thinking locally means thinking critically [19, 20]. Thinking critically is related to the intention to take into account the specificities and implications of the local contexts in which data are created and collected. Some key questions in the field of critical data studies that we should incorporate in the questions addressed in architectural education, especially in relation to the teaching of courses focusing on Big Data and smart cities, are the following: How Big Data are collected? Which is the impact of the local conditions concerning the collection and creation of data on research methods? To what extent how data are collected in dependent on the economic systems within which they are inscribed? Which social groups take advantage of the creation of Big Data? [1] The point of departure of this paper is the intention to investigate how data are collected and instrumentalized when urban scale digital twins are used for urban planning decision-making. Useful for responding to these questions is Christine L. Borgman's remark that "entities become data only when .someone uses them as evidence of a phenomenon, and the same entities can be evidence of multiple phenomena" [21, p 28]. Of great importance for the reflections developed here is Manuel Castells's theory, which would contribute to a better understanding of the relationship between Big Data and urban planning in data-driven society and the new kind of temporality in the so-called 'network-society' [9–13]. Michael Batty and Castells's work are related to the transition from spatial perspectives concerning the investigation of urban data to topological perspectives [9–13, 22].

As Loukissas highlights, in *All Data Are Local: Thinking Critically in a Data-Driven Society*, "[i]f left unchallenged, digital universalism could become a new kind of colonialism in which practitioners at the 'periphery' are made to conform to the expectations of a dominant technological culture" [23]. According to Loukissas, "[a]spiring to the ideology of big data means seeking to collect everything on a subject, downplaying the importance of data's origins, and assuming that data alone can entirely supplant other ways of knowing" [23]. Stefania Milan and Emiliano Treré, call for a "de-Westernization of critical data studies" [24, p 319]. An important shift within the field of smart cities is the shift from technical to socio-technical perspectives [25]. This shift is related to the idea that the concept of smart city should be related to the endeavor to reveal "multiple dimensions beyond an infrastructure-technology focus" [26, p 24]. Useful for understanding that the evangelism that accompanies the discourse around smart cities is not something new, but has a long history is the remark of Benjamin H. Bratton that "[w]ell before smart cities evangelism, the modernist call for a more intense technologization of design's disciplinary doxa, blending urban and cybernetic programs, was a predominant discourse" [27, p 172]. According to Stefania Milan and Emiliano Treré, the myth of 'data universalism' refers to "the tendency to assimilate the cultural diversity of technological developments in the Global South to Silicon Valley's principles". Milan and Treré criticize the "hyperbolic narratives of the 'big data revolution'", arguing that "the main problem with data universalism is that it is asocial and ahistorical, presenting technology [...] as something operating outside of history and of specific sociopolitical, cultural, and economic contexts" [24, p 325]. A key question concerning the myth of "data universalism" is the following: "how does datafication unfold in countries with fragile democracies, flimsy economies, impending poverty?" The uneven access to the technologies and data that make possible smart cities and urban scale digital twins should be seriously taken into account if we wish to go beyond the myth of 'data universalism'. Regarding the uneven access to the technologies and data, Simon Joss, Frans Sengers, Daan Schraven, Federico Caprotti, and Youri Dayot have shed light on the "competitive dynamics created between world cities posited as 'model' smart cities and various second- and third-tier 'follower' cities" [26, p 24].

## 4 Defining Urban Scale Digital Twins: Their Role in Urban Analytics and Computational Social Sciences

The term 'digital twin' refers to the digital representation enabling comprehensive data exchange and can contain models, simulations and algorithms describing their counterpart and its features and behavior in the real world. A 'digital twin' is a digital representation of a physical process, person, place, system or device. The term 'digital twin' firstly emerged in the field of manufacturing sector to refer to digital simulation models that run alongside real-time processes. 'Digital twins' are digital replicas of physical entities. Their creation is based on the use of advanced technological applications, such as sensing, processing, and data transmission. Digital twins are used in the field of urban analytics, as well as in the field of computational social sciences. ABI Research forecasts that urban digital twin deployments will exceed 500 by 2025 [28]. According to Michael Batty, "[t]he idea of the digital twin [...] has emerged from the representation of the city in terms of its physical assets" [22, p 818]. The digital twins are able to get updated following the changes of the physical equivalents thanks to the pairing between the virtual and the physical world. To understand what is the main idea behind the creation of digital twins we should bear in mind that "[a]n ideal digital twin would be identical to its physical counter-part and have a complete, real-time dataset of all information on the object/system" [29].

Recently, within the domain of urban planning and, more particularly, within the field of smart cities, the notion of urban scale digital twin has acquired a central place. Li Deren, Yu Wenbo, and Shao Zhenfeng define the 'digital twin' as a "simulation process that makes full use of physical models, sensors, historical data of operation, etc. to integrate information of multi-discipline, multi-physical quantities, multi-scale, and multi-probability". They also highlight the fact that the current debates concerning the notion of digital twin are characterized by plurality of how this concept is understood. They remark that "a consensus definition has not yet been formed". The common denominator of the different definitions of the term is the shared interest in the "bi-directional mapping relationship that exists between physical space and virtual space". The creation of digital twins is based on the intention to establish "real-time connection[s] between the virtual and the real". In the case of digital twins, the digital models, apart from "observing, recognizing, and understanding" [30] the physical world, they also aim to control and transform it. Martin Mayfield has emphasized the role of urban scale digital twins in providing a holistic approach to urban and infrastructure design [31]. Anah Boyd and Kate Crawford, in "Critical Questions for Big Data: Provocations for a cultural, technological, and scholarly phenomenon", analyze critically the role of Big Data within the current cultural and technological context of data-driven societies [32]. Li Deren, Yu Wenbo, Shao Zhenfeng argue that at the core of the development of urban-scale digital twins is the creation of "a complex giant system between the physical world and the virtual space that can map each other and interact with each other in both directions" [30].

## 5 Exploring How the Opinions of Citizens Could Be Incorporated in Urban Scale Digital Twins: Towards Socio-Technical Perspectives

Among the challenges of data-driven approaches are the measurement errors, the biases, the existence false positives and false negatives, the undesired discrimination effects, the complexity, the network effects, the non-linear dynamics, the wicked problems, and an ensemble of convergence issues. Apart from the aforementioned issues, a problem that should be highlighted is the fact that, in general, how urban scale digital twins are created, often, neglect the importance of the opinions of citizens. Therefore, pedagogies of architecture and urban planning could play a major role in exploring how urban scale digital twins could take into account aspects related to the feedback of citizens during the processes of decision-making, the role of social interactions in urban analytics, the significance of social norms in shaping urban planning strategies, and the role of culture, history, politics, democracy, human rights, ethics in taking decisions concerning the sustainability of our cities. It is, therefore, indispensable to develop approaches that aim to incorporate questions related to the aforementioned aspects in how urban scale digital twins are created and used. The fact that the role of urban scale digital twins in the decision-making processes concerning urban planning will become even more important during the next years makes the incorporation of aspects related to democracy, human dignity, and solidarity in how the urban scale digital twins function even more necessary. Clare Wildfire distinguishes two categories of benefits of the city-scale digital twins: the reactive benefits, on the one hand, and the predictive benefits, on the other. Wildfire relates the first category of benefits to the capacity of enhancing "real-time or near real-time interventions and [to the improvement of] [...] the smooth day-to-day running of the city or asset", and the second category to the use of data for the improvement of "longer-term scenario planning to steer appropriate (and equitable) investment decisions" [33]. Li Deren, Yu Wenbo, and Shao Zhenfeng have examined, in their article entitled "Smart city based on digital twins", the application "Smart City Traffic Brain", shedding light on the fact that this application uses digital twins to collect the Big Data concerning travel trajectories and "real-time dynamic traffic information" [30]. Michael Batty has remarked that "one of the quests in city modelling is to merge social and economic processes with the built environment and to link functional and physical processes to socio-economic representations" [22, p 819].

The shift from technical to socio-technical perspectives goes hand in hand with the effort to construct urban scale digital twins that aim to "reflect the specifics of the urban and socio-political context" [25]. According to Martin Tomko and Stephan Winter, "[t]he term "digital twin" has been applied to representations of buildings and aggregations thereof such as precincts or entire cities—as long as these representations preserved aspects of temporal dynamics and self-updating ("4D")". Tomko and Winter have criticized the term "digital twin" [34]. Their critique departs from Batty's remark that "a computer model of a physical system can never be the basis of a digital twin [i.e., 'mirror'] for many elements of the real system are ignored in any such abstraction" [22]. Tomko and Winter, in contrast with Batty, argue that the notion of "digital twin" should be replaced. They suggest that the notion that should replace the term "digital twin" should be "cyber–physical–social system with coupled properties". They claim that this shift in the description of this phenomenon goes hand in hand with a recognition of the fact that digital models do not function exclusively as a "passive reflection of a mirror", but most importantly as systems serving to establish methods of action. They also mention that "[t]he coupling also implies that the system to describe is not a purely digital one" [34].

## 6 The Cases of Plan.City by AccuCities and the Collaborative Project Between Barcelona and Bologna: Towards Climate-Neutral Smart Cities

A worth-mentioning urban scale digital twin is that of London, which was created by AccuCities. The urban scale digital twin of London is accessible through the application named Plan.City [35]. The datasets of the aforementioned urban scale digital twin of London are publicly available. The users of Plan.City can build a maximum building envelope using a 3D Builder Tool. Unreal Engine 4 (UE4) is used to assist decision-making concerning the urban environment of London [36] (Fig. 1). Among the analyses for which the aforementioned urban scale digital twin of London was used is that concerning wind microclimate. For this analysis, SimScale was used (Fig. 2). In July 2022, the Barcelona Supercomputing Center—Centro Nacional de Supercomputación (BSC-CNS), the CINECA Consortium of Universities, the University of Bologna, the city council of Barcelona and the city council of Bologna signed an agreement concerning the creation of a project focusing on urban governance. This project is based on the development of urban digital twins applications. At the core of this collaborative project between Bologna and Barcelona is the intention to shape evidence-based decision-making strategies concerning not only public policies, but also impact assessment. This collaborative project aims to place particular emphasis on developing urban governance methods concerning parameters related to urban mobility and sustainable environmental design. More specifically, special attention will be paid to urban planning policies intending to eliminate greenhouse-gas emissions (Fig. 3). Moreover, this project aims at the development of urban scale digital twins to be used by the municipality of Bologna as well as the municipality of Barcelona for implementing ecological public policy models in both cities. Aitor Hernández-Morales, in his article entitled "Barcelona bets on 'digital twin' as future of city planning", refers to the MareNostrum supercomputer, which is housed in Torre Girona chapel in Barcelona and is among the most powerful data processors internationally [37, 38]. This supercomputer is used for the aforementioned collaborative project between Barcelona and Bologna. He also mentions that this supercomputer is expected to process data to improve urban policies.

## 7 To Conclude: On How Commoning Practices and Mobility Justice Can Inform Urban Scale Digital Twins

A concept that is useful for addressing the issue of unequal access to urban scale digital twins' data is that of 'mobility justice' [3], that Mimmi Sheller uses to suggest a new way of understanding inequality and uneven accessibility to the mobility commons [39]. This concept is useful for analyzing how urban scale digital twins can be used for traffic simulations. The main idea behind the use of the term 'mobility justice' is the intention to render explicit that while mobility is a fundamental right for everyone, it is experienced unequally along the lines of gender, class, ethnicity, race, religion and age. Sheller, in "Mobility justice in urban studies", argues that "a strong theorization of mobility justice is the best way to bridge these various dimensions of urban inequalities" [40]. She also sheds light on the fact that shaping urban planning methods that aim to promote

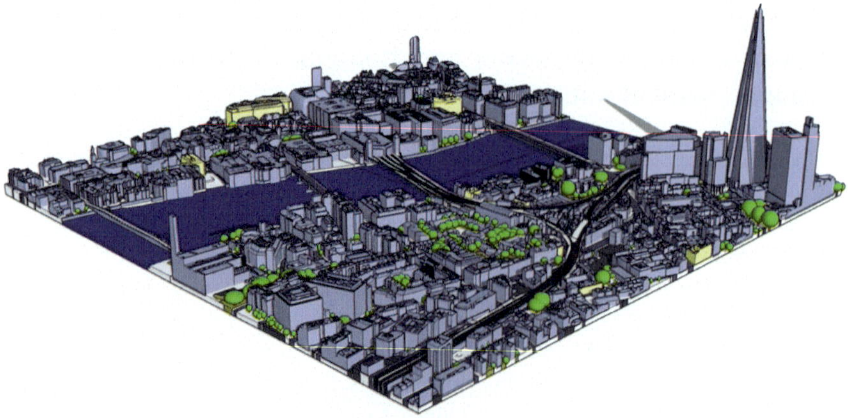

**Fig. 1** Image from 3D London|AccuCities. Licensed under a creative commons attribution-ShareAlike 4.0 international license. *Source* https://www.accucities.com/product/tq3280-free-3d-london-samples/ (accessed 7 January 2023)

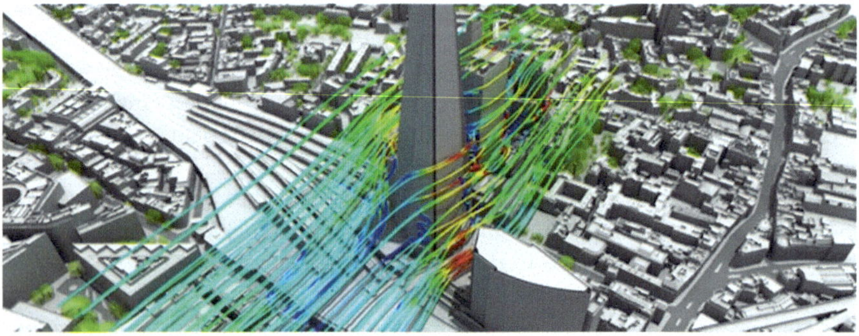

**Fig. 2** Wind microclimate analysis of London by simscale. *Source* https://www.great.gov.uk/international/trade/case-study/3474/3d-model-of-london-in-simscale-wind-comfort-simula/ (accessed 7 March 2023)

a sustainable future for the cities should go hand in hand with using less destructive modes of urban mobility. To shape approaches that promote the use of Big Data for urban analytics without neglecting the social aspects involved in the strategies of formation of urban policies, it is important to bear in mind the weaknesses of 'digital universalism' and the assumptions on which the creation of urban scale digital twins is based. To do so, pivotal is the epistemological shift from technical to socio-technical perspectives.

Of great importance for understanding the connections between the ongoing debates around the concepts of commons and commoning and the history of Advocacy planning movement in the late 1960s are the debates concerning the critiques of urban renewal in Philadelphia during the late 1950s. Denise Scott Brown has commented on advocacy planners' critique of urban renewal program, highlighting that it "derived from the problem that urban renewal had become 'human removal'". She has also "underscored that the main argument of advocacy planners was that architects and urban planners'

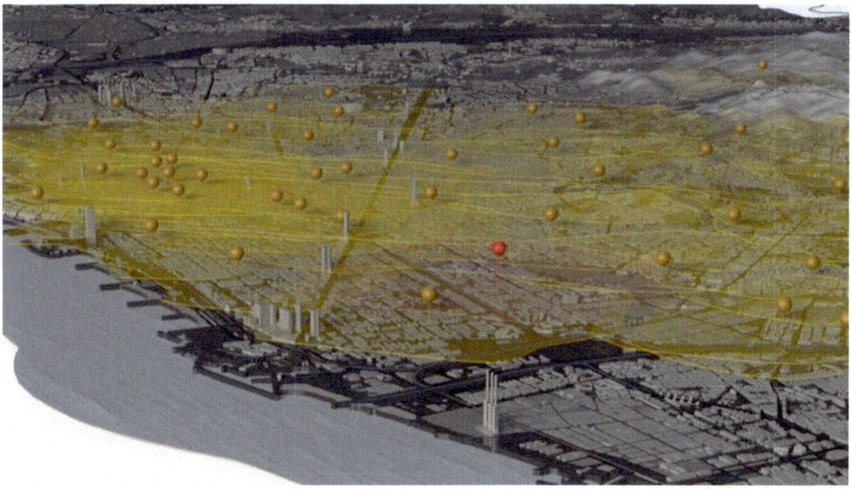

**Fig. 3** An example of accessibility in Barcelona's library network with emphasis on the new Gabriel García Márquez library. *Source* https://www.hpcwire.com/off-the-wire/bsc-and-barcelona-city-council-present-initial-results-of-digital-twin-project/ (accessed 1 Janurary 2023)

leadership had diverted urban renewal from a community support to a socially coercive boondoggle" [41–44]. A distinction that is at the core of the debates around participation-oriented strategies is that between the collaborative and the co-production approaches. The distinction between the collaborative and the co-production approaches is examined by Vanessa Watson, in "Co-production and collaboration in planning: The difference," where she notes that, "co-production" and "collaborative or communicative planning," despite their shared concern "with how state and society can engage in order to improve the quality of life of populations […] with an emphasis on the poor and marginalized," differ in the sense that co-production "works outside (and sometimes against) established rules and procedures of governance in terms of engagement with the state, while this is much less usual (although not impossible) in collaborative and communicative planning processes" [4, 6, 42, 45].

The reflections developed here regarding the role of urban commons in shaping urban planning strategies aimed to render explicit the relevance and importance of establishing methods intending to examine the actual practices of citizens while making decisions. A concept that is of great importance for understanding the challenges and benefits of taking into account the commoning practices of the citizens while taking decisions concerning urban planning is that of 'negotiated planning', which "focuses less on normative expressions of how planning should be (i.e., informed by evidence and participation) and more on the actual practices evident in cities" [46]. As Vanessa Watson has underscored, 'negotiated planning' strategies should be based on a close analysis of "the difficulties of […] [the] processes as well as to the range of contexts and conditions within which participation takes place" [45], which would save them (the 'negotiated planning' strategies) from the traps of an idealized image of collaborative planning based on the Habermasian model [47]. The shift from 'collaborative' toward 'negotiated planning' is related to the intensification of the interest not only in the commoning practices, but, most importantly, in the actual "actors and power dynamics [...] involved," as well as in "the 'virtuous cycle' of planning, infrastructure, and land" [46]. A characteristic of "negotiated planning" that is note-worthy is the attention it pays to the actual "power-laden compromises, contests [...] among various arms of the state, civil society, and the local and international private sector" [46]. More specifically, 'negotiated planning' approaches place particular emphasis on "the actions and agendas of a whole range of stakeholders who together work to configure a fragile system which is constituted through and co-constitutive of each urban context" [46].

Architecture and urban planning education should try to help students understand how urban commons can inform our understanding of the shared codes and conventions characterizing the creation of urban scale digital twins. As David Harvey has highlighted, our understanding of commons should go beyond natural resources given that all resources are socially defined [48]. As John Bingham-Hall has underscored, the notion of common "suggests a community of commoners that actively utilise and upkeep whatever it is that is being commoned" [49]. The role of urban commons in spatial planning decision-making that is at the core of the current debates concerning architecture and urban planning education. At the center of the reflections developed in this paper is the exploration of how we could combine the use of advanced digital tools such as digital twins and civic-oriented participatory design methods. However, despite the aspirations of urban scale digital twins to enhance the participation of citizens in the decision-making processes relayed to urban planning strategies, the fact that they are based on a limited set of variables and processes makes them problematic. During the coming years, given the galloping development of urban scale digital twins applications globally, it will be of pivotal importance to shape methodological tools offering the possibility to develop new forms of social advocacy around Big Data, bringing together the reflections on smart cities and the debates around urban commons.

**Acknowledgements.**

H.F.R.I. Hellenic Foundation for Research & Innovation

The research project was supported by the Hellenic Foundation for Research and Innovation (H.F.R.I.) under the "3rd Call for H.F.R.I. Research Projects to support Post-Doctoral Researchers" (Project Number: 7833)

## References

1. Charitonidou M (2022) Urban scale digital twins in data-driven society: challenging digital universalism in urban planning decision-making. Int J Architect Comput 20(2):238–253. https://doi.org/10.1177/14780771211070005
2. Charitonidou M (2021) Public spaces in our data-driven society: the myths of digital universalism. In: International Latsis symposium 'deep city – climate crisis, democracy and the digital, EPFL Lausanne, Switzerland, March 24–27. https://doi.org/10.3929/ethz-b-000465249

3. Charitonidou M (2021) Mobility and migration as constituting elements of urban society: migration as a gendered process and how to challenge digital universalism. In: Scelsa JA., Tandberg JJ (eds) ACSA teachers conference, curriculum for climate agency: design in action. https://doi.org/10.35483/ACSA.Teach.2021.27
4. Charitonidou M (2022) Urban commons as a bridge between the spatial and the social. In: Brears RC (ed) The Palgrave encyclopedia of urban and regional futures. Palgrave Macmillan, Cham. https://doi.org/10.1007/978-3-030-51812-7_290-2
5. Bresnihan P (2016) The more-than-human commons: from commons to commoning. In: Kirwan S, Dawney L, Brigstocke J (eds) Space, power and the commons: the struggle for alternative futures. Routledge, London, p 96
6. Charitonidou M (2021) Housing programs for the poor in Addis Ababa: urban commons as a bridge between spatial and social. J Urban History 48(6):1345–1364. https://doi.org/10.1177/0096144221989975
7. Charitonidou M (2022) Revisiting civic architecture and advocacy planning in the US & Italy: urban planning as commoning and new theoretical frameworks. In: 110th ACSA annual meeting: empower (ACSA 2022), Los Angeles, CA, USA, 19 May 2022. https://doi.org/10.3929/ethz-b-000528991
8. Stavrides S (2016) Common space: the city as commons. Zed Books, London
9. Castells M (2010) The rise of the network society. Second edition with a new preface. Wiley-Blackwell, Chichester, West Sussex
10. Castells M (1989) The informational city: information technology, economic restructuring and the urban-regional process. Basil Blackwell, Oxford
11. Castells M (2009) Communication Power. Oxford University Press, Oxford
12. Castells M (2010) Globalisation, networking, urbanisation: reflections on the spatial dynamics of the information age. Urban Studies 47(13):2737–2745. https://www.jstor.org/stable/43079956
13. Castells M (2021) From cities to networks: power rules. J Classic Soc 21(3–4):260–262. https://doi.org/10.1177/1468795X211022054
14. Caprari G, Castelli G, Montuori M, Camardelli M, Malvezzi R (2022) Digital twin for urban planning in the green deal era: a state of the art and future perspectives. Sustainability 14(1):6263. https://doi.org/10.3390/su14106263
15. Ferré-Bigorra J, Casals M, Gangolells M (2022) The adoption of urban digital twins. Cities 131:103905. https://doi.org/10.1016/j.cities.2022.103905
16. EU responses to climate change Society. Available at: https://www.europarl.europa.eu/news/en/headlines/society/20180703STO07129/eu-responses-to-climate-change
17. Hatcher J (2022) Digital twins can help sustainability. Smart Build Magaz. Available at: https://smartbuildingsmagazine.com/news/digital-twins-can-help-sustainability
18. Blair GS (2021) Digital twins of the natural environment. Patterns 2(10). https://doi.org/10.1016/j.patter.2021.100359
19. Charitonidou M (2022) Towards a civic approach to urban data: The myths of digital universalism. In: X Congresso Associazione Italiana di Storia Urbana (AISU), Politecnico di Torino, Turin, 8 September 2022. https://doi.org/10.17613/zpmq-1056
20. Charitonidou M (2022) Urban scale digital twins and commoning practices: mobility justice and sharing ground resources. In: 2022 Architectural Humanities Research Association (AHRA) conference 'building ground on climate collectivism: architecture after the Anthropocene, 17–19 November 2022. https://doi.org/10.17613/t3gq-hd32
21. Borgman CL (2015) Big data, little data, no data: scholarship in the networked world. The MIT Press, Cambridge, Mass
22. Batty M (2018) Digital twins. Environ Plann B: Urban Analytics City Sci 45(5): 817–820. https://doi.org/10.1177/2399808318796416

23. Loukissas YA (2019) All data are local: thinking critically in a data-driven society. The MIT Press, Cambridge, Mass
24. Milan S, Treré E (2019) Big data from the south(s): beyond data universalism. Telev New Media 20(4):319–335. https://doi.org/10.1177/1527476419837739
25. Nochta T, Wan L, Schooling JM, Parlikad AK (2021) A socio-technical perspective on urban analytics: the case of city-scale digital twins. J Urban Technol 28(1–2):263–287. https://doi.org/10.1080/10630732.2020.1798177
26. Joss S, Sengers F, Schraven D, Caprotti F, Dayot Y (2019) The smart city as global discourse: storylines and critical junctures across 27 cities. J Urban Technol 26(1):3–34. https://doi.org/10.1080/10630732.2018.1558387
27. Bratton BH (2016) The stack: on software and sovereignty. The MIT Press, Cambridge, Mass
28. ABI Research (2021) New urban use cases drive over 500 cities to adopt digital twins by 2025. 5 January 2021. https://www.abiresearch.com/press/new-urban-use-cases-driveover-500-cities-adopt-digital-twins-2025/. Accessed 3 March 2021
29. White G, Zink A, Codecá L, Clarke S (2021) A digital twin smart city for citizen feedback. Cities 110:103064. https://doi.org/10.1016/j.cities.2020.103064
30. Deren L, Wenbo Y, Zhenfeng S (2021) Smart city based on digital twins. Comput Urban Sci 1(4). https://doi.org/10.1007/s43762-021-00005-y
31. Mayfield M (2020) Can urban scale digital twins address climate adaptation? Buildings and cities. 28 January 2020. https://www.buildingsandcities.org/insights/commentaries/urban-scale-digital-twins-climate-adaptation.html. Accessed 3 April 2021
32. Boyd A, Crawford K (2012) Critical questions for big data: provocations for a cultural, technological, and scholarly phenomenon. Inf Commun Soc 15(5):662–679. https://doi.org/10.1080/1369118X.2012.678878
33. Wildfire C (2021) How can we spearhead city-scale digital twins? Infrastruct Intell 9 May 2018. http://www.infrastructure-intelligence.com/article/may-2018/howcan-we-spearhead-city-scale-digital-twins. Accessed 3 April 2021
34. Tomko M, Winter S (2019) Beyond digital twins—a commentary. Environ Plann B: Urban Analytics City Sci 46(2):395-399. https://doi.org/10.1177/2399808318816992
35. Charitonidou M (2023) Urban scale digital twins vis-à-vis complex phenomena: datafication and social and environmental equity. In: Dokonal W, Hirschberg U, Wurzer G (eds) Digital design reconsidered, proceedings of the 41st eCAADe conference, 20-22 September 2023, Graz University of Technology, Graz, Austria, Volume 2. Education and research in Computer Aided Architectural Design in Europe and Graz University of Technology Faculty of Architecture, Brussels, Graz, 821-830. https://papers.cumincad.org/cgi-bin/works/Show?ecaade2023_472. https://doi.org/10.17613/8ebv-wq50
36. 3D city models overview. https://www.ribacpd.com/accucities/230545/3d-citymodels-overview/411006/movie/. Accessed 3 April 2021
37. Hernández-Morales A (2022) Barcelona bets on 'digital twin' as future of city planning. Politico. https://www.politico.eu/article/barcelona-digital-twin-future-city-planning/. Accessed 1 June 2022
38. Benedetti AC, Costantino C, Gulli R, Predari G (2022) The process of digitalization of the urban environment for the development of sustainable and circular cities: a case study of Bologna, Italy. Sustainability 14(21):13740. https://doi.org/10.3390/su142113740
39. Sheller M (2018) Mobility justice: the politics of movement in an age of extremes. Verso, London
40. Sheller M (2020) Mobility justice in urban studies. In: Jensen OB, Lassen C, Kaufmann V, Freudendal-Pedersen M, Gøtzsche Lange IS (eds) Handbook of urban mobilities. Routledge, London, New York, 13–22
41. Scott Brown D (2009) Towards an 'Active Socioplastics'. In: Idem, Architecture words 4: having words. Architectural Association, London, 32–33

42. Charitonidou M (2021) The 1968 effects and civic responsibility in architecture and urban planning in the USA and Italy: Challenging nuova dimensione and urban renewal. Urban, Plann Trans Res 9(1):549–578. https://doi.org/10.1080/21650020.2021.2001365
43. Charitonidou M (2022) Denise Scott Brown's active socioplastics and urban sociology: from learning from West End to learning from Levittown. Urban, Plann Trans Res 10(1):131–158. https://doi.org/10.1080/21650020.2022.2063939
44. Charitonidou A (2022) Denise Scott brown's nonjudgmental perspective: Cross-fertilization between urban sociology and architecture. In: Grahn F (ed) Denise Scott brown in other eyes: portraits of an architect. Birkhäuser, Berlin, Boston, 98–106. https://doi.org/10.1515/9783035626254-008
45. Watson V (2014) Co-production and collaboration in planning: the difference. Plann Theory Pract 15(1):62–76. https://doi.org/10.1080/14649357.2013
46. Cirolia LR, Berrisford S (2017) Negotiated planning: Diverse trajectories of implementation in Nairobi, Addis Ababa, and Harare. Habitat Int 59(2017):71–79. https://doi.org/10.1016/j.habitatint.2016.11.005
47. Tewdwr-Jones M, Allmendinger P (1998) Deconstructing communicative rationality: a critique of Habermasian collaborative planning. Environ Plan A: Econ Space 30(11):1975–1989. https://doi.org/10.1068/a301975
48. Harvey D (2011) The future of the commons. Radic Hist Rev 109:101–102
49. Bingham-Hall J (2016) Future of cities: commoning and collective approaches to urban space. Government Office for Science, London

# A Pilot Study for a 3D Scanning to VR Workflow for Building Spaces: Exploring Possibilities Through the Use of Different Scanning Hardware

Ian Garcia[1](✉), Mehmet Ozdemir[1], Silvia Van Aken[2], Kristof Overdulve[2], and Jouke Verlinden[1]

[1] University of Antwerp, 2000 Antwerp, Belgium
ian.garcia@uantwerpen.be
[2] AP University of Applied Sciences and Arts Antwerp, 2000 Antwerp, Belgium

**Abstract.** Shaping a more efficient scan-to-BIM (Building Information Modeling) workflow has been an ongoing goal in the Architecture, Engineering, Construction, and Facility Management (AEC/FM) domains. We propose an innovative workflow which processes raw 3D scanning data of a building into usable 3D models for virtual environments that can be directly utilized in VR environments for automating a part of the 3D process. In this case study, two devices with several accompanied tools were explored, namely a high-end laser scanner and a low-cost LiDAR-integrated Tablet. Both hardware solutions differ in accuracy, reach and portability. For reconstruction, we chose a renovated 1800 $m^2$ building, part of the University of Antwerp city campus. This location consists of reference spaces such as offices, classrooms, multifunctional rooms, auditoria, hallways, and stairs. Scanning larger workspace areas bring multiple challenges that are best considered beforehand. In this article, the advantages and disadvantages of both scanning techniques will be compared. On top of this, we will give an overview of the challenges and how to proceed them. Finally, the integration of the 3D scan into the VR workflow will be discussed. Dependencies are textures, file size, reduction capabilities, export configurations and plug-in compatibility. As a result, this case study is the constructional basis for workflows in the industry.

**Keywords:** 3D scanning · Digital twin · Scan-to-BIM · AEC · LiDAR · Case study


M. Blanco Lage et al. (eds.), *Towards a New European Bauhaus—Challenges in Design Education*,
https://doi.org/10.1007/978-3-031-40188-6_21

# 1 Introduction

## 1.1 Problem

Streamlining scan-to-BIM process is invaluable to AEC companies to improve process automation. In this quest, they try to elaborate a workflow through the use of 3D scanning for digitizing existing buildings, eventually creating virtual environments.

Most of the techniques used in these 3D scanning and processing applications are complex for non-tech savvy individuals. Moreover, it requires expert knowledge to execute and process the data of 3D meshes. The core steps of the traditional processes remain manual and require the necessary learning curve to understand and work efficiently. Another part which the person themselves cannot always solve accordingly is the management of redundant data during scanning. Also, BIM conversion becomes cumbersome with increasing levels of detail (LoD) [15].

## 1.2 Knowledge Gap

The architectural field lacks a structured workflow for implementation and facilitation of spatial simulations such as virtual reality environments for training, decision making and communication with external stakeholders [7, 8]. The main challenge is to translate the insights from one mind to others. The 3D scanned models can provide a better visual overview of all the related elements and form a digital overlay of a building or structure.

Besides the explorational and executional tasks on the topic, the knowledge gap is also confirmed by active companies. Through the conducted interviews, the same problem was repeated, namely the lack of experience with 3D scanning [7]. Here follows a quote to highlight the relevance of the problem: 'We need synchronicity between the real and virtual world' -Michael Peeters, digitalcloning.io (2021). However, several companies are actively in the surge for scan-to-BIM solutions. Their own research groups slowly display development progress for specific cases such as virtual reality (VR) training, data visualization for project management and customer experience [4, 23, 37].

The literature also points out that the traditional methods and tools are still widely used but lack the benefits of laser scanning and point clouds to process into multi-faceted 3D models [9].

The research questions we are answering in this article are the following:

- How can we develop a new workflow that streamlines the 3D capturing of buildings, the 3D scan processing and its integration to virtual environments, so that we can build architectural spaces more efficiently in VR?
- What are the challenges during the development of such workflow and which tools need to be used when and how?

### 1.3 Devices

**Leica BLK360**. The accompanied software bundle for the BLK360 are Leica Cyclone packages, namely: FIELD 360, REGISTER 360 and 3DR. The BLK360 scans raw point clouds in formats such as 0.57 and .IMP and separate images in .jpg format with or without HDR [39].

Firstly, FIELD 360 is used in conjunction with the scanner during the scanning process, it provides the interface to operate the BLK360 and gives an overview of the scanned data. Secondly, REGISTER 360 processes the point cloud and prepares the data to be exported as a point cloud. Thirdly, 3DR specializes in the conversion to a 3D mesh and texturing. The Leica ecosystem is therefore an all in-one-solution.

The setup for the Leica BLK360 laser scanner is more elaborate and goes as follows: the BLK360 generates its own WiFi hotspot that can connect to an iPad Pro or desktop computer from up to 50 m. By connecting an iPad Pro to the WiFi hotspot during operation, a user can review scans and change the BLK360's settings on the fly. At the recommended scanning distance of 10 m, the BLK360 has 6 mm accuracy. As distance increases, accuracy will slightly decrease. At 20 m, the accuracy is 8 mm [38].

**iPad Pro**. The alterative mobile scanning method that will be used is the LiDAR scanning technology embedded in the iPad Pro.

The iPad Pro does not have any official LiDAR specifications. However, the available applications for 3D scanning give an idea for its capabilities. For example, the 3D scanner app shows a resolution tolerance of 3 mm at 0.5–20 mm at 5 m [5, 19]. No official apps accompany this hardware, but third-party options such as Matterport allow for processing of the scanned data, which is in .USDZ, .OBJ, .STL, and .PLY file formats. However, no raw point cloud output is possible, and no images are being captured using the third-party apps [28]. Accordingly, there are no official accompanied software packages, but services provided by platforms such as Matterport allow to export the data and create 3D textured meshes [20].

### 1.4 Feedforward - What Will Happen in This Paper

This paper will present experimentation, benchmarking and workflow guidance for 3D scanning enterprise scale buildings. Different applications will first see a round of exploration and potential analysis. The suitable applications will be tested in a benchmark of a classroom at the campus. For the actual scanning strategy for the building two methods were used. The first one is the native scanning with the Leica BLK360 scanner accompanied by the native software bundle. This native method was used to scan the entire building. In the second method, Matterport was used in conjunction with the BLK360 laser scanner, its processing results got forwarded as exported models for further investigation into VR adaptation. Eventually a generalized workflow will be mapped, serving as a guide for future use cases.

## 2 Background

### 2.1 Literature

3D scanning is the method of using photogrammetry or laser scanning to create 3D models. These techniques use digital data, in photogrammetry's case images and in laser scanning digital imagery to obtain (colored) point clouds which then can be converted into texturized 3D objects.

3D scanning has been a trending subject in many fields. For example, in reconstruction of cultural heritage landmarks and archeology [9].

**EAAE related papers**. The teaching process of the design studio model contains a high interaction level between all participants. These include students, teachers and external associates. Within this model, there is a search for the integral application of design methods and techniques in function of the design process. This program has several major challenges, several relevant to highlight are critical thinking and decision making and opportunities for technological advancements [21].

There is a need for desire through experience instead of objects. Finding the craft of effort becomes scarcer and is a sign of authenticity [32]. Although perhaps contradictory, the need for more realism can also be increased using VR. With detailed and realistic scans, we can create and recreate the experiences, where all hope is lost to rebuild or restore the same structures and materials as they were in their original time, e.g. 3D scanning surfaces to use as physical relief, virtual location reconstructions and virtual musea [24].

**AEC publications**. There is an increasing trend of BIM adoption in AECO (Architecture, Engineering, Construction and Operation) industries, it has only been used for some new projects in recent years [33]. Hence, most of the existing buildings do not have a BIM model [15]. In some cases, even blueprints may be unavailable [40]. Therefore, the necessity to 3D scan and provide a scanning workflow accordingly is highly desired for the heritage and real estate market.

Relevancy of BIM and 3D scanning is also covered by trends, a statistic for Swedish companies shows company future investment demands. BIM ends up as most demanding (70%), 3D scanning second (50%) and AI third (40%) [3].

### 2.2 3D Scanning Techniques

**Laser scanning**. Laser scanning systems use different methods to measure distance. "time-of-flight" measurement is done by throwing out light waves that bounce off surfaces and reflect to the sensor. The sensor then calculates how far away the surface is by measuring the time taken for the light beam to complete its journey, this technology is mainly used for long range scanning [29]. Another approach is triangulation, in which the position of a laser and camera determine the distance of an object through geometry, using reflecting angles to calculate the distance [27]. During a single scan, a laser scanner will collect millions of 3D coordinates. When the point clouds from laser scans are processed, they form a digital representation of the scanned surfaces, demonstrating the dimensions and spatial relationships of topographic features and structures [35].

**What is LiDAR**. LiDAR, or light detection and ranging, is a popular remote sensing method used for measuring the exact distance of an object [1]. The Leica scanner uses the same principle, but its calculations are more accurate compared with build-in LiDAR devices such as the iPad Pro 12.9 2021 or iPhone 12/13 Pro/Max [16].

**Clarification**. First of all, 'Laser' is more of a device that uses laser technology to emit light through the optical amplification process. Whereas LiDAR is rather a method for measuring the distance to a predetermined target. In conclusion, LiDAR technology uses laser technology to facilitate the results [18].

On the other hand, the main differences for the hardware solutions are accuracy, range, data type and accompanied software packages. In this case study, we will compare the Leica BLK360 scanner with an iPad Pro 12.9 2021.

**Photogrammetry**. Photogrammetry was not selected as one of the techniques to be included in this case study. This technique uses the capturing of multiple images as input for a reconstruction software that converts these captures into a 3D mesh. The reason to not use this technique was decided after early testing and expert advice, which concludes that this technique will be too time consuming and labor intensive. Especially when going for large scale buildings and outdoor environments [31].

# 3 Methods

As a first step we conducted interviews to obtain insights on current practices in the field. Secondly the choice of the building, defining the type of scanning that will be required based on size and present objects. Then a selection of 3D scanning applications is explored to verify their applicability for the building scan, which are two iPad applications and the BLK360 scanner. Comparison on scanning performance was done using one reference room. The different applications are shown with visual outputs. The doorway of a classroom visualizes the amount of detail and the accuracy in which they scan and translate their respective textures, see reference [11]. The best application was used for the entire building. From here, two main methods were shaped using the BLK360 with native software bundle, and the BLK360 with Matterport support.

## 3.1 Interviews

The participants were partners of the TETRA project [36] namely digitalcloning.io, Immersive lab AP, digitaltwin.be and cyborn.be. The subject of the interviews was mainly about the process of 3D scanning and current workflows. The specific interview questions are to be found in Appendix 1.

## 3.2 Choice of PO Building

The building that was subjected to 3D scanning was the Product Development Campus at the University of Antwerp [25]. The campus on its own is representative for an educational building context where the visual noise of teamwork offices and open workplace environments are present. It also includes the more traditional auditorium spaces and a more practical workplace beneath the ground floor (Fig. 1).

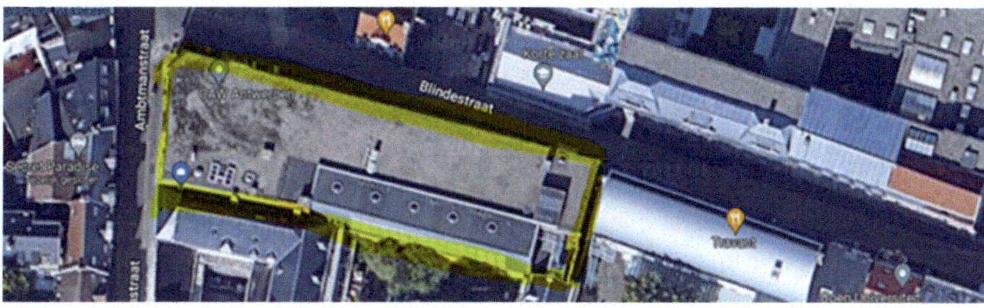


**Fig. 1** Product development campus. Address: Ambtmanstraat 1, 2000 Antwerp, Belgium (Aerodata Internation Surveys Maxar Technologies, Google Maps, 2022)

## 3.3 Choice of Tools

**iPad apps comparison**: Multiple apps on the iPad Pro were tested and benchmarked. This was done using the same area, namely an empty classroom (2.17) with a few props. Each app got monitored for several parameters which are time-to-scan, vertices count and process time on different processing settings varying between low–medium–high [10].

**Leica native and Matterport comparison**: Leica native software consists of Cyclone FIELD 360, with Cyclone REGISTER 360 and Cyclone 3DR for processing.

Alternatively, Matterport also got tested with smaller spaces, but used in conjunction with the BLK360 scanner. This is the only way to form a MatterPak.[1] The most important data from the MatterPak are the textured pointclouds in.xyz and a textured.obj file.

For processing 3D scanned data, a workstation with the following specifications was used: A Dell Alienware Aurora with CPU Intel i9, GPU RTX 2070 and 32 GB of RAM.

Additional tools for improving the scanning process itself were not used due to the size of the project, no anti gloss spray or reference point marking was used.

[1] Service of Matterport that supplies processing results from a 3D data browser upload.

To implement the exported data into VR, the 3D game engine tool Unity was used to facilitate a lightweight VR environment. Crucial due to the amount of data, reduction must be made without lowering the details of the texture and avoid a lack of realism. In this research project, a reduction was made from a native circa 100 million points to circa 10 million points.

Scanning the entire campus took 319 scan points on the Leica Cyclone iPad app and a duration of 40–45 h over 8 days, excluding the preparations. In total there was a point cloud containing over 200 GB of raw data.

## 4 Results and Discussion

The results of this research project are twofold. Firstly, we succeeded into creating a fully implementable and VR ready 3D model from the scanned data of the campus. Secondly, a virtual demo made interaction with the environment possible, thus increasing the realism level of upcoming adaptive user context projects. The green blocks in Fig. 2 are interactable objects, these test the grabbable plug-in [22]. The referenced video showcases the interaction [12]. A realistic environment also results into enhanced recognizability. This phenomenon is noticeable in related practices. For example, when medicine students operate in a room that looks more authentic, the familiarity helps operating more efficiently [30].

**Fig. 2** VR application in unity with the imported mesh of the campus' 3rd floor as an interactable environment

As second result of this article, we constructed a mapped workflow. The workflow is divided into scanning, processing, and VR integration. An overview of the entire process from scan data to importable mesh file to VR application is present in Figs. 4, 5 and 6. The fully detailed approach is given in the appendix. It consists of a combination of the native Leica software and an alternative use through the Matterport ecosystem.

Although the information displayed in these diagrams seems very specific, the processing itself will differ in alternative processing software. But the main steps are valid as a general workflow that would be implementable across the industry: import, setup, alignment, preliminary export, unit checking, potential classification, meshing, texturing, image adjustment and final exporting. Reference companies are Artec, Faro, Trimble, Johnsonlevel, Riegl. They use different processing software but will benefit from the same proposed workflow approach [26].

## 4.1 Lessons Learned and Takeaway

Here we share the lessons learned of using the Native Leica software and alternative Matterport support method. The main takeaway point of this comparison is the larger flexibility that the native Leica software app offers over the Matterport app. Namely, the freedom of scan repositioning and relinking of scans, while also being able to manage and transfer between scanning bundles, offer a big advantage over the Matterport method. Only for processing parts of buildings quickly, Matterport appeared to be a preferable method. But for more manual and specific conversions, the native Leica approach is recommended (Table 1).

Some lessons for 3D scanning in general. Accessibility to the rooms is crucial when scanning, personnel could get informed at which time trespassing was forbidden to avoid unwanted 3D data into the scans. The resulting mesh does not cover densely furnished spaces. Therefore, repositioning and removing clutter was advised before scanning.

## 4.2 Challenges

**Beforehand**. During the planning process, we were cautious about the scanning setup. Scanning with HDR needs to be done across the board when scanning for a coherent space or area. Since otherwise the software will be forced to convert all the HDR images to non-HDR images, which takes multiple hours of processing time when you have a lot of data.

It is important that the amount of point cloud points is manageable for the workstation you are using to run the game engine. The Meta Quest VR headset is lightweight in processing nature and has standalone capabilities. Therefore, to be able to run apps on this device smoothly, a lot of compression is required.

**During**. Attention points while 3D scanning are time of day, lighting condition, reflective/transparent surfaces, and visual noise (intensely crowded areas with obstructing objects). Managing the prevention of unwanted data during the scans went rather smoothly thanks to the organized planning and precautions.

**Table 1** The method overview table provides information about the characteristics of the two methods, displaying their benefits and shortcomings. The last four rows show the common characteristics of the compared methods

| | Leica cyclone field 360 + cyclone 360/3DR | Matterport + cyclone 360/3DR |
|---|---|---|
| *Differentiating characteristics* | | |
| View type | 3D view on the iPad directly in the app | No 3D view, only 2D top view |
| File management | File managing in bundles | File managing with floors |
| Scan relinking | Post relinking scans | Fixed link to previous (with recognition to non-previous but less quality) |
| HDR modes | Modes: HDR on, HDR off | Modes: HDR off + 4 HDR levels (2–5) |
| Scan moveability | Movable individual scans | Non-movable scans |
| Scan indentification | Individual scan naming | Numbering only |
| Processing type | Seamless integration to leica software<br>Manual work to rework point cloud to textured mesh | Pay to upload 1 scan file (through matterpak), gives colorized point cloud (.XYZ) + 3Dmesh (.OBJ) |
| Floor linking | Links floors | No floor links |
| Scan time | Density levels: low (1 min 30) mid (2 min 30) high (4 min 30) | Density levels: low (2 min 00) mid (3 min 00) high (5 min 00) |
| *Shared characteristics* | | |
| Preview options | Preview HDR image of single scan | |
| Scan density modes | 3 scan density modes (low, medium, high) | |
| Alignment requirements | Need for visible link in between scans to pre-aling | |
| Scan continuation | In between floors, bottom up on stairs | |

**Result**. As mentioned, we have two results. The intermediate result which is the scanned data, and the final result which is the importable mesh for the VR application. For the scanned data, the resulting point cloud does not include complete coverage of the to-be-scanned spaces. The placement of each scanning position with the BLK360 is therefore crucial to take the most relevant scanning points. For the final result, the VR integration, it was important to have the right preparation of meshes. The ones that will be enriched with interaction required to be separate from the ones that will be fixed and just serving as background. For example, collision detection asks for applicable mesh groups, if the mesh created beforehand exists of many smaller meshes, it will be a very cumbersome process during the setup to apply all the collision detection to the entire mesh.

### 4.3 Limitations

Although using the main workflow with the native Leica software, it was not meaningful to also scan the entire building with the Matterport workflow. This is mainly due to using the BLK360 laser scanner as main tool for both workflows, and only the interpretation of the data and the amount of service or manual work was the big factor into providing results. The Matterport workflow got a reduced scanned area, but still containing most complexities of the building including hallways, office rooms and stair sets.

The resulting textures were poorly translated into digital textures, coming from high-definition images that the BLK360 took.

Mainly inconsistencies in lighting and usage of non-HDR capture took out detail that adds to the realism.

### 4.4 Bottlenecks

**Computing power**. The entire scan of the building includes over 100 GB of raw data, which is challenging to process into a single mesh, even though using relatively high-end computing power. Therefore, the point cloud data was reduced from over 100 million to 10 million points to enable the mesh generation of the entire building.

**Import restrictions**. You cannot import 3D scanned data in the Matterport environment. If you started performing laser scans with the native Leica software, you need to rescan from scratch [6].

**Capture setup**. Movement of objects needs to be prevented within the duration of scans, but more importantly people need to be notified about the scanning schedule to prevent accidentally register in the scanning data by walking through a scan in process.

The data is usable to the extent of building-sized spaces. It is therefore ideally usable for corporate use, able to scan multiple rooms and floors connected to each other and map them out virtually. Scanning only one building, however, will never suffice to map all kinds of different (office) workspaces.

## 5 Conclusion

This study shows a deep dive in using two different scanning methods with different hardware, namely the LiDAR scanner build-in the iPad Pro and the BLK360 scanner from Leica. Eventually this led to a VR implementation to Unity, making it able to use the scanned environment in feasible use cases. This pilot study shows that making VR-ready environments through 3D scanning have potential to be used in the industry.

Implementation bottlenecks are more related to the VR integration itself. When following a Unity VR template. This eased the process on creating a DIY VR application, capable of demonstrating the functions and realism of the resulting mesh.

## 6 Further Recommendations

Incorporating smaller furniture/interior assets, deleting specific data points, Artificial Intelligence for generating 3D assets, Augmented Reality implementations are subjects that need to be studied more profoundly in relation to 3D scanning [2, 34]. For example, using scanners from the Artec 3D scanning ecosystem will provide more high-fidelity scans for objects that participants will interact with. For even larger buildings and target areas it might be feasible to use drones for scanning setups, these are already being developed but have yet to be testensd properly to their full capabilities [17]. Enhancing the realism factor is a big part of improving the experience and the familiarity of a virtual environment [30]. Therefore, exploration with more accurate texturing becomes interesting. When working towards use cases such as trainings in VR with real environment references it could also come in handy to investigate the modularity of certain objects within the virtual environment, enabling interaction on specific elements or creating distinctive walkthroughs using different layouts [13, 14, 23, 37]. This would all benefit the subject of implementing this processing into VR and make it much more valuable to the industry.

## Appendix

Interview Questions (translated from Dutch)

- Could you briefly introduce yourself and the company(or work)?
- How (or in which context) do you use 3D scanning in your specific case?
- Which 3D scanning methods/technologies do you employ
- How is your planning before performing a 3D scan
  - Are there specific points of attention?
  - Any particular challenge you face?
  - How do you approach for a solution?
- While performing a 3D scan…
  - Are there specific points of attention?
  - Any particular challenge you face?
  - How do you approach for a solution?
- While reconstructing/post-processing the scan…
  - Are there specific points of attention?
  - Any particular challenge you face?
  - How do you approach for a solution?
- Any tips or tricks about the overall process?

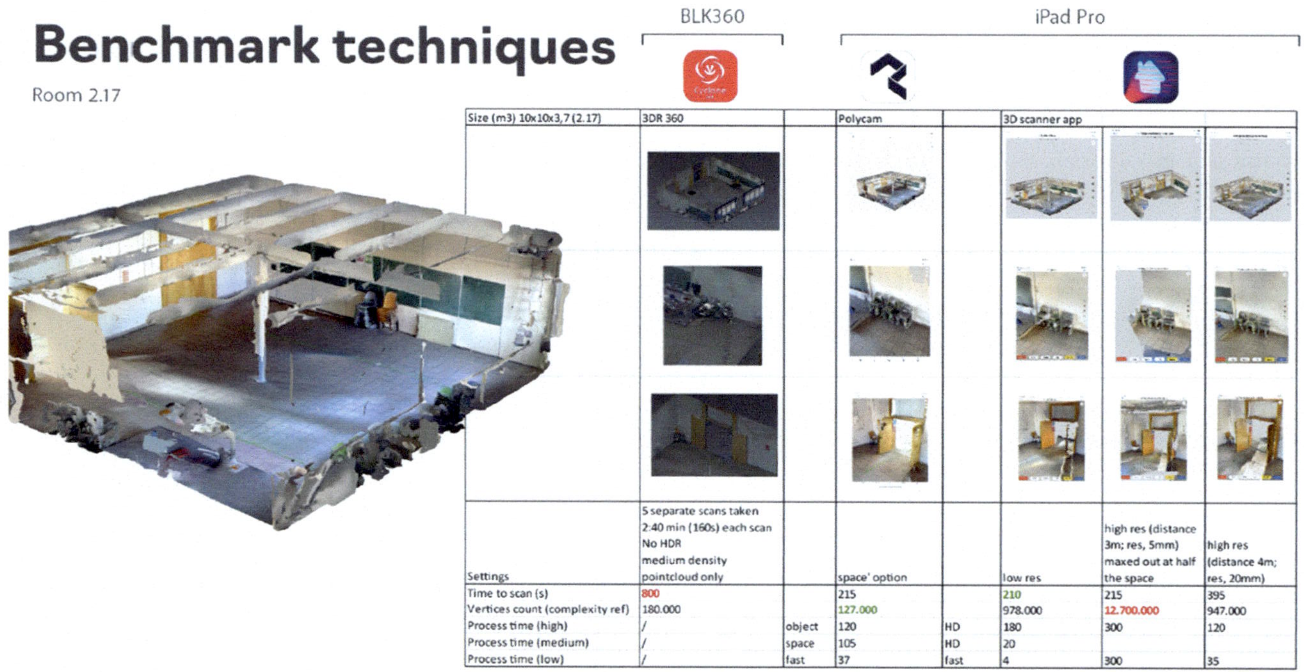


**Fig. 3** Full overview of benchmark

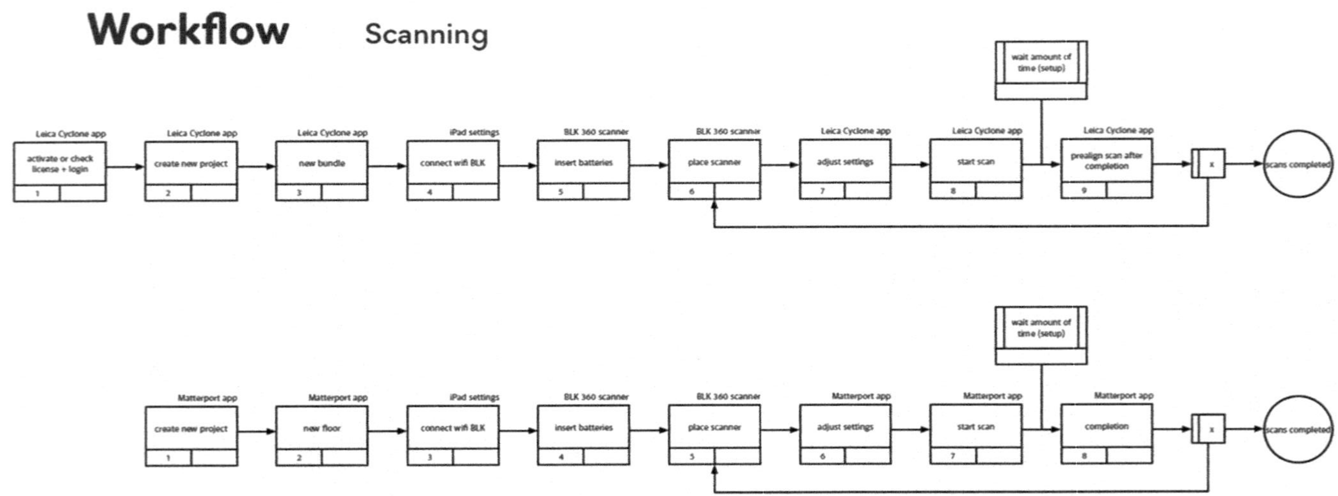


**Fig. 4** IDEF3 visualization of the scanning workflow of both the native Leica process and underneath the alternative workflow using the matterport app, but still the BLK360 as the scanning hardware

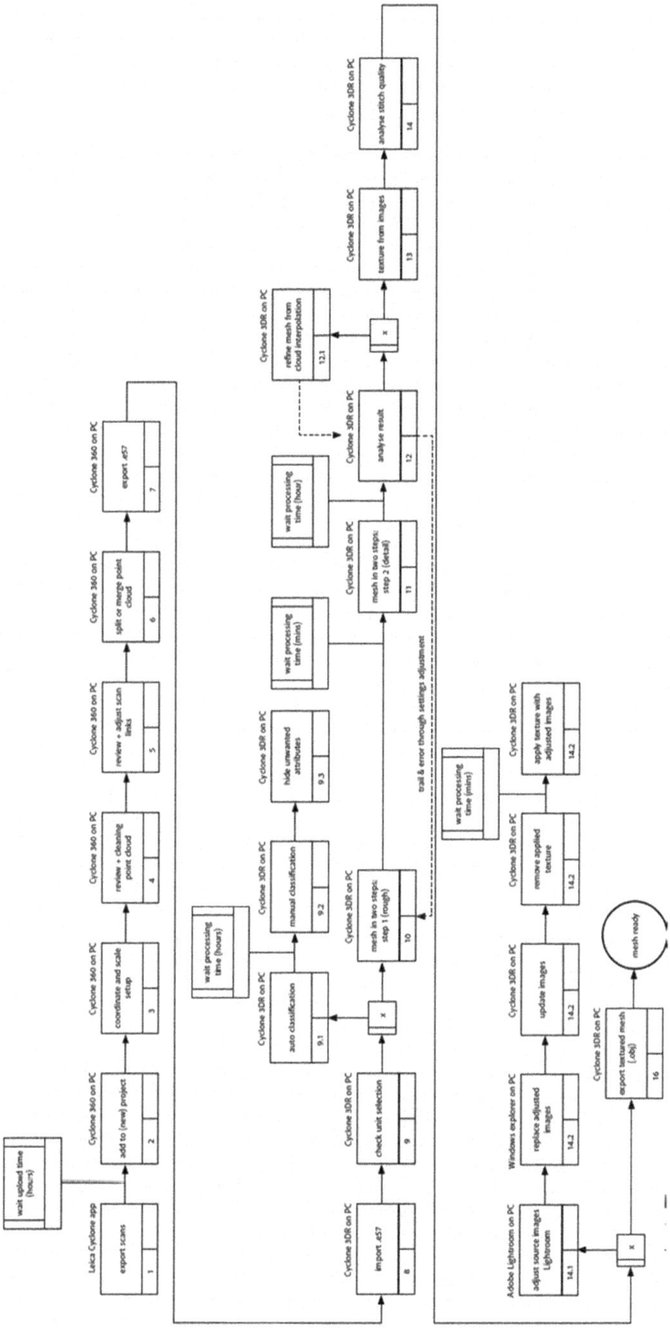


**Fig. 5** IDEF3 visualization of the manual processing workflow in leica cyclone software, displayed only with reference of the leica software

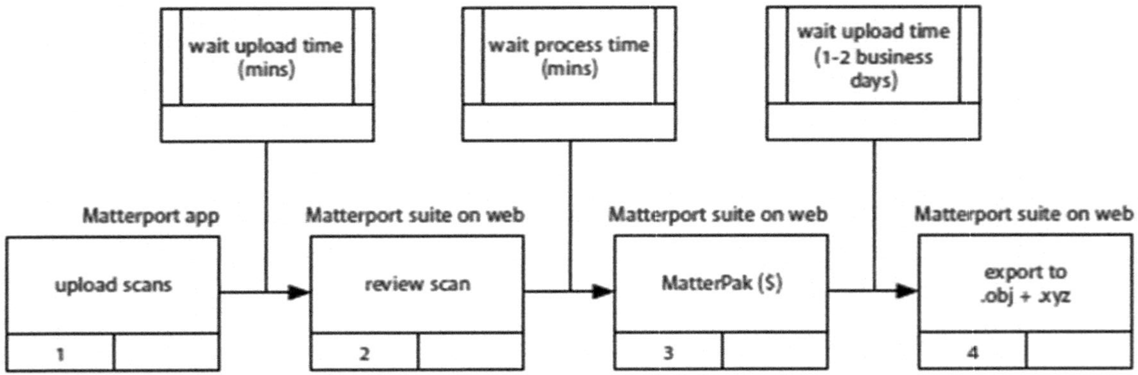


**Fig. 6** IDEF3 visualization of the semi-automatic processing workflow on matterport browser platform

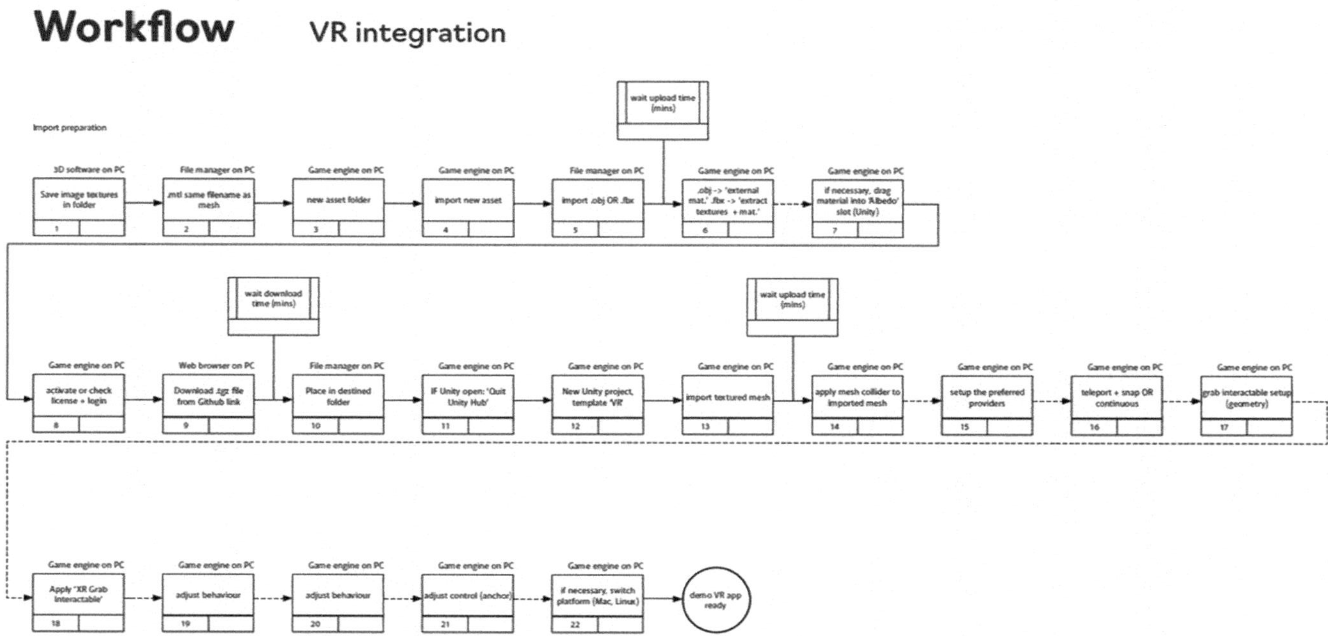


**Fig. 7** IDEF3 visualization of the template workflow using the game engine unity

See Figs. 3, 4, 5, 6 and 7

A workflow map is also made for the integration to VR, this one is based on a downloadable template and shows the setup of the imported mesh and grabbable components, ready to be used as a VR demo application to deploy to the VR headset (Figs. 9 and 10).

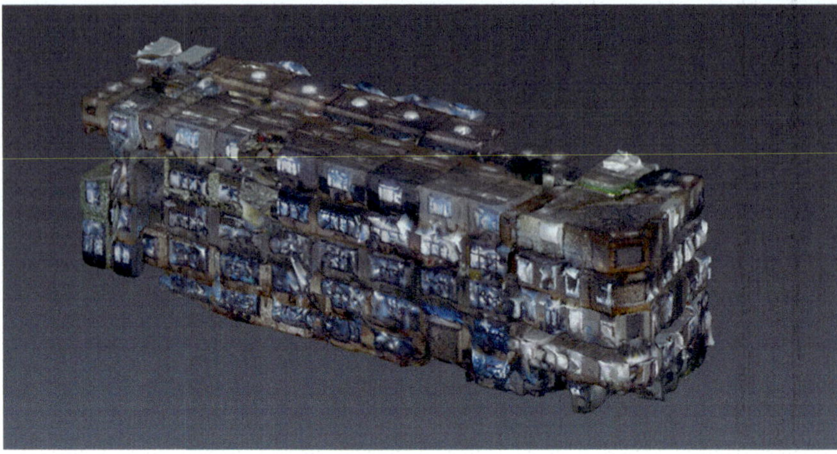

**Fig. 8** Complete watertight mesh of the reduced product development campus building using leica cyclone REGISTER 360 and cyclone 3DR

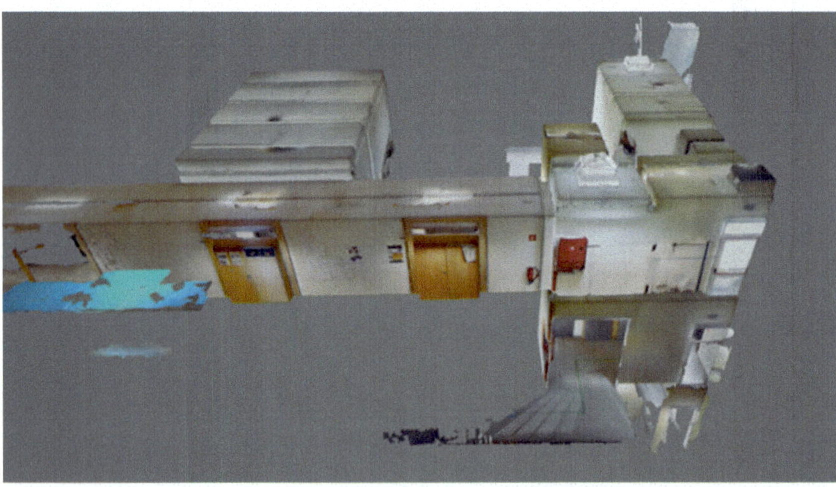

**Fig. 9** Matterport 3D.obj export, scanned with the leica BLK360 laser scanner, shown in windows 3D model viewer

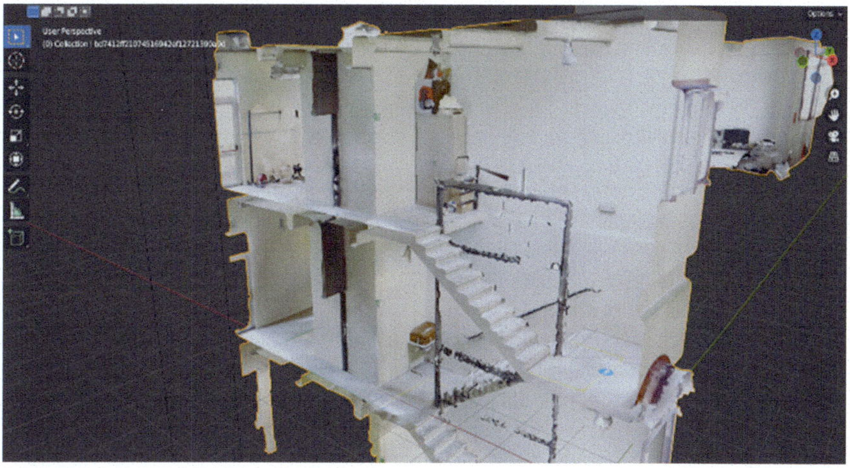

**Fig. 10** The same model as Fig. 8 but imported into blender, to gain a better view of the retrieved data from matterport

We show only a part of the entire campus for the Matterport approach, this is due to the bottle neck of having to rescan the entire campus. It suited better to only take several parts such as some offices, the hallway and a set of stairs to simulate a similar complexity with the original scanned native Leica data (Fig. 11).

**Fig. 11** 3D model export result, imported into unity. To use as a virtual environment in multiple user contexts

## References

1. Bhupendra S (2021, Oct 2) What is LiDAR technology and how does it work? Geospatialworld. https://www.geospatialworld.net/blogs/what-is-lidar-technology-and-how-does-it-work/
2. Bansal M, Kumar M, Kumar M (2020) 2D object recognition techniques: state-of-the-art work. Arch Comput Methods Eng 1:3. https://doi.org/10.1007/s11831-020-09409-1

3. Bosch-Sijtsema P, Claeson-Jonsson C, Johansson M, Roupe M (2021). The hype factor of digital technologies in AEC. https://doi.org/10.1108/CI-01-2020-0002
4. Clifford RMS, Jung S, Hoerrnann S, Billinqhurst M, Lindeman RW (2019) Creating a stressful decision making environment for aerial firefighter training in virtual reality. In: 26th IEEE conference on virtual reality and 3D user interfaces, VR 2019—Proceedings, pp 181–189. https://doi.org/10.1109/VR.2019.8797889
5. DeveloperApple (2022) iPad Pro LiDAR sensor specslapple developer forums. https://developer.apple.com/forums/thread/131013
6. Eric Ernisse (2020) Laser scanning forum. https://www.laserscanningforum.com/forum/viewtopic.php?t=16193#:~:text=The%20BLK360%20can%20also%20integrate,you%20have%20to%20scan%20twice
7. Esmaeili H, Thwaites H, Woods PC (2018) Workflows and challenges involved in creation of realistic immersive virtual museum, heritage, and tourism experiences: a comprehensive reference for 3D asset capturing. In: Proceedings—13th international conference on signal-image technology and internet-based systems, SITIS 2017. 2018-Janua, pp 465–472. https://doi.org/10.1109/SITIS.2017.82
8. Rahimian FP, Goulding J, Pour Rahimian F, Wang X, Editors G, Jack Steven Goulding LH, Research Fellow S (2014) Virtual reality-based cloud BIM platform for integrated AEC projects. J Inf Technol Constr 19. http://www.itcon.org/2014/18
9. Fritsch D, Klein M (2018) 3D preservation of buildings—reconstructing the past. Multimedia Tools Appl 77(7):9153–9170. https://doi.org/10.1007/s11042-017-4654-5
10. Garcia I (2022a) Benchmark table 3D scanning applications. https://doi.org/10.5281/ZENODO.7432996
11. Garcia I (2022b) Benchmark output of different 3D scanner applications. https://doi.org/10.5281/ZENODO.7429133
12. Garcia I (2022c) VR application in unity—3D scanned mesh of campus. https://doi.org/10.5281/ZENODO.7433918
13. Haskins J, Zhu B, Gainer S, Huse W, Eadara S, Boyd B, Laird C, Farantatos JJ, Jerald J (2020) Exploring VR training for first responders. In: Proceedings—2020 IEEE conference on virtual reality and 3D user interfaces. VRW 2020, pp 57–62. https://doi.org/10.1109/VRW50115.2020.00018
14. Heyse J (2019) Self-adaptive technologies for immersive trainings. In: 26th IEEE conference on virtual reality and 3D user interfaces, VR 2019—proceedings, pp 1381–1382. https://doi.org/10.1109/VR.2019.8798207
15. Hossain MA, Yeoh JKW (2018) BIM for existing buildings: potential opportunities and barriers. In: IOP Conference series: materials science and engineering 371(1). https://doi.org/10.1088/1757-899X/371/1/012051
16. Leica Geosystems (2017) blk360_spec_sheet_2_0
17. Leica Geosystems (2022) UAV Oplossingen. https://leica-geosystems.com/nl-be/products/uav-systems
18. Lidarradar (2022) Differences between LiDAR and laser—LiDAR and RADAR information. https://lidarradar.com/info/differences-between-lidar-and-laser
19. Luetzenburg G, Kroon A, Bjørk AA (2021) Evaluation of the Apple iPhone 12 Pro LiDAR for an application in geosciences. Sci Reports 11(1):1–9. https://doi.org/10.1038/s41598-021-01763-9
20. Matterport (2022) Capture, share, and collaborate the built world in immersive 3D. https://matterport.com/
21. Milovanović A, Dragutinović A, Ristić Trajković J, Nikezić A (2020) Towards a methodology for rethinking modernity: between imagined, realized, and lived space. https://publishings.eaae.be/index.php/annual_conference/article/view/54/46

22. Misato (2021, Dec 20) Release v1.0.0 Resistance-Studio/UnityVRTemplate GitHub. https://github.com/Resistance-Studio/UnityVRTemplate/releases/tag/v1.0.0
23. Moore AG, Hu X, Eubanks JC, Aiyaz AA, McMahan RP (2020) A formative evaluation methodology for VR training simulations. In: Proceedings—2020 IEEE conference on virtual reality and 3D user interfaces. VRW 2020, pp 125–132. https://doi.org/10.1109/VRW50115.2020.00027
24. Neslihan İ (2019) Students' approach to participating in informal education In: Zagreb Proceedings 2019. EAAE, pp 100–117. https://publishings.eaae.be/index.php/annual_conference/article/view/57
25. Nulens V (2022) PO_S.MU_Floorplan. https://doi.org/10.5281/ZENODO.7428762
26. Owler (2022) Leica geosystems. https://www.owler.com/company/leica-geosystems
27. Paldivis PG (2006) Laser scanning and triangulation. https://georgepavlides.info/research/LaserScanningAndTriangulation.php
28. Paleoseismicity (2022) Scanning outcrops with the iPad Pro LiDAR scanner—first tests. https://paleoseismicity.org/scanning-outcrops-with-the-ipad-pro-lidar-scanner-first-tests/
29. Pears N, Liu Y, Bunting P (2012) 3D Imaging, analysis and applications. In: Pears N, Liu Y, Bunting P (eds.) 3D imaging, analysis and applications, vol 9781447140. Springer London. https://doi.org/10.1007/978-1-4471-4063-4
30. Radianti J, Majchrzak TA, Fromm J, Wohlgenannt I (2020) A systematic review of immersive virtual reality applications for higher education: design elements, lessons learned, and research agenda. Comput Educ 147. https://doi.org/10.1016/J.COMPEDU.2019.103778
31. Reljić I, Dunđer I, Seljan S (2019) Photogrammetric 3D scanning of physical objects: tools and workflow. TEM J 8(2):383–388. https://doi.org/10.18421/TEM82-09
32. Stone S (2019) Conservation—consumption : preserving the tangible and intangible values. European Association for Architectural Education. https://www.researchgate.net/publication/337008035_Conservation_as_Experience_in_PRESERVING_THE_TANGIBLE_AND_INTANGIBLE_VALUES_CONSERVATION_CONSUMPTION_EAAE_Transactions_on_Architectural_Education_no_66
33. Sustainable Futures Report (2022) Sustainable Futures Report 2022 NBS
34. Taylor C, Mullany C, McNicholas R, Cosker D (2019) VR props: an end-to-end pipeline for transporting real objects into virtual and augmented environments. In: Proceedings—2019 IEEE international symposium on mixed and augmented reality, ISMAR 2019, pp 83–92. https://doi.org/10.1109/ISMAR.2019.00-22
35. Triana J, Pinto F, Jackson P, Qiao G (2020) What is laser scanning and how can it be used?|TopoDOT. TopoDOT. https://blog.topodot.com/what-is-laser-scanning-and-how-can-it-be-used/
36. van Aken S (2021 AI-gedreven VR-training in een adaptieve gebruikerscontext|AP Hogeschool. https://www.ap.be/project/ai-gedreven-vr-training-in-een-adaptieve-gebruikerscontext
37. Villegas A, Perez P, Kachach R, Pereira F, Gonzalez-Sosa E (2020) Realistic training in VR using physical manipulation. In: Proceedings–2020 IEEE conference on virtual reality and 3D user interfaces. VRW 2020, pp 109–118. https://doi.org/10.1109/VRW50115.2020.00025
38. Walker L (2019, Feb 25) Leica BLK360—FAQ. https://shop.leica-geosystems.com/leica-blk/blog/blk360-frequently-asked-questions
39. Walker L (2022) Leica BLK360—FAQ. https://shop.leica-geosystems.com/leica-blk/blog/blk360-frequently-asked-questions
40. Xiong X, Adan A, Akinci B, Huber D (2013) Automatic creation of semantically rich 3D building models from laser scanner data. Autom Constr 31:325–337. https://doi.org/10.1016/J.AUTCON.2012.10.006

# Integration of Collective Knowledge into Simulative Urban Modeling

Kestutis Zaleckis(✉), Laura Jankauskaite-Jureviciene, Jurga Vitkuviene, Indre Grazuleviciute-Vileniske, and Vilma Karvelyte-Balbieriene

Faculty of Civil Engineering and Architecture, Kaunas University of Technology, Kaunas, Lithuania
kestutis.zaleckis@ktu.lt

**Abstract.** The presented research is based on combination of two approaches while working in historical Šančiai neighborhood community in Kaunas. The first approach is based on the principles of tactical urbanism for collection and content analysis of local authentic data about three aspects of neighborhood life: (1) collective memory, (2) present usage of public spaces and (3) interaction with nature. For implementation of the project local inhabitants have collected data in the three online maps: Memory map, Present map, Nature map. The content analysis of collected data was further carried out using qualitative and quantitative approaches for understanding the emotional dimension of people's interaction with the place. The second approach is based on the idea of simulative modeling for complex urban systems, which allows to reflect bottom-up processes essential for urban development both in present and predicted situations. During the project, the location spots and the results of content analysis focused on identification of emotions attached to precise neighborhood spaces, based on data entered by inhabitants were used as an additional input into mathematical graph based simulative models. The results of modeling demonstrated fluctuation in the most important public spaces (e.g., walking routes, intelligible buildings, reachable central locations) depending on various urban life scenarios (e.g., working days, weekends, mornings, evenings, celebrations, etc.) and could be used as a part of decision support system while discussing urban development ideas between interested stakeholders and local community.

**Keywords:** Public participation · Place-making · Tactical urbanism · Simulative urban modelling · Content analysis · Sentiment analysis · Emotional analysis

## 1 Introduction

Both top-down and bottom-up decision-making approaches to the built environment have been used throughout human history. However, the modernistic urbanism treated cities as machines constructed using top-down decision-making ways and caused a significant transformation from city of places to city of spaces [1] which could not be effectively addresses even if New Urbanism or related models are formally employed in urban


M. Blanco Lage et al. (eds.), *Towards a New European Bauhaus—Challenges in Design Education*,
https://doi.org/10.1007/978-3-031-40188-6_22

planning [2]. Currently urban development is facing paradoxical situation: unprecedented opportunities and tools have emerged to democratize urbanization; although this seemingly unprecedented democratization is also accompanied by challenges related to the market-driven creation of space. Decentralization, where central government is giving more and more power to local government, is creating a situation where planning is becoming a market-driven business, where "government no longer controls urban development, but merely facilitates the initiatives of private real estate developers" [3]. Such trends require strengthening citizen participation and co-creation with the community and users, and an understanding of local identity and the ecological vulnerability of a place. Although it is necessary to acknowledge that this is not the case in many projects. One of the reasons for the lack of effective community participation in urban development is the lack of appropriate methodologies and tools. The focus of this study is on the active community of the historical district of Žemieji Šančiai in Kaunas (Lithuania) which served as an urban laboratory for this research.

The history of Šančiai district dates back to the seventeenth to eighteenth centuries, when it was for the first time mentioned in the written sourced as located near Kaunas city, on the bank of the Nemunas River. At the end of the nineteenth century, Šančiai gradually became a suburb of the growing city of Kaunas. Historically, the area has always been linked in one way or another to military infrastructure. The growth of industry also played an important role in the history of Šančiai. The urban structure of Šančiai can be characterized by narrow streets, a semi-regular layout and connections to the Nemunas river and its' ecosystems. The wooden residential architecture is well preserved, as well as the buildings of the historic military town built during the period of rule of Russian Empire [4]. The intensive public activities of Šančiai community started to be visible in a wider context since 2011 with the involvement of artists participating in international projects. Artists V. Gelūnienė and E. Carroll brought together a group of people interested in urban studies. They tried to find out how people feel in places where the wooden houses of former factory workers and restored nineteenth century military buildings have been transformed into current private dwellings, and the links between the old inhabitants and the newcomers. The question of participatory urban planning and place-making in Šančiai acquired particular importance when Kaunas City Municipality raised the need for a two-lane street along the Nemunas riverside with one lane for parking and a pedestrian path. This project was negatively received by the Šančiai community and was seen as an attempt to urbanize the natural environment, the natural riverside of the Nemunas, and to create a barrier between the residents and the recreational and natural area. After discussions with the municipality and understanding the need to educate the community on urban issues, the project "Genius Loci: Urbanisation and Civil Community" was developed with financial support from the European Economic Area and the Norwegian Financial Mechanisms. The idea and objective of the project was to involve community members in spatial planning processes, to provide them with the tools to become active creators of the territory, to share knowledge about urban planning and heritage, thus creating neighborhood identity and contributing to place-making [5]. The activities of this project and the data collected through the period of its implementation allowed searching the ways to integrate local collective knowledge and simulative urban modeling.

The aim of the research was to find an effective way to combine public participation and modeling phases of urban planning and to apply the developed methodology while working in historical Šančiai neighborhood community in Kaunas.

## 2 Methodology of Research

The methodology of the research is based on the integration of two general approaches. The first approach is based on the principles of tactical urbanism [6] for collection of local authentic data from Šančiai neighborhood residents and visitors and its content analysis focusing on the subjective and emotional aspect. The collected data from Šančiai neighborhood residents and visitors reflect three aspects of neighborhood life

(1) heritage of the locality and collective memory,
(2) present usage of public spaces based on modified sociotope approach and
(3) interaction with nature in ecological, recreational and social terms.

The second approach is based on the idea of simulative modelling for complex urban systems [7–11], which allows to reflect bottom-up processes essential for urban development both in present and predicted situations.

## 3 Results

### 3.1 Data Collection and Content Analysis

**Data collection**. Interactive online maps were created in order to involve residents and visitors of the Šančiai neighborhood to actively participate in creating the project data sets and urban vision of the district. The interactive online maps were commissioned and created by professional websites design company desenioras.com as integrated tool available and conveniently accessible on the website of Šanciai community. The website housing these interactive maps (https://sanciubendruomene.lt/lt/) was created in the frame of Genius Loci project [5]. Esri ArcMap 10.6.1 version software was used for analysis of results of mapping activities. Data was collected in three layers: Memory map, Present map, Nature map (Fig. 1). Map data was collected in the period from October 2020 to September 2022. The maps have been publicly available on the community website throughout this period. Information about interactive maps and invitations to participate in project activities were shared in various local media. In order to include as diverse a group of neighborhood members as possible in the project, part of the information was collected during live meetings and creative workshops held with community members. Interactive maps were filled during the events. Due to the different levels of digital literacy and preparation of the participants the necessary information was also collected by other methods of creative mapping. The workshops were organized for different age and social groups (children, youth, adults, people with special needs) as a series of activities completed in groups of 4–5 people supervised by the moderators. During the workshops mental mapping tasks with the elements from sociotop methodology [12] and design-thinking approach [13] were implemented for the better reflection of the topic before the introduction of online interactive maps. Additional information collected during these events was also digitized and uploaded into digital maps by the project researchers. In the spring of 2021, two creative workshops for school children (12–15 years old, 157 participants in total) and workshop for adult community members (39 participants) were held, during which the Memory map was filled. Since the strict quarantine of Covid-19 was announced in the spring of 2021 in Lithuania, the creative workshops were organized remotely using a variety of online creative platforms. 4 creative mapping workshops to 3 different age groups—2 workshops for school children from 11 to 13 years old (66 participants) and youth from 14 to 16 years old (54 participants) and in 2 workshops for adults (75 participants) including one workshop for the adults with special needs and their helpers (10 participants)—were organized as a contact meeting for the Present map in the period of 2021–2022. The Nature map was mainly filled by the active volunteer community members. It is necessary to note that the number of experiences entered in the online interactive maps does not necessarily reflect the number of participants, as there was a possibility to provide as many entries as the participant wants into the maps. During the workshops with youth, children and adult community members, where the functionalities of online maps were presented, the participants were encouraged to enter as many memories or experiences into the maps as they would like to do. Not all the entries in the online interactive maps were collected during workshops; workshops were organized as complementary events to stimulate the active involvement of community members as well as to contribute to the representativeness of the sample of participants, as the workshops were organized for different age and social groups. However, some entries in the maps are by the members of community who were not participating in the workshops and received the information about the maps by other means of communication.

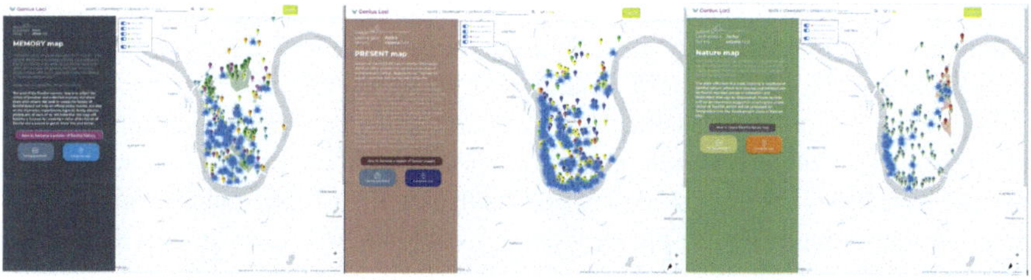

**Fig. 1** Interactive online maps: Memory, Present, Nature [5]

*Memory map.* The aim of this layer is to gather two types of information: official immovable heritage data and Šančiai neighborhood residents and visitors personal and collective memories about the territory and events which happened there: experiences, legends, family albums photos etc. Each user of the map can become a creator of collective memory. On the map, personal memory events and objects are divided into 4 periods according to their date: before 1918, 1918–1940, 1941–1990, after 1990. The map periods are divided according to significant periods of Lithuanian history. The Memory map contains 323 memories. 63 objects are from the list of Lithuania Republic official heritage data collected from the Register of Cultural Values.

*Present map.* The aim of this layer is to identify personal and collective activities, experiences and their locations, to understand the social values of public space. The Present map is based on the sociotop map method developed by Ståhle [12]. The principle of this method is to objectively identify urban public spaces, their social goals and meanings based on people's experiences [12]. On the map, activities are divided into 4 fields: Active activities (for ex. bathing, ball games, fishing, gardening etc.), Passive activities (for ex. rest in a green, admiring flower beds, sitting etc.), Events (for ex. folklore) and other activities. In this map it is important not only to record the experience but as well to identify the activity type and the location. The Present map contains 798 experiences.

*Nature map.* The aim of this layer is to gather two types of information: officially protected species and habitats and their location in Šančiai neighborhood and nature objects, which are important for residents. The Nature map contains 225 entries.

All maps include the possibility to add links to other resources or add images. Total amount of images in all three maps are 4757.

**Content analysis of collected data: emotional and sentiment analysis.** Numerous sentiment analysis studies treat emotions as a binary or scalar value: positive/negative or a value in the range from − 1 to 1; however, natural emotions are much more complex. Some researchers attempted to classify emotions into multidimensional values [14], considering this, the methodology of content analysis of online interactive maps entries integrates qualitative and quantitative approaches. According to Plutchik [15], there are several hundred words in the English language for emotions, which requires some categorization and classification, thus recognizing and identifying emotions in written texts is a complex task requiring appropriate theoretical framework. A qualitative approach based on Plutchik's [15] human emotion classification was used in this research. Plutchik [15] presents a circumplex model of emotion classification analogous to the color wheel, placing similar emotions close to each other and opposite emotions 180 degrees apart. The written comments left by the people filling the online interactive Šančiai maps were analyzed and an emotional label was assigned to each individual comment using the above-mentioned emotion classification. A quantitative sentiment analysis of polarity of emotions was then carried out, assessing whether each comment contained positive or negative emotions. Mohammad [16] and Sathya et al. [17] presented comprehensive reviews on sentiment analysis application and its challenges. According to Sathya et al. [17], because of its relevance to business and society, currently sentiment analysis is widespread in many fields (not to mention computer science), including management and social media marketing. As sentiments in many cases in some ways are attached to particular location, this approach can be beneficial in urban analysis and place-making research. To identify emotions in sentences in this research, a keyword-based sentiment analysis method was used to apply emotional scores to each word [18]. The NLTK VADER sentiment analyzer was selected for this study based on existing similar research experience [18]. The framework for the emotional analysis of the online interactive maps entries is presented in Fig. 2.

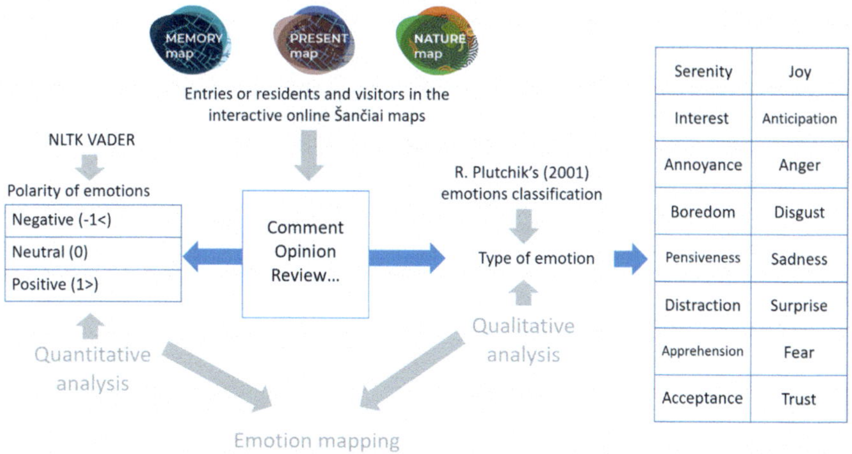


**Fig. 2** Framework for analysis of content of text entries in the interactive online maps of residents and visitors of Šančiai using quantitative (NLTK VADER sentiment analysis tool) and qualitative (Plutchik's [15] classification of emotions) approaches

### 3.2 Šančiai Emotion Maps

Mapping places and emotions is gaining increasing attention from the geoinformation and cartographic science communities [19]. In this context, mention can be made of the studies by Doytsher et al. [20], Misue and Taguchi [14], and Poplin [19]. The terms emotion maps [20], emotion-weather maps [14] have been used in emotion mapping research. Emotion maps are often constructed based on geotagged posts in social media [20]. According to Misue and Taguchi [14], emotion maps or emotion-weather maps, are a kind of thematic maps used to represent emotions. These maps depict the spatial distribution of complex emotions of a large group of people. Most of the work related to emotion maps focuses on simple sentiment analysis and their distributions are plotted on the map [14]. According to Doytsher et al. [20], the purpose of an emotion map is to represent and depict the relationships between emotions and geographical locations. Such maps can help to answer various questions about how people feel in different places [20]. Emotion mapping in urban environments is particularly important as such environments are typically densely populated and heavily used, and present complex emotional interconnections with specific places.

Figure 3 presents six emotion maps generated as the result of content analysis of text entries in the interactive online maps by residents and visitors of Šančiai and localizing of identified emotions on the map. The maps demonstrate both type and polarity of emotions and allow making the conclusions about: (1) diversity of experienced emotions in the area, (2) predominant emotions, (3) the proportions between emotions that can be considered as negative and positive both from the point of view of polarity and type of emotions, (4) the locations that are the most and the least related with emotional experiences, (5) the emotional responses to different aspects of place, like memory, public spaces and nature etc. It is evident that visitors and users experience at least 16 different types of emotions (according to Plutchik's [15] classification) in Šančiai. The widest array of emotions is engendered by the contemporary usage and interaction in public spaces. Positive emotions predominate in experiencing all three aspects of the neighborhood—memory, public spaces and nature. Predominant positive emotions are interest and joy. As the content analysis has demonstrated, negative emotions (anger, annoyance) are most often related with the events of destruction or abandonment of cultural heritage, cutting trees and destruction or threat to other ecosystem features and the actions of municipality that community members consider against the interests of community. The concentration of emotional experiences in particular locations, like the main Jouzapavičiaus prospect and Nemunas riverside reveal the importance and potential of these places for neighborhood development.

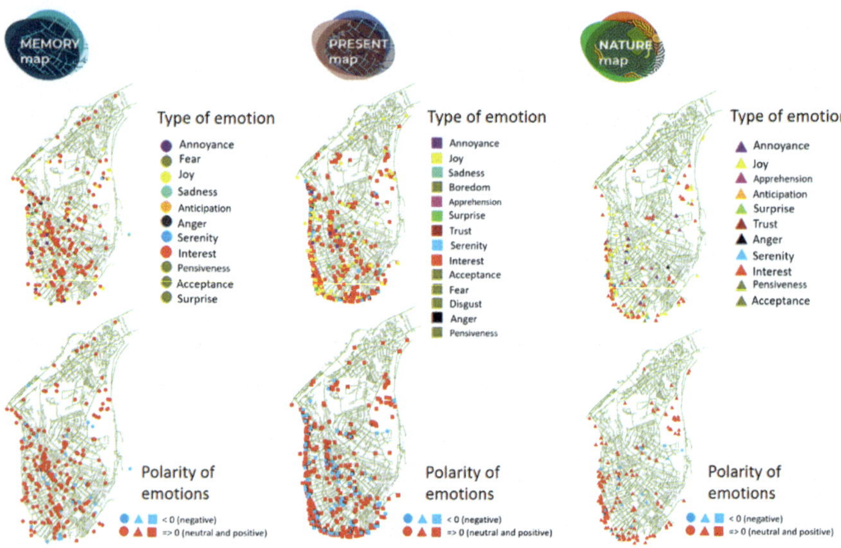


**Fig. 3** Results of emotional and sentiment analysis of Šančiai using the content analysis of data obtained from text entries in the interactive online maps by residents and visitors of Šančiai. Maps in an upper row show the types of place-related emotions identified using Plutchik's [15] emotions classification related with memory of the place, interactions in public spaces and interactions with nature. Maps in the lower row demonstrate the polarity of experienced emotions respectively. NLTK VADER sentiment analysis tool was applied for determining the polarity of emotions

### 3.3 Simulative Modeling for Complex Urban Systems

The essence of graph model, if applied to a city, is the following: a whole urban structure is seen as an integral network made of nodes and links. Node, depending on the model and approach, can represent street axis, street segment [21], part of visual space [22], so called convex space [23], intersection of streets [24] or building [11]. In all cases, connectivity and configuration of a network is analyzed as it is seen much more important for urban functionality based on the idea of complexity, than properties of graph nodes themselves.

The mathematical concept of centralities of graph nodes and its first calculations were introduced by Freeman [25] with three types of centralities offered to calculate:

- Degree centrality as a sum of connection of a node. More connections mean bigger importance in terms of connectivity to neighbors on a network.
- Closeness centrality as a sum of distances from a calculated node to the rest of network. A smaller number shows nodes which are the most reachable and, in urban context, represent objects which potentially can become destinations for many journeys in a city.
- Betweenness centrality as a count of transit journeys which cross a calculated node while simulating all possible shortest journeys between all pairs of nodes in a network.

Because of the small size of the investigated area and need for as high precision as possible Urban Network Analysis Toolbox (UNAT) [11] as walkability modeling oriented software was chosen. UNAT investigates a building graph, where initially graph is constructed on the base of the streets and after that—expanded with buildings as nodes. Simulation in essence is based on few simple rules:

- People move in the shortest ways between origins and destinations;
- Each journey has origin and destination which are presented by a graph nodes or buildings;
- Nodes can have different weights which reflect real urban situation, e.g.: area of commercial buildings, number of inhabitants in a building, etc.

Movement economy principle and idea of the symmetry by Hillier [7] could be named as additional concepts for interpretation of graph centralities. Movement economy idea is "…built on the notion of natural movement, proposes that evolving space organization in settlements first generates the distribution pattern of busier and quieter movement pattern flows, which then influence land use choices, and these in turn generate multiplier effects on movement with further feedback on land use choices and the local grid as it adapts itself to more intensive development" [7]. Symmetry which is described as "… the property that if A is a neighbor of B, then B is neighbor of A" [26]. Symmetrical relations between node means that if they as space or buildings have symmetrical relations to each other created by short distance and direct links, they bring people, activities, functions together and create more multi-functional, diverse zones in a city.

Despite the fact the above-mentioned models are many times validated and could presented as working ones, including case of Kaunas Walkability Compass modeling and analysis [27] it still should be agreed that, as based on similar rules, they are very generic. In order to make the simulative model more reflective to local, subjective aspect of urban life and community, it was decided to add data from Memory, Present and Nature maps described earlier, as weights to the graph and use it for better understanding and prediction of effects of various infill interventions on neighborhood functioning in the future.

The used UNAT allows to calculate various centralities of nodes in a graph. Three centralities which represent three aspects of functioning of urban structure as kind of "gravity fields" or attractive power of an area created by travel destinations; kind of intelligibility or possibilities to see nodes of a network on straight visual axis and transit flows of pedestrians were chosen for the presented research.

Gravity measures accessibility or closeness to a node of the graph and assumes that it is proportional to weights of destinations nodes surrounding node for which calculation is performed and inversely proportional to the distances between those nodes [11, 28]. Depending on available data as the weight could be used area or volume of buildings based on the premise: "bigger volume—more inhabitants" or "more attractive commercial object"; presence of function—e.g.: commercial function, school, kindergarten, etc. The following formula is used for calculation:

$$\text{Gravity }(i) = \sum_{j \subset G-i; d[i,j] \leq r} W(j)/\exp(\beta * d[i,j])$$

where Gravity ($i$) is gravity centrality of node $i$; $W(j)$ is a weight of any node $j$ in the graph $G$; $G$ is a graph within which reached buildings are counted except building $i$; $d[i,j]$ is the shortest path from building $i$ to any other building $j$ which is reachable on the shortest distance equal or smaller to radius $r$; *exp* means natural exponential function which describes distance decay effect on gravity when it is increased; coefficient "$\beta = 0.001$ and represents pedestrians willing to walk substantial distances in pleasant urban environments" [11]. In the presented research gravity demonstrates attractive power of the area as the final destination of all journeys within Šančiai neighborhood.

Straightness illustrates the extent to which the shortest paths from a node of interest to all other nodes in the system resemble straight Euclidian paths [8]. In essence straightness shows higher probability that a precise node will appear on straight paths and correspondingly—straight visual axes along streets so it could be compared to intelligibility of nodes within a street network. Straightness is calculated according to the following formula [11]

$$\text{Straightness }(i) = \sum_{j \subset G-i; d[i,j] \leq r} \frac{\sigma[i,j]}{d[i,j]} W(j)$$

where Straightness ($i$) means straightness of the node $i$ the graph $G$; $G$ is a graph within which reached nodes are counted except node $i$; $d[i,j]$ is the shortest path from building $i$ to any other building $j$ which is reachable on the shortest distance equal or smaller to radius $r$; $\sigma[i,j]$ is the straight-line Euclidian distance between the nodes $i$ and $j$.

Betweenness simply counts all the shortest journeys within a graph which pass by a node for which calculation is conducted. Bigger betweenness values mean bigger transit flows besides or through a node. It is calculated according to the following formula [11]

$$\text{Betweenness }(i) = \sum_{j,k \subset G-i; d[j,k] \leq r} n_{jk}[i]/n_{jk} * W[j]$$

where Betweenness ($i$) is the betweenness of building $i$ within the Search Radius $r$, $n_{jk}[i]$ is the number of shortest paths from node $j$ to node $k$ that pass by node $i$; $n_{jk}$ is the total number of shortest paths from $j$ to $k$; $W(j)$ is a weight of any building $j$ in the graph $G$.

The OSM were used for the modeling. The obtained objective quantitative data on buildings (functions, volume) and subjective data (valuable for community memory places, visited recreational areas, activities in public spaces) were used as specific travel destinations in the graph.

### 3.4 Simulative Maps of Šančiai with Integrated Collective Knowledge

The mathematical graph simulative modeling was conducted in five steps while consistently adding more different data for weighting of the nodes. Such sequence of the process while moving from more generic to more specific models allow to compare results between them and evaluate changes in sensitivity levels of the models. Modeling was conducted with 1000 m radius thus reflecting focus of the research on pedestrian distances within very walkable neighborhood. The first UNAT calculation was performed without any weighting (Fig. 4)—it means that every building (except temporary buildings and contractions which were filtered out based on OSM data), despite its function or size was used as both origin and destination of journeys between all possible pairs of buildings within radius 1000 m. No special weights were used so the weight of every building was equal to 1. The results of calculation (gravity, straightness, betweenness) in this case are affected just by street network and density of buildings. The model (Fig. 4) has few features, which should be treated cautiously because generic nature of the model, but are quite interesting:

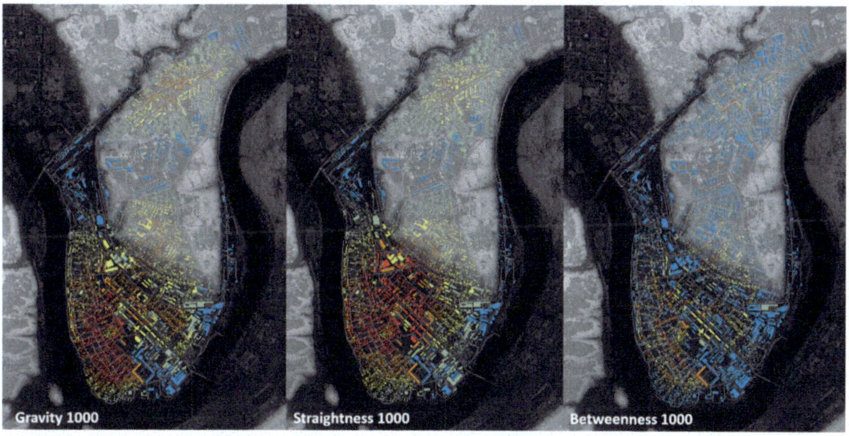


**Fig. 4** Calculation of three centralities without any weighting. Red color shows the biggest numerical values, blue—the lowest ones

- Gravity center of the neighborhood, where, according to the movement economy concept, we should expect to have highest densities and multifunctionalities is located not on the main urban corridor of Šančiai—Juozapavičiaus prospect, but in the old housing area closer to the river. Such result from one side reflects the importance of the river for organic urban structure. From another side it demonstrates that expansion of Šančiai with construction of military barracks and new street in the second half of the nineteenth century created a kind of double structure where new part is intelligbale according to straighness, but possibly not attractive in terms of street culture and other fomrs of social interaction.
- Straighness demonstartes intelligibility of both the riverfront with organic housing structure around and the barracks besides Juozapavičiaus prospect.

- Betweenes demonstrates that the most intensive pedestrian transit is probably happening within the organic housing zone and close to the river as well.

The sensitivity of the model in essence is low as it would not react significantly if a new building or a small group of buildings would be added. Significant change of street network of building densities would be needed to cause noticeable changes in the model.

The second modeling attempt was conducted while modeling just journey from all buildings to specific destinations—commercial, administrative, educational, recreational, etc., which were identified based on OSM data. Each target destination was weighted by a volume of a building calculated based on polygon area and Earth surface Lidar scanning data received from geoportal.lt [29]. The results of the calculations (Fig. 5) show significant differences with the previous not weighted model (Fig. 4): the zone of high gravity is concentrated along Juozapavičiaus prospect thus reflecting its multifunctionality; intelligible zone identified by the highest values of straightness is concentrated around the high gravity zone but includes some big parts of organic urban structure with dominant housing functions thus revealing importance of this zone for visual identity of the area; betweenness clearly show the highest pedestrian transit in Juozapavičiaus prospect. The precision and sensitivity of the model with the second weighted calculation was increased—now, at least theoretically, it can react even to significant changes of the volume of a single building if it attracts journeys. Despite the increased accuracy of the model, it still should be seen as a generic one as subjective preferences and local way of life is not reflected in it.

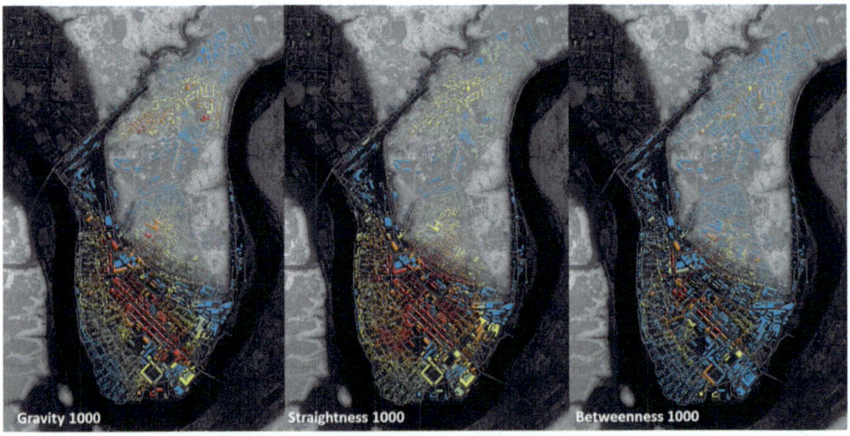


**Fig. 5** Calculation of three centralities with volume and function of buildings used as a weight. Red color shows the biggest numerical values, blue—the lowest ones

The third UNAT calculation was performed with the data from the Memory map: all buildings were considered to be origins of journeys while collective memory spots were counted as destinations. At the moment each destination has a weight of 1, while each building has a weight of 0. In this case not only allocation of memory spots but also their clustering or density becomes important. The calculation results (Fig. 6) show few interesting features:

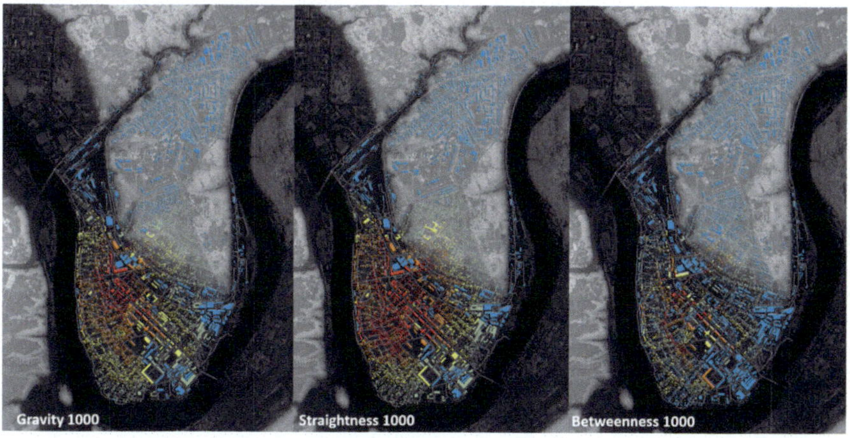


**Fig. 6** Calculation of three centralities data from Memory map used as a weight. Red color shows the biggest numerical values, blue—the lowest ones

- The high gravity zone is concentrated in the north-western part of Juozapavičiaus prospect where a historical center with church, former market and shops before WW2 was located. The nineteenth century barracks, which make up the majority of the official immovable cultural heritage are not reflected very much in the gravity map.
- Intelligibility or straightness map concentrates around the gravity center and includes the bigger part of historical housing zone with the river front.
- Betweenness shows probable pedestrian transit inside the historical housing zone as well.

The model could be sensitive even to the appearance or disappearance of a single point of attraction depending on its position in the network. The third calculation results are interesting because they demonstrate significant shift of the zones of high centralities if the journeys of people are simulated not based on generic rules focused on utilitarian and functional priorities, but on looking at every person as *flaneur* involved in journey of secular pilgrimage by Benjamin [30, 31]. The zones of high intelligibility-straightness or gravity could be seen not as ordinary core of urban neighborhood but more as collection of collective memory entangled in urban space or concentration zone of outer collective consciousness artefacts according to the concepts of Cole [32]. The same is true if we speak about the fourth and fifth UNAT calculations: they represent orientations of *flaneur* focused not on history and memory, but on experience of nature (Fig. 7) and street culture (Fig. 8).

Figure 7 represents UNAT calculation with destinations modeled according to the Nature map in the same way as it was performed with the Memory data in the previous calculation. Shift of height centrality zones towards riverbank and low importance of Juozapavičiaus prospect reveal big importance of natural riverfront as the main attractive recreational area of the neighborhood and its high significance for functioning of urban structure in terms of attraction of people, generation of pedestrian transit flows and intelligibility of the area.

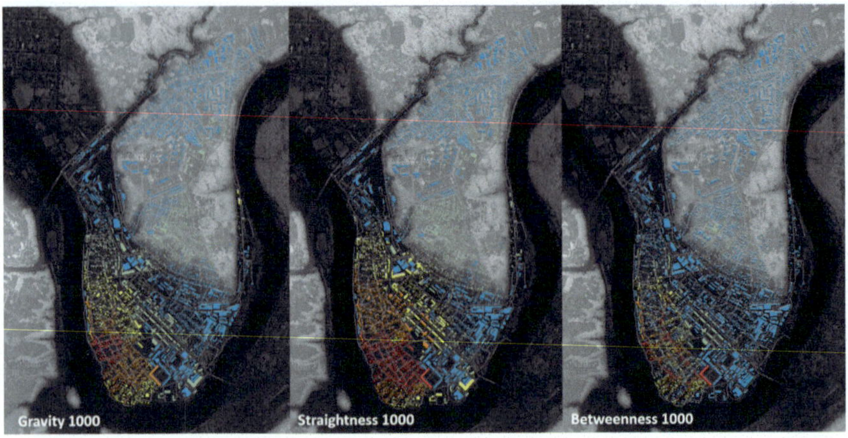


**Fig. 7** Calculation of three centralities data from Nature map used as a weight. Red color shows the biggest numerical values, blue—the lowest ones

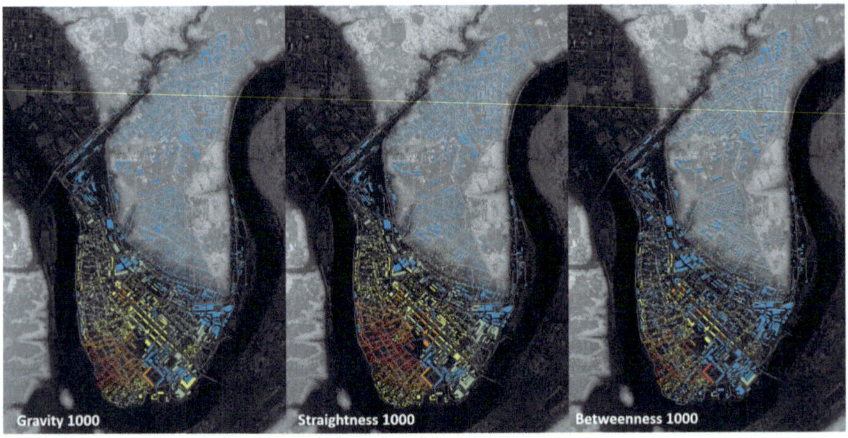


**Fig. 8** Calculation of three centralities data from Present map used as a weight. Red color shows the biggest numerical values, blue—the lowest ones

Figure 8 represents the last calculation focused on various activities of street culture based sociotop or Present map where various activities in public spaces are modeled as targets of all journeys. The model reveals unexpectedly high significance of the waterfront which, together with historical organic part of urban structure, performs, according to the model, function of the node of not only recreational activities, but social interaction as well and thus confirms importance of this area for social-cultural-urban identity of Šančiai.

## 4 Discussion and Conclusions

General System Theory by Von Bertalanffy creates a fundamental background for all science-based models as it states that the World is a big system, made of smaller subsystems, which are organized together in a fractal way—by intersecting, overlapping, nesting within each other at different scales. Each system is made of specialized parts which are kept together by flows of energy, materials and information. Complex system theory describes a specific type of systems which have a very big number of elements and connections, are constantly self-developing based on self-organization forces and are hardly understandable if "traditional" scientific analysis based on observation, classification, search for top-down organizing forces is applied. According to many scholars, complex systems can be well reflected and understood in simulative models only. While classical scientific models are deterministic, focused on classification, static, having limited number of parameters, focused on identification of top-down relations within a modeled system, the simulative models reflect opposite—probabilistic nature, dynamic, numerosity, non-linearity, connectivity, adaptation, bottom-up self-organization, etc. Because a city could be seen and is often seen as a complex system, then simulative modeling could be very helpful while understanding logic and identifying patterns of urban interactions, predicting results of planned changes, etc. The mathematical graph-based model, besides agent based and cellular automata, is often used as a background for simulative modeling. Space Syntax could be mentioned as one of the well know examples of graph-based simulation offered by Hillier [7], but it is not the only one as similar mathematical method is applied by Porta [24], Sevtusk [10], Batty [9] and many others.

Active community of the historical district of Šančiai served as an urban laboratory for searching the ways to integrate local collective knowledge and simulative urban modeling. Developed methodology includes the following approaches: (1) principles of tactical urbanism for collection of local authentic data using workshops and interactive online maps and content analysis of collected data focusing on the aspect of emotional experience and (2) simulative modeling for complex urban systems.

The content analysis identifying type and polarity of emotions of text entries in the interactive online maps by residents and visitors of Šančiai and the localization of identified emotional aspects on the maps allowed developing emotional maps of Šančiai. The emotional maps show diversity of experienced emotions in the area. The analysis allowed determining predominant emotions, the proportions between emotions that can be considered as negative and positive both from the point of view of polarity and type of emotions, the locations that are the most and the least related with emotional experiences, the emotional responses to different aspects of place, like memory, public spaces and nature. The positive emotional portrait of Šančiai has emerged as a result of content analysis and mapping, revealing the emotional significance of particular locations, such as Nemunas riverside.

The combination of the data from public participation and mathematical modeling helps to create "a living" simulative model of the urban structure which reflects not only its objective features but subjective contents as well. It could be seen as a complex and flexible decision support tool for various stakeholders involved in urban development and planning. The combined model (set of models) creates potential connections with such fundamental theories, still not very much reflected in urban planning, as cityness by Mumford [33], outer consciousness by Cole [32], architectural-urban patterns by Alexander [34], secular pilgrimage by Benjamin [30], etc.

The outcomes of the modeling all together demonstrate results which could be called a "fluctuating graph" which allows to see how different scenarios of walking activate different zones of the neighborhood. It does not mean that the generic models should be rejected or neglected, but that the simulative models based on subjective data collected by inhabitants, help to understand local situation much deeper and create more complex view on the investigated area. At the same time, they allow to model predictive results of urban interventions with higher sensitivity level which reflects preferences of local people.

The model allows us to identify the most actual zones and optimal points for various "activations" or "transformations" related to urban memory, detailed social content of spaces, nature attraction, etc. The model has predictive power and could be made even more detailed based on the results of content analysis or emotional analysis of inhabitants' data, e.g.: the "fluctuating graph" might focus on travel destinations related to certain emotions, just specific type of sociotop activities, memories related just to specific historical period, etc.

The limitations and biases of the presented models can be related with the reliability, accuracy and fluctuation of data presented by research participants, since such data related to the perception of and experiences in urban spaces as well as personal and collective memories is highly dependent on the age, social class, origin and other sociodemographic characteristics of the research participants as well as on their moods and attitudes ant the specific moment than data was provided. In order to mitigate these biases, the data must be collected from representative samples of individuals and accumulated for a longer period. Another limitation may be related to standardized classifications and Artificial Intelligence based evaluation of human emotions, as this represents the simplified picture of human emotional experience while interacting with the place. However, such standardized operations are more convenient for dealing with large volumes of data.

**Acknowledgements.** The research was accomplished in the frame of the project "GENIUS LOCI: urbanization and civil community" funded by the European Economic Area and Norwegian Financial Mechanisms and organized by Žemieji Šančiai Community in collaboration with Kaunas University of Technology. We express gratitude for the New European Bauhaus Prizes 2022 committee for distinguishing and awarding Genius loci—digital mapping tool in the category of Regaining a sense of belonging.

## References


1. Tuan YF (1979) Space and place: humanistic perspective. Philosophy in geography. Springer, Dordrecht, pp 387–427
2. Zaleckis K, Tranavičiūtė B, Grunskis T, Gražulevičiūtė-Villeniškė I, Vitkuvienė J, Sinkienė J, Doğan HA (2021)Modernization of public spaces in Lithuanian Cities. Sciendo
3. Bureau Urbanisme. Why is urban sustainability so important? http://sustainabilitycompass.eu/context/. Last accessed 08 Nov 2022
4. Inytė V (2019) Ypatinga atmosfera alsuojančios Kauno gatvės gyventojai pokyčių vengia dėl svarbios priežasties, https://www.lrytas.lt/bustas/architektura/2019/03/03/news/ypatinga-atmosfera-alsuojancios-kauno-gatves-gyventojai-pokyciu-vengia-del-svarbios-priezasties-9342840. Last accessed 18 Mar 2022
5. Genius Loci. https://sanciubendruomene.lt/en/genius-loci/. Last accessed 18 Nov 2022
6. Silva P (2016) Tactical urbanism: towards an evolutionary cities' approach? Environ Plann B Plann Des 43(6):1040–1051
7. Hillier B (2007) Space is the machine: a configurational theory of architecture. Space Syntax, London
8. Porta S, Crucitti P, Latora V (2005) The network analysis of urban streets: a primal approach. Environ Plann B 35(5):705–725
9. Batty M (2013) The new science of cities. MIT Press
10. Sevtsuk A (2013) Networks of the built environment. In: Ofenhuber D, Ratti C (eds) Decoding the city – how big data can change urbanism. Birkhäuser
11. Sevtsuk A (2022) Urban network analysis for Rhinoceros 3d tools for modeling pedestrian and bicycle trips in cities. City Form Lab, Massachusetts Institute of Technology. https://cityform.mit.edu/projects/una-rhino-toolbox. Last accessed 15 Nov 2022
12. Ståhle A (2013) Sociotope mapping-exploring public open space and its multiple use values in urban and landscape planning practice. Nordic J Archit Res 19(4):59–71
13. Diethelm J, Flannery JE, Giuliani NR, Flournoy JC, Pfeifer JH (2016) De-colonizing design thinking. She Ji: J Design, Econ, Innov 2(2):166–172
14. Misue K, Taguchi K (2015) Development of emotion-weather maps. Proc Comput Sci 60:292–301
15. Plutchik R (2001) The nature of emotions: human emotions have deep evolutionary roots, a fact that may explain their complexity and provide tools for clinical practice. Am Sci 89(4):344–350
16. Mohammad SM (2017) Challenges in sentiment analysis. A practical guide to sentiment analysis. Springer, Cham, pp 61–83
17. Sathya V, Venkataramanan A, Tiwari A (2019) As certaining public opinion through sentiment analysis. In: 2019 3rd international conference on computing methodologies and communication (ICCMC), pp 1139–1143
18. Park SH, Byung-Chull B, Yun-Gyung Ch (2019) Emotion recognition from text stories using an emotion embedding model. In: 2020 IEEE international conference on big data and smart computing (BigComp)
19. Poplin A (2017) Cartographies of fuzziness: mapping places and emotions. Cartogr J 54(4):291–300
20. Doytsher Y, Galon B, Kanza Y (2017) Emotion maps based on geotagged posts in the social media. In: Proceedings of the 1st ACM SIGSPATIAL workshop on geospatial humanities, pp 39–46
21. Turner A (2004) DepthMap4: a researcher's handbook. UCL
22. Turner A, Doxa M, O'Sullivan D, Penn A (2001) From isovists to visibility graphs: a methodology for the analysis of arcthiectural space. Environ Plann B 28(103–121):108–109

23. Van Nes A, López M (2013) Spatio-socio classification of deprived neighborhoods in the Netherlands. Sejong University, Seoul, Strategies for neighborhood revitalisation. In Ninth International Space Syntax Symposium
24. Porta S, Latora V, Strano E (2010) Networks in urban design. Six years of research in multiple centrality assessment. Netw Sci 107–129
25. Freeman L (1979) Centrality in social networks conceptual clarification. Soc Netw 3(1):215–239
26. Hillier B, Hanson J (1984) The social logic of space. Cambridge University Press, Cambridge
27. Zaleckis K, Chmielewski S, Kamičaitytė J, Gražulevičiūtė-Vileniškė I, Lipińska H (2021) Walkability compass—a space syntax solution for comparative studies. Sustainability 14(4):1–25
28. Hansen WG (1959) How accessibility shapes land use. J Am Plann Assoc 25(2):73–76
29. Lidar_DR—Lietuvos Respublikos apskričių centrų skaitmeniniai erdviniai lazerinio skenavimo taškų duomenys. https://www.geoportal.lt/geoportal/paieska#savedSearchId={B2C739CB-4F91-4FC0-8CDB-EBA6B0A46F79}&collapsed=true. Last accessed 15 Nov 2022
30. Benjamin W (2002) The arcades project. An Imprint of Harvard University Press, Belknap Press
31. Parker S (2004) Urban theory and the urban experience: encountering the city. Routledge, London and New York
32. Cole M (1996) Cultural psychology: a once and future discipline. Harvard University Press
33. Mumford L (1961) The city in history: its origins, its transformations, and its prospects. Houghton Mifflin Harcourt
34. Alexander C (1999) The origins of pattern theory: the future of the theory, and the generation of a living world. IEEE Softw 16(5):71–82

# The Skopje Project—Building the "World City"

Ognen Marina(✉) and Teodora Mihajlovska

Faculty of Architecture Skopje, Ss. Cyril and Methodius University in Skopje, Skopje, North Macedonia
ognen.marina@arh.ukim.edu.mk

**Abstract.** In 1966, the United Nations completed the large scale programme of assistance to the City of Skopje, the purpose of which was to assist the local authorities in drawing up a new city plan for its future development after the devastating earthquake in 1963 that left 80% of the city in ruins. Following this idea, a number of world architects were invited in Skopje, including Kenzo Tange from Japan, Jaap Bakema and J.H. Van den Broek from Holland and Constantinos Doxiadis from Greece, to contribute to this effort with their ideas, plans and designs. It was anticipated that once the reconstruction would be completed, and given the engagement of such eminent designers, the city would provide solutions to the contemporary "urban crisis". However, sixty years after the adoption of urban plan for the City of Skopje, the legacy of this international project led by UN remains relatively unattained. Revisiting the idea of the power of architecture and urban design as the drivers of reconstruction of the societies and cities the paper explores the contexts and networks of political and institutional actors that enabled and led to such an ambitious international project, mapping the theories and urban concepts that were in the foreground of the main project proposals and draws conclusions that will lead toward better understanding of the capacity of contemporary architecture to exercise and work with concepts of solidarity, just society and urban reconstruction.

**Keywords:** Skopje · Earthquake · Reconstruction · Urban plan · "World city" · Solidarity

## 1 Introduction

The large-scale emergency response and support are essential in the aftermath of the natural disasters and man-made destructions for saving human lives and rebuilding communities and cities that have been affected. The process of organizing and executing such a large scale and often international operations of support and solidarity are rather complex processes that require political and expert knowledge and experience that is beyond the capacity of a single organization, nation or state.

In the morning of July 26, 1963, Skopje was struck by a devastating earthquake that resulted in 1.070 casualties; more than 3.300 people were injured and 75–80% of the city's-built structures were destroyed or damaged beyond repair; 150.000 of Skopje's inhabitants left were homeless (Fig. 1) [10]. This unfortunate event was followed by


M. Blanco Lage et al. (eds.), *Towards a New European Bauhaus—Challenges in Design Education*,
https://doi.org/10.1007/978-3-031-40188-6_23

unprecedented effort by United Nations to reconstruct and redesign the city. By the nature of emergency which prompted it, the Skopje Project was unlike any other operation of its kind ever undertaken by the United Nations Special Fund. It was a unique opportunity to demonstrate the ability of the International community led by the United Nations for a much larger endeavor of constructing the role model of the future city—the "World city" [12].

**Fig. 1** The devastation of the earthquake in Skopje and the citizens leaving the city

The paper explores the contexts and the networks of political and institutional actors that enabled and led to such an ambitious international project, mapping the actors, organizations, processes, theories and urban concepts through the actor-network theory and draws conclusions that will lead toward better understanding of the capacity of contemporary architecture to create a better world based on solidarity.

## 2 The "World City"

In the months after the earthquake and amid the dramatic political confrontation of the bi-polar world, Skopje has become a place of collaboration and coexistence that reversed the negative preconception and the confrontation of the Cold War era. This message was further elaborated and the city, once rebuilt, was supposed to show the governments worldwide that only through unconditioned collaboration and through the sharing of knowledge would it become possible to create an environment suitable for the entire humankind. The reconstruction of Skopje has become a political project aimed at providing a model of social justice through the process of redesigning the city [12]. On October 14th 1963, with a resolution from the General Assembly, the United Nations complied with the request from the Yugoslav government and agreed to be the coordinator of the most important activities concerning the renewal of Skopje, thus becoming a crucial part of the expert network.

United Nations have recognized and understand the potential of Skopje's situation embodied in the figure of Ernest Weissmann, a Croatian born pioneer of Modernism, member of Zagreb Work Group and CIAM, participant of the 4th Congress of CIAM in Athens and active contributor to the debate about the Athens's Charter. He belonged to the "leftist group" of CIAM members led by Jose Luis Sert, promoting the ideas that would "bridge the gap between the non-political, inherent professionalism of architects, urban planners and civil engineers, and their actual social and political importance in societies with different economic systems and different levels of technology and industrialization" [2].

At the time of the Skopje earthquake Weissmann was holding a position of vice-director of at the United Nation's Economic and Social Council and who later become the key figure in the process of reconstruction of Skopje chairing the International Board of Consultants (IBC). He committed himself to transforming the city into a place where it would be possible to bring together the most successful international experts and that this would show that the only alternative to the destruction was in collaborative planning. The main idea was to modernize Skopje and to turn it into the "world city" in order to show to the international community how planners can achieve a better life for the many. The idea of the "world city" has been brought to Weissmann attention during his work in the late 20's of the 20th Century at the Le Corbusier and Pierre Jeannerette office in Paris on the avant-garde proposal for the competition for the League of Nations in 1926. It coincided with the Le Corbusier's collaboration with Paul Otlet, Belgian peace activists and lawyer and their unrealized project for *Mundaneum* and the *cite mondiale*. This was the city that should become the center of universal knowledge through which the humanity will ensure the world's peace and wellbeing [12]. Weissman saw the opportunity in the post-earthquake reconstruction of Skopje to create the "world city", the city that should be seen as the epicenter of knowledge that would promote peace, understanding and collaboration. Following this idea, a number of world architects were invited in Skopje, including Kenzo Tange from Japan, Jaap Bakema and an den Broek from The Netherlands and Constantinos Doxiadis from Greece, to contribute to this effort with their ideas, plans and designs. The reconstruction of Skopje was a great opportunity to demonstrate the real potential of Tanges' Metabolist architecture, Bakema's Open Society and Doxiadis's Ekistics theory (Fig. 2).

The Skopje project demonstrated the importance of architecture as a collaborative effort in the divided world promoting the belief that the construction of the human habitat is a humane way to solve socio-political and economic obstacles of the contemporary world.

## 3 The Actor-Network

Actor-network theory (ANT) was developed by Michael Callon and Bruno Latour [3, 4, 6–8] in the field of sociology of science and technology. ANT is a theoretical and methodological approach to social theory where everything in the social and natural worlds exists in constantly shifting networks of relationships. In ANT the connections between actors can lead to the creation of new entities that are not simply the sum of characteristics of the constituent entities. An example of such fusion of entities into another entity is a "gunman", an actor-network formed from two separate entities, a man and a gun, which are connected in a third entity: the gunman [8]. A "gunman" is

**Fig. 2** The UN and international experts involved in post-earthquake reconstruction of Skopje

different from both a man and a gun in the sense that a gunman is able to shoot someone whereas both the man and the gun cannot do this separately. It is important to underline that the term actor is, everything that in some way affects the network, not only human beings, but also, for instance, technology, methodologies, techniques, social rules and institutions, knowledge, and last but not least, external objects.

In order to understand the actor-network formed around the disaster in Skopje, first, we must identify its problematization (Fig. 3). In order to do this, we need to go back to our example of the gunman. Just like in that example, we have two actors that fuse to form the very first entity of our network, Skopje and the earthquake merge to form Disaster-stricken Skopje, a new entity that has very unique characteristics different from its components. Now we can treat Disaster-stricken Skopje as an actor, rather than a network, that has very distinct and urgent interests.

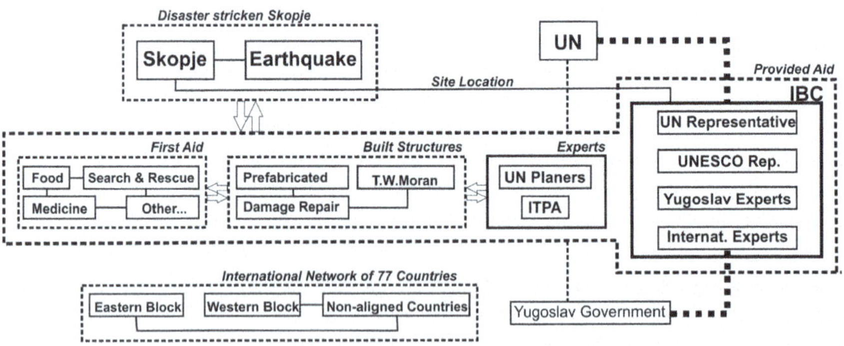


**Fig. 3** The problematization of actor-network in Skopje

This new actor has attributes that Skopje alone doesn't because natural disasters are powerful catalysts that can make other actors forgo their interest and align themselves with the actor in need. In the midst of the Cold War, Yugoslavia remained one of the few non-aligned countries in Europe, even though it was a desirable ally for both the Eastern and Western bloc. For the situation in Skopje this resulted in an outpour of help from all sides of the political spectrum, giving it the appropriate nickname City of Solidarity.

The second network, the built structures network, at this point is focused on the provision of short-term shelter for the survivors. This category mainly includes materials for the reparation of buildings and prefabricated structures, as well as experts and workers that can help with construction (Fig. 4).

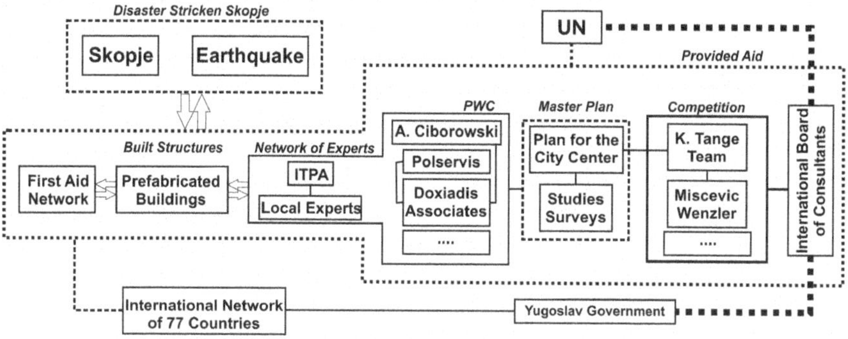


**Fig. 4** The second translation—the detour of the actor-network in Skopje

The third network, the expert network, is comprised of experts, both local and international, whose primary objective is to coordinate and plan the efforts. As the international effort was gathering an unprecedented amount of aid, it became apparent to the city administration that the United Nations and its family of agencies could provide a crucial contribution in the form of coordination, scientific and technical assistance.

## 4 The Skopje Project

The third sub-network, the expert network, at this stage, was the least developed because their primary focus would be the long-term reconstruction of the city. It has become clear that if the complex and international led planning and reconstruction process was to be delivered the involvement of international experts that will closely collaborate with local experts was imminent. Major roles were paired—an international expert with the local expert and further international involvement was requested by the IBC led by Weissmann.

The methodology used in different parts of the project was the result of the high-level negotiated modes of cooperation and involvement of UN and foreign experts, but in the same time exhibited the strategies and interest of each of the individual experts. E. Weissmann's background and proactive engagement in CIAM has shaped the overall idea of the Skopje Project, M. Rotival strategy called "the principle of parts and counterparts" provided a ground for a balanced presence of foreign planners, states and institutions in the process and later was a source of frustration for winning architects for the urban plan for the center of the city because it enforced a continues series of compromises and adaptation as a working method in defining the final version of the urban plan [9, 12].

In the case of C. Doxiadis most of the methods, tools, and techniques used in his Doxiadis Associates (DA) office and the success of every project was not only based on the value of the scientific methods used but also depended on the rhetoric supporting these methods [13].

Now, Skopje is established as the focal actor of a significant actor-network in which the actors have a clear aligned interest, reconstructing the city. The network is generating a huge amount of material recourses for the cause, but another crucial translation of the network emerges.

The second part of the planning was the preparation of a detailed city center plan. First, a restricted international competition was held based on an outline of the Master plan which was not yet completed, inviting four Yugoslav and four international teams to part-take. Two entries were chosen; the Kenzo Tange team from Japan, because of the high quality of its overall design composition and detailed ensemble layouts and the plan of Miscevic and Wenzler from Croatia because their proposals made a valuable contribution to the efficient and practical realization of the program [11].

The contribution from J.H. van den Broek and J. Bakema from The Netherlands represented the concept of the open society in the urban plan for future development of the city center. Bakema described his understanding of the open society as the "hidden potential of our new social structure of society", as the new reality of a "changed social pattern" [1].

In Tange's vision, Skopje remained a planned city under an architect's complete control, an ultimate form for the whole is designed on virtually constitutional basis and all development is made to agree with this form;". This method would "make it possible to produce a total image" (Fig. 5). Tange chose this approach because he felt that the Skopje project was less about stimulating the growth and redevelopment of a living city than it was about establishing a total image around which a devastated city could be resurrected [9].

According to Jenks [5] K. Tange is positioned within the Neo-Expressionism together with J. Bakema, but also in close proximity of the Brutalist architecture, and not far away from the Bureaucratic Neo-International Style where Doxiadis is located [5].

The origins of their theoretical positions are clearly emerging from the pre-WWII Modernist Architecture and Functionalism, and even De Stijl in the case of van den Broek and Bakema, but probably the most dominant resource is the work and declarations of CIAM where some of the protagonist of the Skopje Project like E. Weissmann actively participated and contributed. Local architects and urban planners from Yugoslavia in vast majority in the terms of the theoretical background and their conceptual positions belong to the legacy of the Modernist Architecture, CIAM and International Style like R. MisceviĆ, F. Wenzler and A. ĐjorđjeviĆ; with some of the authors demonstrating a strong sensibility for the local interpretations of the Modernist legacy in Skopje and in Yugoslavia like the work of S. Brezovski and E. Ravnikar (Fig. 6).

With the completion of the plans, the detour was over and Skopje was back on track. The actor-network of experts completed its task and provided the by-pass route to Skopje's reconstruction (Fig. 7).

For twenty-odd years, the city continues being built and a plethora of different actors become tied with the realization of the plan. Construction carried on with various degrees

**Fig. 5** The model of Kenzo Tange team project proposal for Skopje city center (Photo credit: Peter Sägesser, www.ostarchitektur.com)

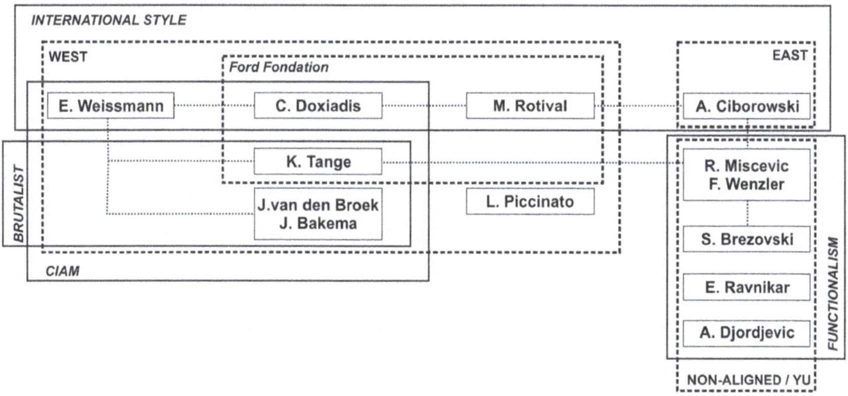


**Fig. 6** The network of actors, theoretical background and affiliations

of speed during the 1970s, however by the 1980s it slows down to a halt. The costs of the plan had begun to exhaust the political momentum created by the earthquake. Criticism of the plan became more prominent, mostly citing the cost and lack of public involvement during the design process. This marks the beginning of the end for the actor-network. In 1985 a new master plan is put forward, but it is just a slight revision of the plan after the earthquake and it does little to further its agenda. The situation completely deteriorates in 1992, with the break-up of Yugoslavia. The Republic of Macedonia is formed as an independent country, and transitioned from a socialist to capitalist system. This brings change to the structure of the whole political system, including the local authorities in Skopje, and the local actor- network tasked with the reconstruction dissolves. The first plan for the city center after independence was made official in 1997. The plan that was

**Fig. 7** Comparison of modern-day Skopje with the plan for the city center form 1963. **a** Skopje today **b** The ninth variant of the plan **c** Overlap of the planned elements with their realization

made is the work of M. Grčev, V.P. Korobar and M. Penčić. This plan came after a gap of more than thirty years in the planning processes in the city. At its core, this plan is a reaction to the rigidity of the previous modernist plan, and makes significant changes, permanently neutralizing the previous urban plans.

## 5 Conclusion

The Skopje Project, the origin of which could be traced back to dominant paradigm of the CIAM, was an attempt to shape the process that will extrapolate the added value of the synergetic acts of the politics and urban design in the midst of the divided world by the Cold War. What made the difference in Skopje is the specific role of the actor-network of experts and transfer of knowledge among them.

The universality of the Skopje Project was strongly embedded in the complex process of negotiations between the institutional positions, different theoretical and urban planning paradigms, design concepts; and experience and knowledge of architects and planners. It was led by the local and international organizations and networks of experts in order to rebuild the city and the society that will create a favorable condition for social and cultural progress. This synergy of politics and knowledge of how to build better and more just cities were unique feature of the Skopje project that demonstrates the importance of the capacity of architects and planners to solve the problems of the cities and societies. It suggests that in order to create a better and more sustainable world the role of the architects in the contemporary society must be redefine. It points out toward the Latourian "Gunman"—a new entity of a man and the tool—in this case, the architect that has in possession of the tools of power exercised through the best practices of politics in a complex process of negotiation and inclusivity. Hence, we must embrace the proactive role of the architect who practices the tools of power and influence, the knowledge, expertise and experience in creating policies and projects for a better world for the communities in need around the world, so we can yet again start to build the "world cities".

## References


1. van den Heuvel (ed.) (2017) Jaap Bakema and the open society" Jaap Bakema, 1960–2000. In: Post box for the development of the habitat (B.P.H.) (5) 27-1-1961, collection Het Nieuwe Instituut, Rotterdam in D. Archis Publishers, p 4
2. Bjazic Klarin T (2015) Ernest weissmann—socially engaged architecture, 1926–1939. Croatian Academy of Sciences and Arts, Zagreb, pp 200–214
3. Callon M (1986) Some elements of a sociology of translation: domestication of the scallops and fishermen of St. Brieuc Bay in Power. In: Action and belief: a new sociology of knowledge? Law J (ed), vol 32. Routledge, London
4. Callon M (1986a) The sociology of an actor-network: the case of the electric vehicle. In: Mapping the dynamics of science and technology: sociology of science in the real world
5. Jenks C (2000) The century is over, evolutionary tree of twentieth century architecture. Architectural Record
6. Latour B (1985) Give me a laboratory and i will raise the world, science observed: perspectives on the social study of science. In: Knorr-Cetain KD, Mulkay MJ (eds) London, Sage
7. Latour B (1987) Science in Action: How to Follow Scientists and Engineers through Society. Open University Press, Milton Keynes
8. Latour B (1999) On recalling ANT, actor network and after. In: Law J, Hassard J (eds) Oxford, blackwell and the sociological review, pp 15–25
9. Lin Z (2010) Kenzo Tange and the Metabolist Movement. Routledge, London, pp 188–195
10. Nikolovski A (1975) Posledicite na zemjotresot vo Skopje. Skopje: Izdanie na Pravniot fakultet na Univerzitetot "Kiril I Metodij vo Skopje
11. Senior D (1970) Skopje Resurgent: The Story of a United Nations Special Fund Town Planning Project. UN Development Programme, New York
12. Tolic I (2017) Ernest Weissmann's World City-The Reconstruction of Skopje within the Cold War Context, Southeastern. Europe 41:171–199
13. Tsiambaos K (2018) Designing on a moving terrain: doxiadis associates and the reconstruction of Skopje. In: The future as a project—doxiadis in Skopje. Amygdalou, Tsiambos K (eds) Hellenic Institute of Architecture, pp 50–88

# Inclusive Permeable School Environments

Alessandro Massarente[1](✉), Michela De Poli[2], Mariagrazia Marcarini[3], and Alessandro Tessari[1]

[1] Dipartimento Di Architettura, Università Di Ferrara, Ferrara, Italy
msslsn1@unife.it
[2] MADE Associati, Treviso, Italy
[3] Alma Mater Studiorum, Università Di Bologna, Bologna, Italy

**Abstract.** Is it possible to conceive a school as an urban ecosystem? We developed a research by design experience, focused on a architectural project of a complex for primary school, kindergarten and nursery school, conceived as an urban ecosystem interacting with the multi-ethnic community characterizing the Lunetta neighborhood in the Italian city of Mantua. Innovative teaching methodologies are related to 3 main strategies at an architectural level: extension of teaching activities towards the exterior spaces; visual intertwining and proximity between quiet area and classroom spaces; wide opening between common spaces and patios. This design is able to promote a naturalistic intelligence through the concept of "biophilia", so that the "outside" represents a "third educator", where to experiment innovative teaching methodologies, didactic application of ecological principles and promote an "authentically ecological pedagogical concept".

**Keywords:** Urban ecosystem · Research by design · Permeable platform · Third e-ducator · Quiet area · Inclusive school

## 1 Theme and Site

### 1.1 Outside Spaces as Third Educator

As the 3 pillars of NEB New European Bauhaus are Sustainability, Inclusiveness and Aesthetics, in the session "Equality through Design" we have tried to explore and measure in a design process the space and the distance between the two terms Inclusiveness and Equality.

As an ecosystem is formed by all the organisms and the physical environment with which they interact, an urban ecosystem is represented by a dynamic spatial structure created and maintained by humans.

In 1826 Fröbel [6], with a forward-looking "pre-vision", had already laid the foundations of outdoor education, Pedagogy of Nature and of Environment.

In various pedagogical researches and texts, school spaces are considered as a "third teacher" [17, 20] due to their ability to influence thoughts, emotions and actions, and have a direct impact on well-being [12], behavior [15, 19], learning [1] and inclusion of pupils, without discrimination or ghettoization [2].


M. Blanco Lage et al. (eds.), *Towards a New European Bauhaus—Challenges in Design Education*,
https://doi.org/10.1007/978-3-031-40188-6_24

Well-being is a fundamental element of the project of a school space, and the World Health Organization believes that health is not only the absence of disease but also a state of physical, mental and social well-being.

The use of biophilic design, which connects and includes people with their environment [7, 23], allows an "in–out school" continuity through direct access from interior spaces (for example classrooms), to the respective outside spaces (porches, patios, gardens).

We have started our work from this hypothesis: is it possible to conceive a school as an urban ecosystem?

We have tried to find an answer to this question, developing a research by design experience, following the concrete approach suggested by NEB.

This research experience was developed in the frame of an "idea consultation" involving 20 Italian schools of architecture and aiming at the renovation of 18 existing educational complexes in the city of Mantua. Each school has formed one or more research groups, with architects, urban planners, engineers, pedagogists, Ph.Ds.

All the groups participated to a main conference, seminar and site surveys dedicated to the presentation of themes and areas of consultation, within the context of a Research Project of National Interest, financed by the Italian Ministry of University.

## 1.2 Lunetta

Our research group designed a complex consisting of primary school, kindergarten and nursery school, conceived as an urban ecosystem interacting with the multi-ethnic community characterizing the Lunetta neighborhood.

The satellite district Lunetta was built in the 1970s on the eastern shore of the Lower Lake as an organic space for modern development in Mantua, a city where still remains strong limits on contemporary development, and are rather suggested replacement of abandoned areas and building recovery.

The Urban Plan of the mid-1970s defined its layout, set on a square-based grid that proposed a sort of "carpet" urbanization of buildings with large four-storey courtyards, able to include the already built-up parts in the area. The urban design repeated that of most of the new post-war neighborhoods set on vehicular/pedestrian streets with apartment buildings.

The main axis of the district is flanked by a seven-storey high continuous curtain wall: a megastructure that represents the backbone of this settlement, conceived as an opportunity to explore the possibilities inherent in the separation of vehicular traffic (at ground level) from pedestrian paths (elevated), commercial spaces on several levels, which sets a platform for public services and collective spaces.

The Lunetta district is one of the many cases where the daily living in an area that has arisen almost out of nowhere has produced ghettoization, partially due to the assignment of residences to social classes with a low economic and social profile.

The nursery school Peter Pan is now located on the west side of the quartier, immediately at the east of the former there is the kindergarten school Berni, and at the east of the abovementioned main axis, the primary school Allende (Fig. 1 left).

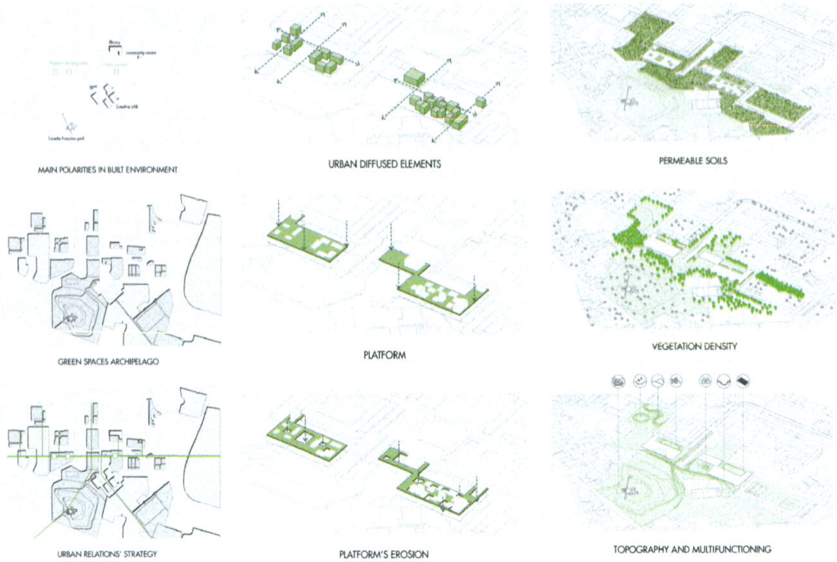


**Fig. 1** From left: urban studies and strategy, platform's morphology, landscape strategy

## 2 Urban and Landscape Strategies

### 2.1 Permeable Platform

In this site existing permeability represented in Fig. 1 (left) by green spaces archipelago was interpreted as part of a possible strategy to gain a new ecosystemic relation between the 3 schools and the quartier.

We have explored the possibility to evolve from existing urban diffused elements (actual schools as isolated and single buildings), to a platform system able to connect the 3 schools and the related civic and community center, and then to a permeable platform (Fig. 1 middle), through which is possible to reach different relations with the soil.

Existing permeable soils, different densities coming from trees and vegetations, and topograpy as support for various activities (Fig. 1 right) become 3 possible landscape strategies introduced in the site that are able to define a new permeable system acting on an urban scale, involving with various activities different inhabitants.

Through its open form this permeable system multiplies different active environments available to the school and neighborhood communities in order to involve local people and promote social interaction between citizens.

This new system at an urban level is designed at architectural level as a thick vegetated band, articulated in cultivated soils and botanical spaces outside and inside the new school (in the patios): the nursery school with a central patio; the kindergarten with 2 patios with classrooms and school services surrounding them; the primary school as a sequence of patios, related with classrooms and common spaces (Fig. 2).

This system is characterized by a permeable platform formed by a two-fold surface, through which is possible to connect the 3 schools and the civic and community center

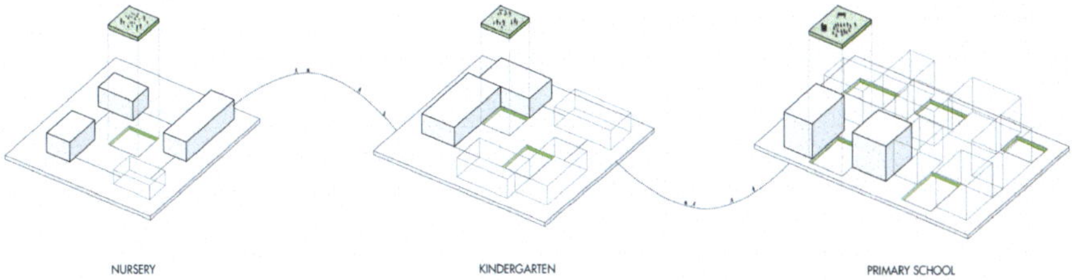


**Fig. 2** Architectural relations between schools and patios

at ground floor level with the district playroom at west near the nursery school and the laboratory classroom for children, parents and elderly people at east near primary school (Fig. 3).

**Fig. 3** View of the model at urban scale

This permeable platform formed by a two-fold surface, at first floor level become a system of hanging gardens, which are conceived to let the neighborhood "breathe", but also to foster playful, educational and social functions, using landscape natural resources to stimulate new relationships between people, places, spaces. The platform plays as a system able to register the topographic variations related to the different inside and outside spaces and in the meantime to be used as a hanging garden by students and teachers during school time, and by citizens during other daytime.

Viewing from south (Fig. 4), primary school appears as a double permeable platform: at ground level, internal patios are located near the border, aiming to favour a wide visual relation between garden, classrooms and common spaces; at first floor level, laboratory classroom and hanging gardens form a new urban connection system, open to different users during daytime.

**Fig. 4** View from south of the primary school

This school conceived at urban scale as a permeable inclusive ecosystem is formed by welcoming spaces open to the community even after school hours: a school in the school, where students, teachers and inhabitants of all ages can meet and share common spaces where to carry out various educational as well as recreational activities, promoting inclusion and personalization of students' educational experience. Orchards, didactic gardens, greenhouse, laboratory classrooms form a continuous pattern through which is possible to reach various spaces and play different activities.

Furthermore, in view of the Universal Design for Learning [3] the school was designed in order to stimulate the affective-emotional sphere and the motivation to learn through the possibility of proposing an active didactic centered on the students [13], as well as differentiated and inclusive [22].

## 2.2 Naturalistic Intelligence

In this new school students can use the green spaces in moments of relaxation or breaks in which it is possible to promote the "multiple intelligence" [8], in particular Naturalistic Intelligence and the concept of "biophilia", so that the "outside" represents a "third educator", where to experiment innovative teaching methodologies, didactic application of ecological principles and promote an "authentically ecological pedagogical concept" [4].

Spaces in which relationship between inside and outside, classrooms, common space, and external spaces of patios, in the different declinations of these 3 new schools, forms a "school for the five senses" [16].

As shown in Fig. 5, «didactic orchards», where students can cultivate vegetables for educational purposes and for the school canteen; «jungle gardens» in the school patios where it is possible to organize classroom group playing activity; «garden in the box» where aromatic plants and flowers are cultivated in great vases along linear sitting elements; «fantasygarden» in the school patios connected with classrooms where are possible external didactic activities; «vegetal patterns» that cover the upper surfaces of the vase in the hanging garden to increase biodiversity and favour the creation of ecological niches, which also intentionally involves Motor Intelligence [8].

**Fig. 5** Plan of the primary school roof and its hanging gardens

The school space is divided, through a "learning street" [11] with areas that have a specific educational value and allow activities in which reflection and creativity are encouraged. The spaces structured in this way offer a set of "affordances" [9] which depend on the educational potential inherent in the environment and in the organization of the elements that compose it.

## 2.3 Innovative Teaching Methodologies and Architectural Strategies

In this project ecological principles are applied in a didactic and innovative way to spaces for education.

The spaces designed are flexible and configurable in relation to didactic proposals for the different learning profiles of the students [5] so that they can be placed in the best conditions for learning, also taking into consideration the emotional dimension as there is a close distance between cognition, emotions and learning [10] which has been given the name of "warm cognition" [14] (Fig. 6).

Innovative teaching methodologies are related to 3 main strategies at an architectural level (Fig. 7): extension of teaching activities (in classrooms and laboratories) towards the exterior spaces (patios); visual intertwining and proximity between quiet area and classroom spaces; wide opening between common spaces and patios.

As the classrooms conceived to be organized through different settings (for example working group, individual test, assembly, exploration, frontal lesson) and to be widely related to patios and common spaces for playing and relax.

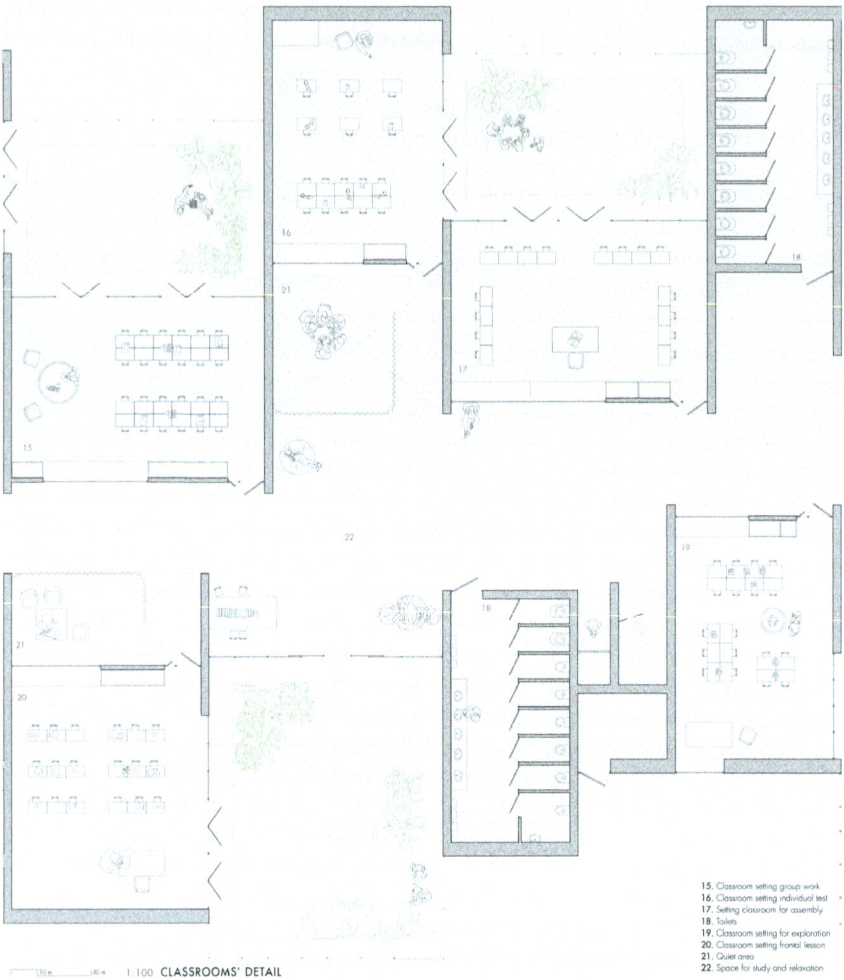


**Fig. 6** Ground floor plan of primary school: six types of setting and quiet areas

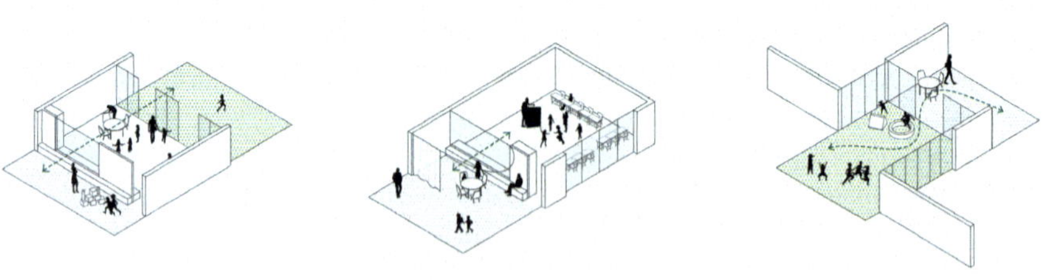


**Fig. 7** Main strategies at architectural level

As the "quiet area", visually connected with classrooms, where all children, not just pupils with deficits or severe autistic syndrome, can relax, reflect or engage in alternative activities that are more motivating for them, without being excluded or margina-lized in separate spaces (Fig. 8).

**Fig. 8** View of a patio from the model

The "Quiet Area" [21] or Breakout Area [18] characterized by an environment connected, in a visible way, with the "classroom-space", allows control of a teacher and acquires a particular meaning in terms of personalization and inclusion of the learning paths of each student.

To build a school is always an opportunity to create new relationship between people and places. This building is also an opportunity to generate a new landscape, a new ecosystem and an organism as a social stimulator. A school as a porous organism with spaces available to create connections in order to regenerate the neighborhood. Each classroom corresponds to a green space usable to study in open air. Each space has a particular characteristic in order to invite students to know landscape natural evolution.

Among various examples of landscape and outside spaces integration in school complexes there are primary school in Vila Nova da Barquinha, Portugal (Aires Mateus), kindergarten school in Lugano, Switzerland (Fioretti Marquez), kindergarten school in Bolzano, Italy (Modus Architetti), primary school HCampus in Roncade, Italy (Zanon and RSHP Rogers Stirk Harbour and Partners), kindergarten school Benetton in Ponzano, Italy (Campo Baeza).

This school is completely integrated in the place, there are not physical limits to divide different areas. Some natural "devices" create soft boundaries, like elevation, depression, shrubs to insert harmoniously the new structure. The new design frees space by making it for everyone, without the insertion of specific aids. This because mobility is a requirement of freedom that must be understood as a basic emotional experience, as a fundamental human and social need, as a manifestation of autonomy and freedom.

This research by design experience aims to define a possible protocol to explore the new ecosistemic role that schools and related community services can play in housing complexes of second half of twentieth century, comparing different case studies at an European level.

## Appendix

Research project credits

Alessandro Massarente (research project coordinator), Alessandro Tessari, Elena Verzella, Alessandro Ramini, Marina Servidei, Martina Ulbar (architecture), Michele Bottarelli (technical plant), Michela De Poli (landscape), Mariagrazia Marcarini (pedagogy), Eleonora Masia, Riccardo Besio (interiors model).

## References

1. Barrett PS, Zhang Y, Davies F, Barrett LC (2015) Clever Classroom. University of Salford, Manchester
2. Canevaro A (2013) Scuola inclusiva e mondo più giusto. Erickson, Trento
3. CAST (2011) Universal design for learning guidelines. Version 2.0. Wakefield, Ma
4. D'Antone A (2018) Epistemologia e prassi della pedagogia della natura. In: D'Antone A, Parricchi M (eds) Pedagogia della natura. Epistemologia, prassi e ricerca. Zeroseiup, Bergamo
5. Dunn R, Dunn K (1978) Teaching students through their individual learning styles: a pratical approach. Reston Publishing Company Inc., Reston
6. Fröbel FWA (1860) L'educazione dell'uomo. la Nuova Italia, Firenze. Original title: (1826) Die Menschenerziehung, die Erziehungs-, Unterrichtsund Lehrkunst, angestrebt in der allgemeinen deutschen Erziehungsanstalt zu Keilhau. Verlag der allgemeinen deutschen Erziehungsanstalt, in Kommission bei A. Wienbrad, Leipzig
7. Fromm E (1964) The heart of man: its genius for good and evil. Harper and Row, New York
8. Gardner H (1983) Frames of the mind: the theory of multiple intelligences. Basic Books, New York
9. Gibson JJ (1979) The ecological approach to visual perception. Houghton Mifflin, Boston
10. Goleman D (1995) Emotional intelligence: why it can matter more than IQ. Bantam Group, New York
11. Hertzberger H (2008) Space and learning. Lessons in Architecture. 010 Publisher, Rotterdam
12. Hughes H, Franz J, Willis J (eds) (2019) School Spaces for students wellbeing and learning. In: Insight from research and practice. Springer, Singapore
13. Jonassen D, Land S (eds) (2012) Theoretical foundation of learning environments. Routledge, New York and London
14. Lucangeli D (2019) Cinque lezioni leggere sull'emozione di apprendere. Erickson, Trento
15. Mallgrave HF (2013) Architecture and embodiment. In: The implications of new sciences and humanities for design. Routledge, New York and London
16. MIUR (2022) Futura. Progettare, costruire e abitare la scuola. Linee Guida per l'Edilizia scolastica. Roma
17. Montessori M (1935) Manuale di Pedagogia Scientifica. Alberto Morano Editore, Napoli

18. Nair P, Fielding R, Lackney J (2009) The language of school design: design patterns for 21st century school. Designshare.com, The International Forum for Innovative Schools, Minneapolis
19. Robinson S, Pallasmaa J (eds) (2015) Mind in architecture. In: Neuroscience, embodiment, and future of design. Massachusetts Institute of Technology (MIT), Boston
20. Rousseau JJ (1965) Émile. Editrice La Scuola, Brescia. Original title: (1762) Émile, ou De l'éducation. L'Aia, Jean Néaulme
21. Sandri P, Marcarini M (2019) Ambienti di apprendimento, inclusione e benessere: dati di ricerca. In: Gherardi V (ed) Spazi ed Educazione. Aracne, Roma
22. Tomlinson CA (1999) The differentiated classroom. In: Responding to the needs of all learners. ASCD, Alexandria (VA)
23. Wilson EO (1993) The Biophilia Hypothesis. Island Press, Washington

# Digital Craft: A Contemporary Bauhaus Model from Design Through Build

Michelle Pannone(✉) and Rebecca Dolgas

Marywood University, Scranton, PA 18509, USA
mpannone@marywood.edu

**Abstract.** Emerging technologies and digital fabrication have revolutionized architectural education and practice, transforming the methods and tools of making in the design process that progress an idea from creative artistic vision into the fabricated built environment. Contemporary architectural education, pedagogically rooted in Bauhaus principles, must therefore utilize the advantages of these modes of making, fundamentally aligning teaching methodologies to the technologies that forge this process. As stated by Gropius (1919), "The Bauhaus believes the machine to be our modern medium of design and seeks to come to terms with it." Today, more than a century later, the intersection of art and design remains essential to making. With technological advancements in the profession, it therefore has become imperative for academia to direct focus towards digital craft. Digital fabrication has enabled the pursuit of limitless making through more comprehensive and efficient tools, equipping the designer to fabricate any creative idea, rather than limiting design potential by the ability to fabricate an idea. Gropius (1919) elaborates on the need for a greater connection between the drawing board and the construction process, stating that "The ultimate goal of all visual artistic activity is construction!… Let us establish a new guild of craftsmen without the presumption of class distinctions building a wall of arrogance between craftsmen and artists." To better position students to see design through the lens of construction and with the contemporary digital tools that define making in the field of design, this paper examines a case study community design build program, and the transformative role of linking the profession and academia through making at full-scale. In connecting the principles of the Bauhaus to the technological tools and processes of digital fabrication, while investigating the interplay between various materials and assembly, design build represents a teaching methodology pertinent to contemporary design education.

**Keywords:** Digital craft · Pedagogy · Community sharing · Fabrication · Engagement · Design

## 1 Introduction

### 1.1 A Contemporary Bauhaus Methodology

Making is fundamental to the Bauhaus ideology. It is through the act of making that students are able to develop the problem-solving skills that translate to the construction site. As Chad Kraus writes, "Hands-on values were central to the pedagogy of the Walter


M. Blanco Lage et al. (eds.), *Towards a New European Bauhaus—Challenges in Design Education*,
https://doi.org/10.1007/978-3-031-40188-6_25

Gropius-led Bauhaus. Gropius implored architects to mend the artificial divide that had opened up between the head and the hand" [1]. Based upon this Bauhaus pedagogical model with an emphasis on making and in an attempt to bridge the gap between academia and the profession, one of the primary teaching objectives of the coursework is to have students demonstrate bringing their design ideas through the entire build process to construct a full-scale implementation.

Architectural practice is rooted in the development of a design for construction, projects often taking years from conceptualization to realization. In contrast, architectural education is situated at the intersection of concept and craft constrained by the academic calendar. This format, condensing months or years into days or weeks, often leaves fabrication considerations and the act of construction hypothetical, a condition to more comprehensively investigate when an individual enters the profession. As Claire Nicholas and Arlene Oak write, "The hypothetical aspect of standard architectural education, and its emphasis on imaginative experimentation, challenges students to invent what might be possible, rather than to actually make it" [2]. This commentary highlights a critical opportunity in the current pedagogical system for students to work at full scale. Through contemporary implementation of the Bauhaus' aspirations, methodology, and mission, community design build projects such as the case study outlined here draw critical connections between society, craft, and technology in the pursuit of a built project.

### 1.2 Context

Located in Scranton, Pennsylvania, Marywood University has been conducting a community-focused design build program since 2018. The coursework, rooted in digital craft, accelerates students from design conception through fabrication in ~ 4 months. Mirroring the ideology advocated through the Bauhaus, the Community Design Build program empowers a group of students to embark on the process to design, test, and build an inhabitable structure. The Community Design Build program engages students alongside community members to test assumptions with concentrated research, theorizes design concepts at a constructible scale, establishes the expectations for a finalized product, and concludes with implementation of the final full-scale structure. Each phase of the Community Design Build process considers society, craft, and technology.

## 2 Methodology

A case study project will demonstrate the methodology of the Community Design Build program; the project Noodles, completed in 2021, was aptly named for its flexible wrapping element which occupies the structure (Fig. 1). Through interactions with the local community, the Community Design Build team recognized that there was a lack of opportunities for relaxation in public space. The students also discovered the recent addition of a middle school in the neighborhood devoid of any playscape. These findings informed the goal to design a playful gathering space, offering equal parts energetic fun as well as casual relaxation to provide areas of interaction that would engage a wide range of age groups.

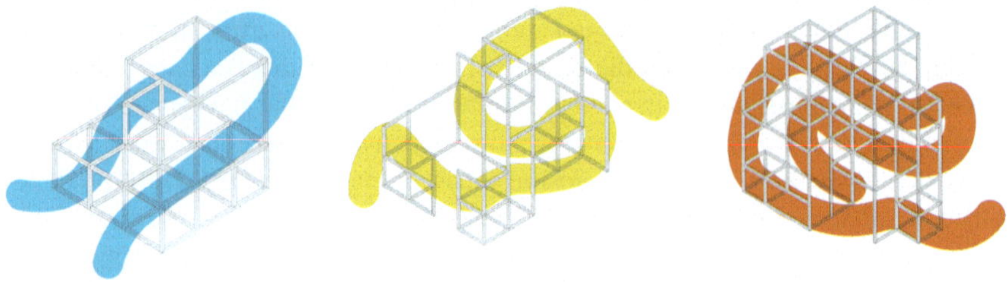

**Fig. 1** Concept of Noodles structure and wrapping elements

## 2.1 Society: Community Engagement

While design build projects do not inherently necessitate a community-centric approach, this program was developed with community participation as an integral component, aiming to design with rather than for the community. This approach was integral to position implementations to address the needs of community members, improve public space, and connect the project to a specific, tangible user group. Chad Kraus writes about the intersection between participatory community engagement, which he terms as "public interest design," and design build:

> Designbuild pedagogies are often associated with participatory design movements, such as public interest design, characterized by a renewed sense of social responsibility and community engagement in design and construction processes.... Public interest design (PID) represents an ideological shift in the way architects and allied professionals approach the design process. While PID and designbuild are two distinct frameworks, there is a very real intersection between the two [3].

In context, participatory planning and design build are often conducted together (as they are in this case study's *community* design build framework) to provide the basis for a design build project to be grounded in creating a meaningful impact on public space for a community, rather than designing and building merely as an exercise in craft. Therefore, rooted in co-design, the course begins with the context, defined as the community in which the intervention will interact. The students are then tasked with designing methods of engagement to frame their approach to the project. While traditional design projects utilize a hypothetical user to outline the needs of the project, the challenge with this approach is that a fictional role drives the intent of the project in the form of an intangible essence; the students shape the theoretical user's needs to match their own preconceived notions rather than shift their own perceptions, and the resultant design, to address the needs of a real user. Therefore, through engaging with the community, Community Design Build instead works to remove the barrier of the intangible by placing the project in the hands of local users.

To determine the site, students explore with community members a public space to identify a specific location for a design intervention. Students then continue the exploration of both the spatial qualities and the community members who interact with it, implementing surveys, interviews, and modeling methods. From this research, students gather information and feedback to determine the desires of the community. The students are breaking down the notion of designer as an outside entity by immersing themselves as facilitators in the conversation, bringing together designers and users. Integral to this type of process is the network and system of collaborators, beginning with student facilitators, who bring their research and response to the design process. Next are the community collaborators, members of the community with a passion to improve their public space. Also imperative are curious students; these are students who are not directly facilitating but are inspired by the engagement in the design process and want to be involved. These additional hands not only further root the process in the academic community but also offer the lessons of craft and construction to a greater range of students, and in some instances these volunteers have later taken leadership roles in consequent build projects. The final component of the network and system of collaborators are industry professionals which play a critical role in linking professional practice with the academic setting, along with providing technical assistance and in certain cases funding. These are individuals from a wide range of technical backgrounds offering their expertise to support students as they transition their ideas from representation to reality.

Collaboration at the scale of the community and the profession is integral to the Community Design Build process from conceptualization to implementation. The Bauhaus guiding principle of society is evident in the approach to designing with rather than for the community through co-design to achieve a tangible outcome.

## 2.2 Craft

Craft is the lens through which ideas are visualized and materialized to transition from concept to physical object. In architectural practice, craft is integral not only in the final construction, but also in determining potential assembly methods and materials; craft informs all design considerations if concepts are to be brought from paper to the built environment. Höweler and Yoon describe how craft and building trades, among other considerations, permeate through many aspects of the design process, ultimately defining how craft is positioned at the transition between concept and construction, in stating:

> Contemporary processes by which architecture is conceived, modeled, and mocked up illustrate the messy process of translation from concept to construction…. If architecture is to have greater agency in addressing the many challenges encompassed by the built environment then questions of material procurement and reuse, embodied energy, and building trades and craft cannot be separated from design [4].

In Community Design Build, craft is also the catalyst that allows students to explore through testing the means by which their conceptual ideas are brought into spatial constructs. Simultaneous explorations of craft are imperative at varying scales throughout the design process; in developing the overarching concept, prototyping mockups for feedback, and exploring material qualities and potential fabrication methods.

In the Noodles project, once the team had established initial design ideas it was then essential to test the early concepts and solicit feedback. The students utilized small scale models during the early phase of the design. These conceptual models established critical interaction points both for community members and designers. With the community members, these models provided valuable feedback on the concept and usability. From a construction vantage point, students worked with collaborators to evaluate connections, materiality, and buildability. The final design culminated in a series of three steel grid structures that support noodles to form sittable, climbable, and colorful elements. Upon first glance the white steel structure, wrapped with colorful elements, presents itself as a bright, enticing, and inviting spatial structure. Each aspect, from location to construction method to color, was a part of the team's carefully curated facilitation process; designed and tested alongside the community, their involvement was integral to the project's iterative testing and decision-making processes.

### 2.3 Technology

From small scale physical models, the team shifted from the conceptual design and feedback process to developing the plan for full-scale construction. Through the use of three-dimensional modeling software, the students represented their ideas by forming them into digital models. The implementation of digital modeling tools were then utilized to demonstrate the viability of the project; by creating a digital model the team could quickly assess weak points in the design as well as diagram potential fabrication methods both within the design team and with the community. As Malcolm McCullough describes this interplay between the physical and the digital as "a means for combining the skillful hand with the reasoning mind" [5]. In addition to the benefits to the design process, the digital models also provided advantages in communication and representation of the project which allowed for the efficient creation of visualizations to then generate dialogue with the community.

To further communication methodologies between team members and with the community, detailed assembly guides were created from the digital models following the various physical tests that were performed. These legible guides (examples shown in Fig. 2) streamlined and simplified the job for constructors, even allowing for community members and collaborators unfamiliar with the design to become a part of the fabrication process. Additionally, creating this level of detail in the design documentation through the assembly guides verified that construction methods, materials, and connections were resolved prior to the start of construction. These detailed drawings were a result of the explorations of assembly and craft pursued during the testing phase of the project. For example, after the students had gained experience working with steel, many of them working with this material for the first time, they translated the physical findings of the material qualities and connections necessary to achieve the design intent. The fabrication and documentation phases of the project provided a significant amount of dialogue regarding the relationship between digital precision and built craft. Digital craft is therefore the ultimate outcome of an education environment that requires students to work fluidly between the digital and physical world. For example, students discovered instances in the assembly process where it was necessary to verify the dimensions and angles in the field to account for onsite conditions such as tolerance. These findings highlight the knowledge gained through the process of translating a concept from design documentation to a built implementation at full scale.

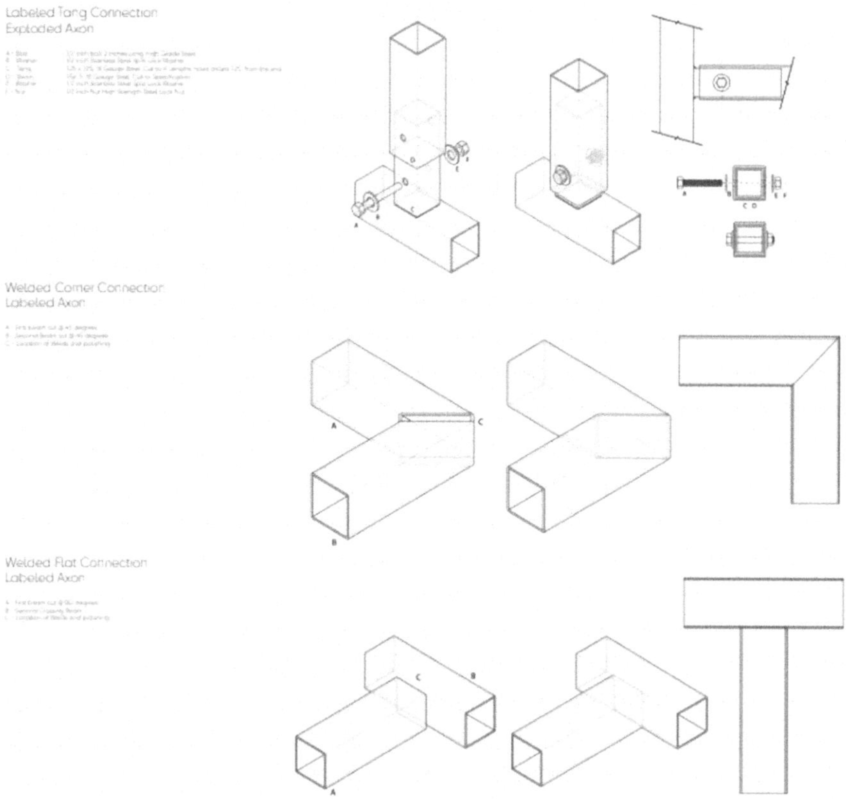


**Fig. 2** Excerpt from assembly guide documentation

# 3 Conclusions

Opportunities to build at full scale allow students to explore the transition between the development of concepts and the execution of built work that occurs as the principles of society, craft, and technology converge. As described by Bruce Lonnman "the Bauhaus re-established the critical relationship between the designer and the medium: the materials of construction, the processes of forming and fabrication, and the constraints these place on the design" [6]. Community Design Build provides more than an opportunity for students to experiment with construction methods; this program provides a framework in which design, physical and material limitations, and community engagement converge. Beyond individual considerations of society, craft, and technology that shape the conceptual framework of these types of projects, the overarching process requires these principles to work in tandem with each other. Community Design Build provides a methodology to connect these principles through the exploration of digital craft, a pedagogical approach to deliver learning outcomes more closely aligned with architectural practice.

When building at full scale students gain knowledge of fabrication processes in addition to the practical knowledge associated with material consumption and labor resources necessary to complete a built project. Examples of these takeaways are seen in the Noodle case study. One takeaway is evident in the use of community resource sharing. The recycled noodle filling is a response to the environmental demand to rethink the way that discarded objects such as trash are transformed into a building material. Figure 3 shows an image of students filling the fabric noodle, after several tests to determine the composition of used plastic bags, bottles, and foam packing materials that would make the optimal seating cushion. In the final form, these noodles filled with recycled objects serve to create spaces of gathering and play. Another takeaway is a result of the process of construction, which after undergoing several rounds of testing, included acquired skills and techniques such as welding, painting, sewing, and filling the shell. Unique skills and techniques were learned from the iterative process of testing and research to achieve a final assembly process. A comprehensive understanding of digital craft was necessary in order to realize the final product at full scale. These material findings yielded the opportunity to share skills among collaborators which in turn expanded the knowledge base of the group for processes such as textile manipulation, sewing at an industrial scale, and welding. Each part of the assembly process relied on multiple individual knowledge bases coming together to create an actualizable plan of execution.

**Fig. 3** Photograph of recycled materials being used to fill one of the fabric noodles

The Noodles occupy the public space of the Art Field, with both children and adults using the structure in various ways ranging from relaxing to studying to play. In addition to the didactic outcomes of Community Design Build, the final implementation of this case study included three structures and noodles at varying scales one of which is pictured in Fig. 4. As the pedagogical discourse of design education continues to progress, it is necessary to continue to distill the lessons, methodologies, and benefits of building at full scale and further integrate these opportunities to explore digital craft.

**Fig. 4** Photograph of Noodles in use. Image by Paige Bechtle

## References


1. Kraus C (2017) Designbuild education. Routledge, New York, p 3
2. Nicholas C, Oak A (2020) Make and break details: The architecture of design build education. In: Design studies 66, pp 35–53. https://digitalcommons.unl.edu/cgi/viewcontent.cgi?article=1050&context=textiles_facpub. Accessed 30 Nov 2022
3. Kraus C (2017) Designbuild education. Routledge, New York, pp 2–3
4. Höweler E, Yoon J (2021) Verify in field. Park Books, Zurich, p 70
5. McCullough M (1996) Abstracting craft: the practiced digital hand. The MIT Press, Massachusetts, p 81
6. Lonnman B (2010) Constructing design in the studio: projects that include making. In: Kinnard, Goodwin (eds) Re.Building, pp 67

# Participatory Planning for Post Industrial Sites in Sibiu

Oana Paval[1] (✉) and Maria Cristina Gavozdea[2]

[1] Ion Mincu University of Architecture and Urbanism, ATU-Association for Urban Transition, Bucharest, Romania
sooana.paval@gmail.com

[2] A.Plan Architecture Studio, Bucharest, Romania

**Abstract.** In Romania, the abandoned post-industrial buildings or sites are usually valued by developers based on a tabula-rasa, top-down approach, by demolishing the existing and building commercial areas or housing units at maximum capacity, even above. We propose a different approach based on participatory planning, using the Urban Design Management (UDM) methodology, developed at the University of Helsinki and brought to us through a Social Challenges EU project. UDM is based on the concept of integrating as much value as possible into the development process, for most stakeholders. It is an iterative approach, aiming to generate a shared vision, development strategy and/or a masterplan solution. It is a participative process including both professional and non-professional diverse perspectives, that far outreaches the requirements of formal planning. The first two applications of the UDM in Romania were the former Amylon factory and the Independenta industrial site. Both sites are situated at the verges of the intra-muros historical center of Sibiu, being part of the XIXth century industrial belt that developed on the banks of Cibin river. The participatory planning processes unfolded in 2019 and 2022, leading to balanced, mixed-use development solutions including apartments, offices, commerce, public facilities and spaces. This represents a novelty in recent Romanian planning practice. Both experiences have proven extremely valuable and all stakeholders have well appreciated them. Thus we promote this methodology in brownfield regeneration and other complex projects, as it can be an active instrument in the formation of the New European Bauhaus culture in Romania.

**Keywords:** Participatory planning · Stakeholder engagement · Post-industrial · Brownfield regeneration · Urban development · Public

## 1 Participatory Planning in Romania

### 1.1 Context

Central and Eastern Europe evolved in a different way than the Western states, entering a process of industrialization specific to the communist era. This included a distinct industrial architecture and led to the construction of numerous factories, compounds and


M. Blanco Lage et al. (eds.), *Towards a New European Bauhaus—Challenges in Design Education*,
https://doi.org/10.1007/978-3-031-40188-6_26

even mono-industrial towns (most of which are now in decay). Abandoned industrial sites/areas, as resources for urban regeneration, already have some advantages regarding their position within the city and the relation they have with the infrastructure. This should be a means to balance the chaotic urban sprawl in Romanian magnet cities. But the countries of the former communist states relate to the recent industrial past, and implicitly to the industrial heritage, in a way closely linked to the past political regime and the trauma it left behind. This association, overlayed with the general attitude towards urban development and built environment, led to a certain approach regarding abandoned factories and industrial areas. [1]

The reluctance of local councils to regulate private property areas or to require public services, as well as the ownership culture—over 95% of Romanians being home owners [2] are part of the communist trauma. This mostly led to greenfield and, more recently, brownfield developments that are based on a tabula-rasa approach. Both types of properties are valued through building housing or commercial areas at maximum capacity or even above, with low mixed use ratio or none at all. For developers this approach is a recipe for fast profit. This is also the common approach in Sibiu, a magnet city where housing demand well exceeds the offer. People, coming generally from villages or smaller cities, have low expectations considering housing standards and community involvement, so everything goes. Here too, the old industrial sites were abandoned until recently, when new housing areas were built, as in the case of Balanța Residence (see Fig. 1).

**Fig. 1** Left—Balanta Scales Factory (Google Earth caption, 2007), right—Balanta Residence Sibiu (Google Earth caption, 2022)

## 1.2 Legal Frame

Urban planning of areas within the city is required in the case of urban areas with a high degree of complexity or accentuated urban dynamics, in the case of central urban/rural areas, of built protected areas and areas for monuments' protection, for areas that are subject to urban restructuring or regeneration [3], such as post industrial abandoned sites in accordance with the Romanian legal frame. Though public authorities have the main responsibility in planning city development, they usually elaborate strategies and general urban plans at the scale of a whole city, metropolitan area or region, leaving the more detailed urban planning to private initiatives.

Public–private partnerships are taboo due to a flawed legal frame. This leads to a great number of documentations that are elaborated by private urban planning offices and processed by several institutions. In some cases the same area is planned several times in the span of a few years, by different owners, each focusing on his own interests. The consequences are an unbalanced, poorly coordinated urban development, high investment costs especially for small initiatives, low public administration leverage for large investments, process delays due to institutional bureaucratic overload. The need to conform to EU directives led to one main consultation procedure that the local authority is responsible for in all urban planning processes; some also require an extended Environmental Agency procedure that also involves public consultations.

Order 2701/2010, approving the methodology for public consultations within urban planning processes, compels the Public Administration to inform and consult the public in three key moments at least: when initiating the process, when founding studies are delivered and when a first draft of the urban or territorial development plan is delivered. Unfortunately the methodology lacks sanctions for non-compliance, so most local administrations conduct a sort of consultation only when plans are almost final, through unfriendly and hard to access platforms, presenting urban planning technical documentations that are incomprehensive to non-professionals.

Only recently Sibiu has put in place an interactive map [4] —that is actually hard to reach, but it is used only to show areas that have been recently regulated or that are being processed; one can also upload some scanned documents. However, the public consultation processes are not communicated to a broad audience in a highly visible way. If anyone is interested in certain development projects regarding the city, they have to intentionally, regularly follow the city hall website and a certain newspaper. The same goes for most local administrations—open governance is making very small steps, slowly.

### 1.3 Urboteca Programme and the First Community Led Neighborhood Development Plan

In this context, URBOTECA programme, coordinated by ATU-The Association for Urban Transition, started in 2014 with a vision for a more educated and better informed community. URBOTECA, etymologically defined by the Latin terms "urbs" = city and "theca" (derived, in turn, from the Greek "theke" = box, chest, receptacle), is a neutral place and accumulator of resources, shelter and generator of pedagogic actions aimed at increasing the degree of information and interest in city themes, aimed at providing support for urban actors who are thus encouraged to approach the complex theme of participatory planning [5].

The first project aimed to translate the specific urban planning language to the broad audience, so that the citizens in a community could get more involved throughout all the stages of an urban development project. At that time, there were only a few community organized civic initiative groups, mostly in Bucharest [6]. Usually, these civic initiative groups represent areas or neighborhoods and are the fuel for grassroots participatory initiatives but also the core that mobilizes the citizens to react if development projects affect their communities quality of life.

After trying to be a translator, URBOTECA settled down in one place for 3 years (2017–2021), during the Urban Education Live (UEL) Project. The main objective of the project was to create and test models of collaboration between universities and urban communities and the community partner was the Bucurestii Noi Civic Group. The result was the Community Led Development Plan (CDP) for the Bucureștii Noi Urban Area, [7] realised through a participatory planning process. It was set out to regenerate the identity of the neighborhoods in Bucureștii Noi area, taking into account both the field reality and the wishes expressed by the citizens. The premise for a healthy neighborhood is to both meet the basic needs of a community (accessibility, primary facilities, mobility, sanitation) and social needs (sense of belonging, equal opportunities or self-realization).

The CDP is organized in four sections that are defined around important themes for that urban area (their relevance being established after conclusions were drawn from interviews and questionnaires conducted between January 2018 and August 2019). The four sections are: AMENITIES/equipment for the public (spaces for activities with an audience), ENVIRONMENT (problems and opportunities), MOBILITY (public transport users, drivers, pedestrians, cyclists) and NEW-OLD TENSIONS (protection regime, urban design, notifying and authorization). Important formulated objectives, on all four sections, were also related to encouraging and including small and medium sized local business and also to creating the opportunity for residents to carry out activities with multiple values in their own neighborhood.

The CDP is an instrument through which different stakeholders can further collaborate because it clearly states development objectives for a neighborhood. The built environment specialists have an important role in using this tool, because their involvement consists also in translating the ideas and aspirations expressed in the CDP. The ideas stated in the Bucureștii Noi urban area CDP were meant to be the starting point for the City Hall to organize an architecture competition, in partnership with the Bucharest Chamber of Architects, for the core of the area, but the pandemic hit, also elections took place during it, and now it is on standby.

Drawing on earlier lessons learned, UEL developed a set of innovative approaches for inclusive, vibrant and accessible urban communities. Collective Networks for Everyday Community Resilience and Ecological Transition (CoNECT) in the newest project that URBOTECA is a part of [8]. CoNECT will catalyze existing networks for collective action in six EU countries—Romania, France, Netherlands, Sweden, Norway, Spain—aiming to boost community organizing capacity by recognizing, mapping, connecting, and strengthening everyday practices of community resilience toward ecological transition [9]. The project will foster collaboration between citizens, thinkers, doers, planners from the partner countries, who will co-design and conduct parallel synergistic Labs implementing joint tools and methodologies.

These initiatives, the PCD and CoNECT are very well suited for certain contexts that era based on bottom-up approaches, Professionals of the built environment are aggregators and translators for the community and of the community's needs and expectations towards local administrations. The particularity of the two case studies of participatory planning for post industrial sites in Sibiu is that the processes have been initiated top-down. The urban design management methodology was a means to include the community in the development processes of two large urban areas right in the vicinity of the historic city and turn their opinions into regulations through urban zone plans.

### 1.4 Urban Design Management as an Instrument for Participatory Planning in Sibiu

Sibiu is one of Romania's magnet cities. Its population tripled in the last 10 years and expanded mainly in greenfields, usually preserving the agricultural property divisions that are not suitable for healthy, sustainable urban developments. In 2019 questions were raised within the civil society about a consistent greenfield area that had somehow escaped the real estate rush, though it was like a pizza slice on the city map. Fear of chaotic development lacking urban commons [10] was fed by the experience of the last decade of urban development.

What was there to be done? Authorities should be able to mediate owners and developers into lot restructuring at neighborhood scale, but nobody knew how to do that. Through the Social Challenges call a solution that included the Urban Design Management (UDM) methodology for participatory planning was funded. UDM is an iterative planning approach, aiming to generate a shared vision, development strategy and/or a masterplan solution. It is a participative process including both professional and non-professional diverse perspectives, that far outreaches the requirements of formal planning. It can be applied in different stages of planning and design, so it can even be used for defining the solution of a complex architectural project (ex: community /center, cultural center etc.) or for designing a public space. It suits both one-owner and multiple-owner situations, thus large scale properties and highly fragmented property areas are equally served [11].

This methodology has an innovative approach, building on the premises that the value of a project is the sum of the benefits of all those involved or influenced, not just the financial profit of the owner and investor. Customers are not only the beneficiaries defined in the contract, but all future users of the created product and also those who will be influenced or affected by it in any way. In the case of a neighborhood, the customers include the future residents, workers, students, business owners and neighbors, sometimes the city as a whole (citizens, business environment, NGOs, authorities) and more, depending on the scale, location and context.

Any development has much more value for more people when it is founded on their values and principles, their needs and desires, in an inclusive approach. Thus it becomes the project of many, not only of the developer, which makes it a stronger case within the community as a whole. The developer, whether public or private, is also gaining time and money by answering the needs and desires of the community in a more focused and efficient way. So rolling UDM in the planning process is a small upfront investment for an extensive long term benefit.

The UDM process is well structured into four stages that develop the solution from an overview level of the subject at hand, zooming in. Each stage includes a workshop that generally follows on a monthly basis, that can be enriched with other public engagement tools. The time between workshops helps ideas mature both for the UDM facilitating team and for the participating stakeholders. (see Fig. 2) The first and last workshops gather only primary stakeholders. The other two include secondary stakeholders too. Participants are invited as a result of a stakeholder analysis that aims to capture a relevant picture of the community. These workshops are also an opportunity for diverse stakeholders to meet in person, to better understand each other's perspectives, to embark on the project journey as service or cultural operators, co-investors, business owners and so on [12].

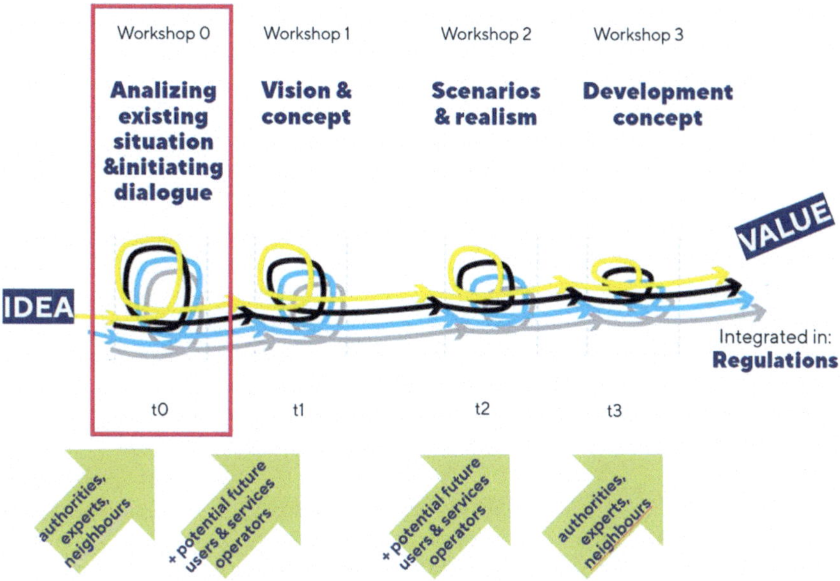


**Fig. 2** The stages of the UDM process throughout the 4 workshops

UDM builds on the concept of place creation. Urban planning and architecture generate contexts of life and mold life habits of individuals and communities at large. This connection between the build environment form and lifestyle is rarely really understood. So the facilitator's task is to bring into discussion lifestyle aspects, that the architectural team then translates into the urban plan or architectural project. In the case of urban planning, it is very important that, apart from the major legal constraints and public strategies, discussions start on a clean slate. Showing people even a sketch of a project inhibits opinions and feeds the conviction that cards are already dealt. So the solution iterations should always build after the discussions within the participatory process, not beforehand.

## 2 Two Brownfields in Sibiu as Case Studies

A belt of industrial sites developed in the nineteenth century along the Cibin river valley, beyond the defensive walls of Sibiu. Most of these factories also operated during the communist regime, with some modifications, re-engineering, modernization or expansions, but later they fell into disrepair.

The city expanded beyond the Cibin valley and the railway, through neighborhoods dominated by single-family dwellings (houses) and blocks of flats, benefiting from reduced public services and amenities. The connection of these neighborhoods with the city center is generally poor due to the presence of these closed, non-functional industrial premises, which stretch over large areas, respectively due to the barrier represented by the railway, which is rarely crossed by road or pedestrian traffic routes. All these sites need regeneration and weaving with the neighboring urban tissue.

### 2.1 The Pilot. Amylon 2030

The first application of the UDM methodology in Romania was in the context of the Amylon neighborhood development in 2019. The 4.2 ha site of the former Amylon factory and of the former Abattoir had been demolished more than 10 years before (see Fig. 3).

**Fig. 3** Aerial perspective of the Amylon site, Google Earth caption

The client had a vision—a complete neighborhood, for people to lead a good life—that needed refinement and a more in-depth definition. Public authorities, neighbors, the developer, potential future residents and business owners participated in the process and discussed certain aspects of the future life of the neighborhood. Discussions also took into account the city strategies and regulations, physical site considerations, certain concepts were explained and exemplified.

The approach is largely based on the following principles: the place as a resource, accessibility, attractiveness and sustainability. For example, attracting talents and highly qualified people to a city involves providing an attractive living environment for them, corresponding to their values and needs (alternative lifestyle, diversity, gentle mobility, sociability, culture, education, sports, etc.) that should intertwine with the local community interests.

While most ideas that were discussed within the participatory process only confirmed the architectural team expectations, some ideas came fresh. Everybody agreed on mixed use and diversity, superior quality of life, public and green space variation, soft mobility and connectivity, sustainability and so on. Certain aspects were refined:

- visual permeability and accessibility of inner courtyards at street level,
- gradual transition from public to private, with medium permeability,
- mix of types of apartments for a healthy mix of categories of inhabitants, adapted to the real needs of society.

Fresh ideas that emerged from the UDM process included a jogging trail within the neighborhood and volumetric accents to mark points of attraction/perspective ends.

Two first customers started concrete negotiations with the developer after having participated in UDM: a sports and wellness center operator and a private school and kindergarten. The masterplan needed to include these two functions in the first stage of development, which covers about half the site surface, along with commercial and office spaces, apartments, a two level garage, a restaurant and skybar. It also comprises public green space, the streets and neighborhood utilities (see Fig. 4).

Following the UDM, the zone urban plan documentation integrated its conclusions. As a result, the consultation procedure led by authorities had only two observations, one of which was just to welcome and encourage the project.

## 2.2 Abandoned Independenţa Post Industrial Site

Independenţa is the largest central brownfield in Sibiu, with a surface of almost 9 ha (see Fig. 5). It has long been a barrier between the historic city and the nearby neighborhoods, especially since its closure. The site contains some listed buildings, both individually and as an ensemble. Several city strategies and the masterplan, commissioned and adopted by the Sibiu City Hall, pay particular attention to this former industrial platform. However, it's regeneration is not a priority on the local administration agenda.

Independenţa is privately owned by a Romanian investment fund and has long been the subject of legal disputes, especially on built heritage issues, as the owners and some specialists consider that the heritage listing has been excessive. Recently, the board members decided to regenerate the site. Profit is the main driver, but they understand that the place has its specificities that need to be addressed with hand gloves. They initiated a zone planning documentation that got stuck in the approval process, then searched for another team of urban planners. Several recommendations led them to A.PLAN architecture studio, which accepted the job on the condition of first going through a participatory planning process based on the UDM methodology. It is a delicate subject for Sibiu and the Amylon experience had proven the positive impact of this participatory planning methodology in the community.

Within the UDM for the regeneration of this site, again, everybody agreed on aspects such as valuing the heritage both within the site and surrounding, connecting to the historical center, open green and public spaces, soft mobility, the opening of the main

**Fig. 4** Amylon masterplan proposal

street of the site that connects the city center with the river bank and the neighborhood beyond it, making the premises permeable etc. More definition was gained on the social diversity of future users, mixed functions with a significant arts and culture component, activity density and public spaces. A demand, which participants agreed upon, was to maintain the possibility to carry out small production activities, light industry and/or crafts within the zone urban planning regulation limits, which also included some small neighboring properties.

As general interest for the future of Independenţa was very high and workshops could only involve a limited number of people, an open questionnaire was launched between the second and the third UDM workshop. It collected 1237 answers in three weeks, mostly from citizens of Sibiu, with a superior education profile. The questionnaire explored five themes: functions that the center of Sibiu lacks for free time activities, the public–private character of open spaces, materiality, mobility and functions within the Independenţa site. It revealed a strong desire for more promenade space, cultural and sports facilities, a preference for open versatile public spaces, with a consistent tall vegetation presence, soft mobility, a very mixed use with a strong public character. The subsequent iterations of the UDM included the questionnaire conclusions. The conceptual volumetric solutions

**Fig. 5** Drone photographed aerial perspective of Independenta by Sergiu Brega

had a smaller variety that the ones for Amylon, as a large part of the Independenţa site is listed as historical monument (see Fig. 6).

Although high interest is generally expressed on the subject, the rate of participation at workshops did not sustain it. Very few people participated at all four workshops, mostly non-professionals, which sometimes made discussions sterile, fragmented the process and deprived it from the benefit of ideas maturation. As far as we know, no negotiation with future operators started as a result of the process. Authorities had a very low and inconsistent participation in both cases—Amylon and Independenţa.

## 3 Instead of Conclusion

Community engagement in the decision-making process is growing in Romania, mostly in the largest cities starting with Bucharest, but also in smaller ones like Sibiu. Bringing all stakeholders to one table is a great challenge, as daily routines are challenging enough for most people, regardless of their roles. Complementing such meetings with open surveys or other community involvement tools may be needed in the post-communist context more than in older democracies.

UDM has several competitive advantages as a participatory planning tool. It has a predictable and reasonable timeframe—4 to 6 months, which also means a reasonable cost. The result is a plausible solution that has already integrated several professional and inclusive considerations. It can be rolled at several development stages: strategy, territorial and/or urban planning, urban and/or architectural design. It engages stakeholders

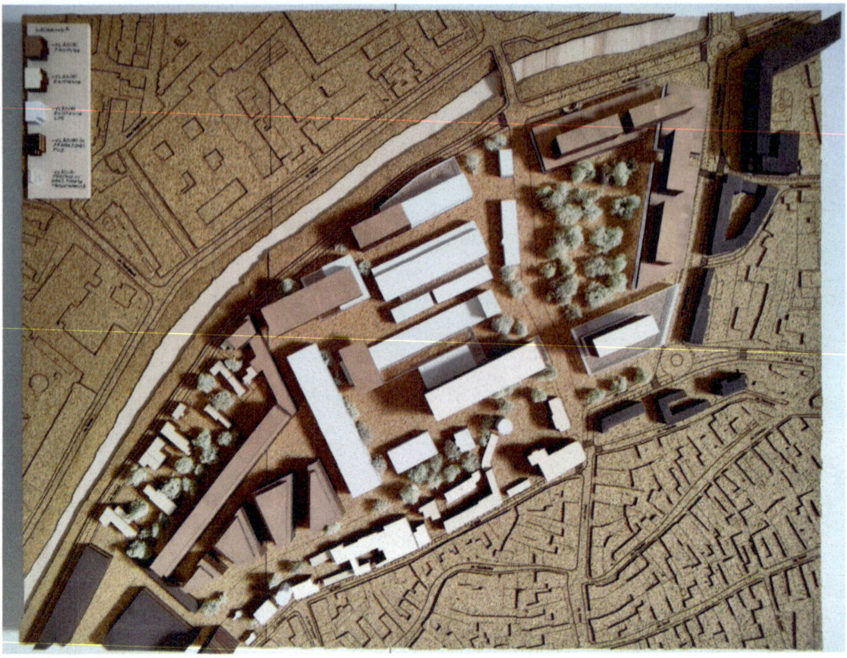

**Fig. 6** Model of the volumetric concept solution for Independenta: white—monuments, brown—new proposed volumes, nature wood—existing buildings

at a personal level. The UDM has proven its positive impact in the case of Amylon. We hope the Independența experience will confirm it too, despite its setbacks. A more engaged administration is very important, even in private development cases; it's a recent promise for some cities that showed interest in UDM.

Meanwhile, many administrations make public investments under the radar, without urban planning procedures or open competitions to ensure transparency, high professional quality and social inclusiveness. More and more people are upset by the poor results. Various community initiatives emerge, among which For Sibiu that deals exactly with public interest projects that need to be discussed in the public realm. CoNECT also aims to contribute to building a more resilient community [13] in Sibiu, starting with 2023.

### 3.1 Sibiu Starts to Grow a Civic Muscle

Sibiu has been attracting workforce from all over the country for many years, and urban development has occurred to the greatest extent beyond the administrative limits of the city, on the territories of neighboring towns, where it was easier to build low-quality housing, lacking public facilities. The extremely high demand made it possible to trade them regardless of their quality. Among the effects of this phenomenon are the overloading of the city's public services in the absence of contribution to the local budget, traffic problems, a high level of real estate prices and the lack of a good and superior housing stock.

The conversion of industrial spaces along the Cibin valley, close to the historical center of Sibiu, represents an extraordinary opportunity for the city to revitalize itself and define itself as a creative and attractive city not only for tourists, but also for residents with a higher level of training and with higher living standards. There is a strong group of different professional entities actively involved in the education and engagement of the local community in urban development processes [14] and the two initiatives presented before just led to a more aware and active local community. For example, FOR Sibiu is an initiative led by the Sibiu-Valcea Branch of the Romanian Chamber of Architects and although it only started in 2022, it is bringing the community together around important urban themes. This leads the way to future projects building a resilient community in Sibiu, through the CoNECT Lab that will start in 2023. The local Partners of the project are the Sibiu-Valcea Branch of the Romanian Chamber of Architects and the Sibiu Community Foundation, and thus FOR Sibiu will actually pave the way to a more receptive and engaged local community in 2023, already familiarized with some of the topics that CoNECT will put on the table and try to work with.

## References


1. Paval O (2019) Abandoned industrial sites in urban context. The re-use capacity of industrial heritage. PhD thesis, Ion Mincu University of Architecture and Urban Planning
2. Eurostat. https://ec.europa.eu/eurostat/databrowser/. Last accessed 28 Nov 2022
3. Law no. 350/2001 regarding territorial development and urban planning (Legea 350/2001 Privind amenajarea teritoriului si urbanismul)
4. Sibiu City homepage. https://www.sibiu.ro/. Last accessed 28 Nov 2022
5. Marin V, Calciu D (2018) Urban pedagogy in Bucharest: URBOTECA by ATU. In: Revista Arhitectura nr.2–3/2018, pp 150–153
6. Andrei M, Voicu O (2015) The activity of informal civic groups in Romania. Research report (Activitatea grupurilor civice informale in Romania. Raport de cercetare), Open Society Foundation
7. Marin V, Calciu D, Bădescu G, Dumitru A, Mocanu R (2019) Community based development plan for the Bucurestii Noi Urban Area. Available at http://atu.org.ro/wp-content/uploads/2021/04/ENG-Copy-of-PLAN-COMUNITAR-_compressed.pdf
8. Urban Europe Homepage. https://jpi-urbaneurope.eu/project/conect/. Last accessed 25 Aug 2022
9. Petrescu D, Petcou C, Gibson K, Safri M (2020) Calculating the value of the commons: generating resilient urban futures. Environ Policy Governance J
10. Petrescu D, Petcou C (2020) Resilience value in the face of climate change. Archit Des 90(4)
11. Ahlava A, Edelman H (2014) Urban design management a guide to good practice. Taylor&Francis
12. Kaner S, Lind L, Toldi C, Fisk S, Berger D (2014) Facilitator's guide to participatory decision making, 3rd edn. Jossey-Bass
13. The Young Foundation (2012) Adapting to change: the role of community resilience. The Young Foundation, London
14. Trogal K, Bauman I, Laurence R, Petrescu D (2018) Architecture and resilience: interdisciplinary dialogues. Routledge, London

# Transferring Data from a Factory Heritage Site into a Building Information Modeling (BIM) Which Integrates Sustainable Development Indicators

Sandra Rihs(✉), Betty Baud, and Roubini Makridou

HES-SO HEIA, Fribourg, Switzerland
sandra.rihs@hes-so.ch

**Abstract.** This research project looks at how to integrate indicators for sustainable development into a building information model. The idea is to combine the fields of sustainability, digital construction, and renovation. The case study for the ongoing research is the listed building "Halle Blanche (white hall)", which is part of a factory site and will be transformed into the faculty of architecture of the University of Applied Sciences and Arts Switzerland. The data requisitioned from the existing building is used to create a digital twin. It will undergo several iterations to arrive at an optimal version for reuse and simulation of possible designs, to carry out a precise analysis of the state of the building, its structure, its components, and its materials to optimise the choices for a renovation that considers the values of sustainable development. To create a reliable and accurate decision model, we use various tools such as Light Detection and Ranging (LIDAR), drone, and photogrammetry combined with conventional data such as plans, documentation, and knowledge of renovation and history specialists as well as feedback from former and current occupants. To build a model that can be evolving and participative for future users and visitors to come. This research aims to provide an innovative approach to integrating a precise, reliable, and applicable methodology for buildings with heritage value.

**Keywords:** BIM for renovation · Sustainable digitalisation · Compatibility of models · Data interoperability

## 1 Establishment of Procedures

### 1.1 Context

Globally, the construction industry is the most significant contributor to carbon emissions and global warming [1], yet it is one of the fastest-growing industries in Europe and most other countries. In Switzerland, according to the Swiss Federal Office of Energy (SFOE), the building stock is responsible for about one-third of $CO_2$ emissions and CHF 60 billion annually is invested in the construction sector [2]. Of this investment, between 3 and 5 billion is spent on error correction on construction sites in Switzerland per year, according to the president of Bauen Digital Schweiz, Markus Weber [3].


M. Blanco Lage et al. (eds.), *Towards a New European Bauhaus—Challenges in Design Education*,
https://doi.org/10.1007/978-3-031-40188-6_27

In this context, it is surprising how timidly the industry is integrating automation [4]. The interest in leading the construction industry towards a sector that reduces greenhouse gases aims at a circular economy of materials used to create a construction process and a building that integrates the values of sustainable development and is, therefore, a priority to achieve energy and climate transformation.

The renovation of buildings is named a priority issue for the Swiss building stock by the SFOE for its 2050 energy strategy [2].

Considering these factors and recommendations, this research project develops a method for surveying a building with heritage value, allowing digital modelling, and integrating ecological, economic, and social indicators.

While in Switzerland, the interest in applying the BIM methodology is mainly focused on the construction process of new buildings, this research aims at the renovation process.

Traditionally, renovation projects are carried out based on conventional surveys and plans of the existing building, which are often dated and imprecise in terms of the changes they undergo during the life cycle of the building. The requisition of data linked to incomplete, erroneous, or outdated documents can lead to additional costs and delays during the renovation process.

To meet these challenges, this project proposes to optimise the pooling of requisitioned data of a building to be renovated to allow easy management of the renovation process for all involved parties.

The targeted reduction of data and the agreement on a common language between the different parameters will facilitate the process of finding a common denominator and thus allow the partners of the team involved in the construction process to work together efficiently.

### 1.2 Research Methods

The project employs an applied research methodology for a scan-to-BIM approach as suggested by Rocha et al. [5], providing conclusions and recommendations directly applicable to other construction actors seeking to renovate a property with heritage value.

The methodology is applied to a case study of a former beer brewery industry site, "Halle Blanche", which will house the architecture faculty.

The first data acquisition was made in a specific room of the building, using an iterative process, and then the method was refined and applied to the whole building.

The research uses a combination of capture tools such as laser scan, drone, and conventional methods. These allow the integration of different aspects into the virtual model beyond purely geometrical information and facilitate the development of optimised solutions.

Interdisciplinary scientific support has been provided by Smart Living Lab (SLL) of the Swiss Federal Institute of Technology in Lausanne (EPFL), the Chair of Circular Engineering for Architecture (CEA) of the Swiss Federal Institute of Technology in Zurich (ETHZ), external experts and an architectural historian of the city.

**Project planning**. The research was subdivided into five work packages (Fig. 1). In the first step, work package 1 (WP1), a metadata model was defined, and modelling standards were established. Communication protocols and collaboration objectives for the multidisciplinary work team were defined. A critical look was taken at a targeted selection of data to avoid an excessive and unmanageable amount of data.

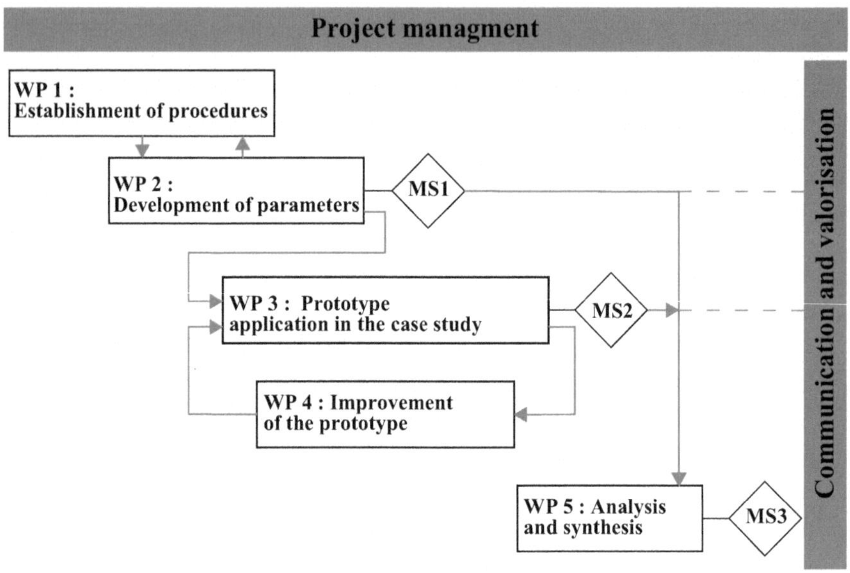


**Fig. 1** Workpackages 1–5

In parallel and continuous exchange with the progress made in WP1, a detailed model was defined in work package 2 (WP2), which was used to develop the parameters. This allowed identifying software solutions, data exchange possibilities, and network environment specifications. The main issue of this step was to make a judicious choice of tools and to base the selection of these on reliable values.

In the third work package (WP3), data collection, data pooling and modelling of a virtual model were primary tasks. This WP3 worked in continuous exchange with the fourth work package (WP4). WP4 consisted in making adaptations and modifications of the models. The first model was limited to one room of the "Halle Blanche", tested, improved, and then applied to the whole building.

In the last step fifth work package (WP5) it is planned to analyse the process and synthesise it. The methodology will be documented, a workflow manual will be developed, and the virtual model of the case study will be finalised.

**Research Topics**. Throughout the research, four topics (Fig. 2) were addressed in different experiments. In the Scan 3D topic, different approaches with laser scans were conducted. Various paths to integrate sustainable development indicators were looked at in the SD indicators topic. The interoperability topic focused on testing data exchanges between platforms and multiple programs. Different ways to make the included data visible and accessible were explored for the visualisation topic.

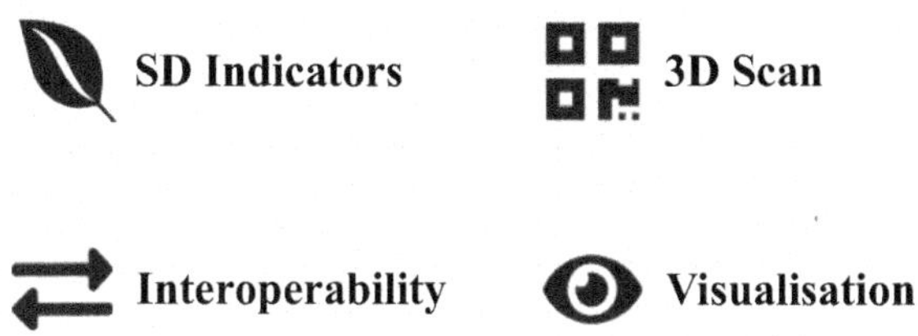


**Fig. 2** Research topics addressed in various experiments

## 2 Development of Parameters

### 2.1 Sustainable Development Indicators

The first experiment for integrating sustainable development indicators into the BIM model consisted in finding the standard for sustainability most appropriate, extracting one example and integrating all necessary information into the model. The Swiss Sustainable Development Standard, SNBS [6] was chosen as it considers criteria comparable to most common Swiss and worldwide standards.

Table 1 shows the list of various indicators and what information is necessary to integrate them into the BIM. Some indicators may only require a view template of a plan as input while others need to be calculated and therefore have values such as currency, energy, mass, or surface to be added to the model.

A more detailed look is taken at the natural light indicator in the visual and acoustic comfort category of societal indicators. Table 2 shows the enumeration of information for this indicator. The data type varies from text, surface, distance, percentage, angle, and lists in different IFC building components such as ifcProject, ifcBuilding, ifcSpace and ifcWindow.

Regarding the different programs used to enable such a data transfer, this research used Revit [7] with Speckle [8] to import pieces of information into Excel sheets, for example. Therefore, it was necessary to acquire precise information on the size, position, and number of windows in the building, which was achieved through various LIDAR scans as shown in the following chapter.

**Table 1** Examples of how to imbed sustainable development indicators into the BIM

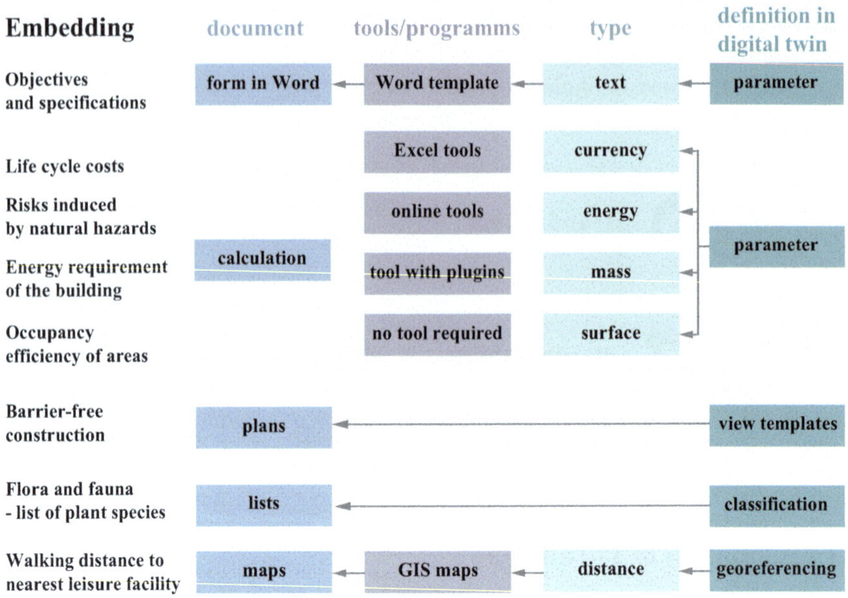


**Table 2** Excerpt of values for natural light indicator as inputs in Revit

| Category | Name | Data type | Unit | Enum | Applicability |
|---|---|---|---|---|---|
| Object | Project name | Text | NA | | ifcProject |
| | Owner | Text | NA | New com | ifcProject |
| | Architect | Text | NA | | ifcProject |
| | Date | Text | NA | | ifcProject |
| Summary | Total net area | Surface | $m^2$ | | ifcBuilding |
| | Main surface | Surface | $m^2$ | | ifcBuilding |
| | Main surface in | Surface | $m^2$ | | ifcBuilding |
| Types of rooms | Name premise | Text | NA | | ifcSpace |
| | Allocation | Text | NA | List | ifcSpace |
| Dimensions | Length facade | Length | m | | ifcSpace |
| | Height | Length | m | | ifcSpace |
| Use of daylight | Glass surface | Piece | NA | | ifcWindow |
| | Glazing transmission | | | Clear, normal | ifcWindow |

# 3 Application in the Case Study

## 3.1 3D Scan

In the first step, the 3D Scan focused on one designated room of the building to test and then implemented the methodology found [5]. A Light Detection and Ranging (LIDAR) scan with a FARO Focus S and scans using mobile phones for smaller remarkable objects were used for the interior of the building [9]. Furthermore, using a drone for the envelope and a test with a drone for inside photogrammetry helped improve the digital twin. The data combined with conventional methods of data requisition informed the BIM model of the "Halle Blanche" and allowed correcting any deviation (Fig. 3). The LIDAR scan allowed an adjustment of the original plans as seen below the red arrows in Fig. 3.

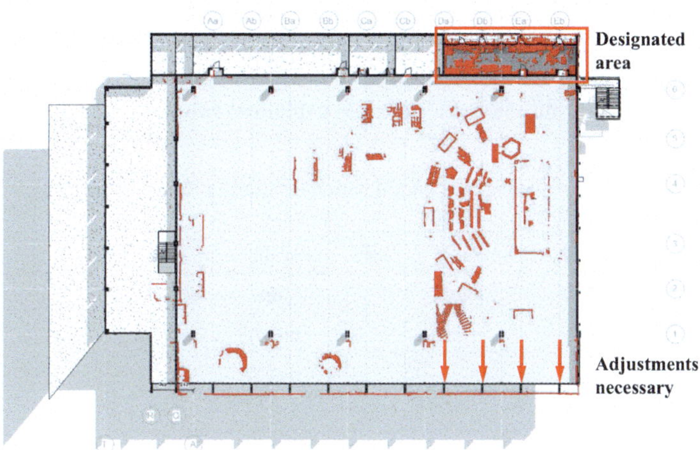


**Fig. 3** First-floor point cloud overlapping Revit plan to check for improvements

The density of points in the designated area as well as for smaller objects was heightened, to allow a more precise reconstruction of the elements this research focused on. The different densities can be seen in Fig. 4.

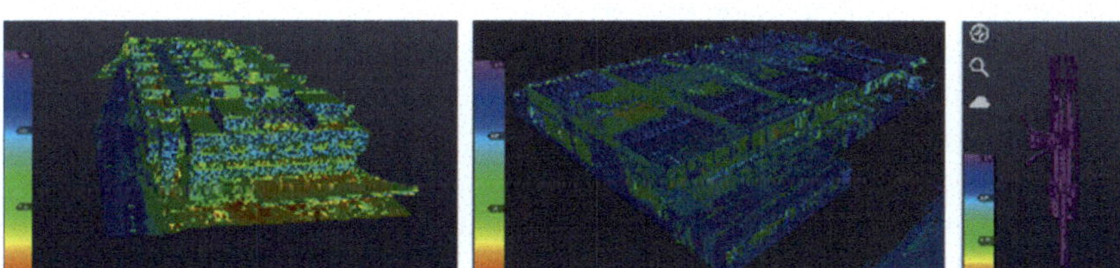

**Fig. 4** Density comparison left: drone outside point-cloud, middle: Faro first floor point-cloud, right: iPhone scan of crank handle for blinds

The various densities allowed an adjustment of the modelled elements of the test room and gave insight into the inaccurate databases of original analogue plans. To further inform the model, building-specific parameters have been added and data collected as further explained in Sect. 4.

## 4 Improvement of the Prototype

### 4.1 Interoperability

Concerning interoperability, different experiments were conducted for extracting data for the sustainable development indicators and for exchanging the BIM model between Revit and other programs. Therefore, the georeferenced Industry Foundation Classes (IFC) model had to be set up to make the data transfer efficient and seamless [10]. A unique Parameter Set (PSet) for the relevant cultural heritage values was created as seen in Table 3 to include deterioration status, heritage significance, a link to pictures of the element, and materialisation. This allowed the model to be further refined and made ready to be exported for different end-users as explained below.

**Table 3** Parameter set for cultural heritage values of the listed building

| Revit family | Wall |
|---|---|
| Parameter | Value |
| Link to photo | https://… |
| Photo of material | http://… |
| Deterioration status | 1 |
| Heritage significance | A |

### 4.2 Visualisation

Various tests were conducted on the visualisation of the building so different users could access different views of the digital twin. The idea is to allow digital twin users to visualise relevant aspects for them (Fig. 5).

The visitor model is exported so it can be viewed without specialised programs, a 3D walkthrough of the building with basic information about floors, rooms etc. The owner model has access to more information, for example, used or unused surfaces, heated or non-heated parts of the building, and similar relevant links. The participatory user can be an architect or engineer and needs work-related information, such as the dimensions of a load-bearing beam, the connections between elements etc. A model for technical maintenance gives the user a link to the deterioration state of an element to be able to schedule necessary rehabilitation work. The different models are shown in Fig. 6a–d.

The original BIM model has different declinations accessible through various programs or platforms; using IFC standards and building specific PSets is a prerequisite.

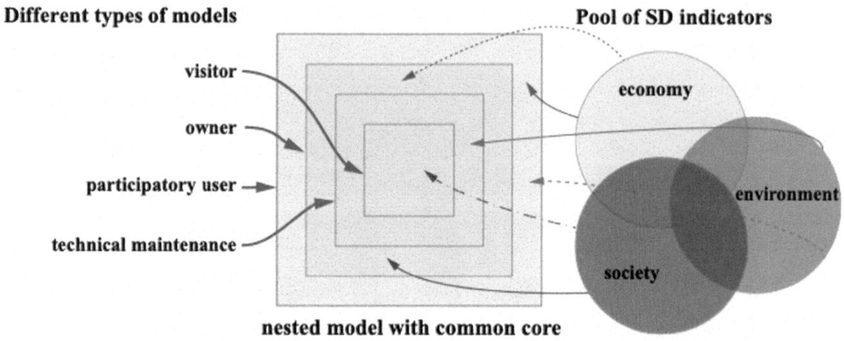


**Fig. 5** BIM model scheme, different models, include various indicators and targeted information for each user

The different users can therefore select the information that is most useful to them. The visitor model (Fig. 6a) is accessible through a Vectorworks viewer [11], the owner (Fig. 6b) has direct access through Speckle, from which he can choose to highlight heated or non-heated areas of the building or extract an excel sheet with all net-areas of the building, the participatory user (e.g. engineer) looks at the structure (Fig. 6c) through BIMvision [12], and the technical maintenance person has a direct link from an object in the model to a picture of its current state (Fig. 6d). The different ways in which the original model can be used and implemented into the various workflows of the users is expected to facilitate rehabilitation, communication, and maintenance of the heritage building.

# 5 Analysis and Synthesis

## 5.1 Expected Impacts in Scientific, Technological, and Socio-economic Terms

Currently, the data from the surveys of buildings to be renovated often need to be coordinated, leading to incomplete results. The research aims to provide an exemplary and complete method for managing data flows efficiently and accurately between the various players in the construction field.

Having all the information about the existing building as a virtual twin allows for undertaking an informed renovation and saving costs, as variants can already be studied with the help of the virtual model. The advantages of an informed virtual model compared to traditionally used plans are numerous. The renovation will be plannable with constructional and design choices that can be tested on the virtual model before construction, thus avoiding changes on-site and saving time and costs in the construction phase. The possibility of optimizing the rehabilitation virtually and before the beginning of the construction process is expected to improve the sustainability factors, as different scenarios can be tested, compared, and sound choices made. Implementing a BIM process in rehabilitating a heritage building can optimize the construction process. The most carbon–neutral version throughout the building's lifecycle can be chosen based on thorough simulations in the digital twin.

a

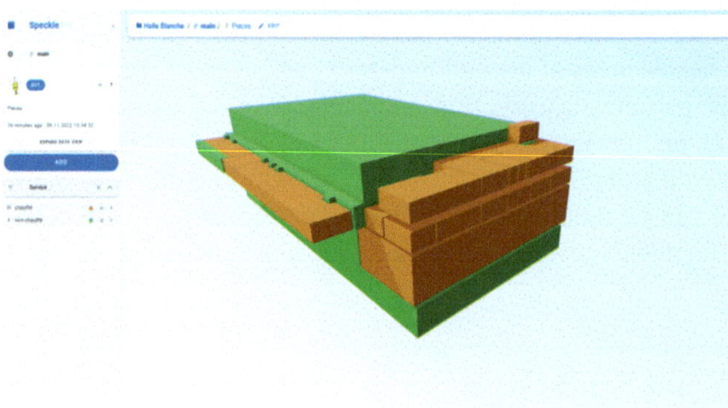


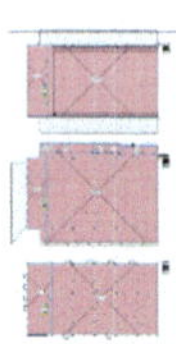

| SIA - Surface Nette Subdivision> | | | | |
|---|---|---|---|---|
| A | B | C | D | E |
| Numéro | Nom | Surface nette SN - Extrne nette SEN | Subdivision de la surface nette | Surface |
| N2-03 | Escalier | SN : Surface nette | SD : Surface de dégagement | 15 m² |
| N0-03 | Escalier | SN : Surface nette | SD : Surface de dégagement | 15 m² |
| N0-04 | Escalier | SN : Surface nette | SD : Surface de dégagement | 14 m² |
| SD : Surface de dégagement: 3 | | | | 44 m² |
| N1-05 | Balcon | SEN : Surface externe nette | SEU : Surface externe utile | 25 m² |
| SEU : Surface externe utile: 1 | | | | 25 m² |
| N0-01 | Halle Livraison | SN : Surface nette | SUP : Surface utile principale | 1913 m² |
| N1-01 | Halle Embouteillage | SN : Surface nette | SUP : Surface utile principale | 2538 m² |
| N1-02 | Pièce test | SN : Surface nette | SUP : Surface utile principale | 79 m² |
| N1-03 | Bureau | SN : Surface nette | SUP : Surface utile principale | 119 m² |
| N1-04 | Bureau | SN : Surface nette | SUP : Surface utile principale | 373 m² |
| N2-02 | Bureau | SN : Surface nette | SUP : Surface utile principale | 354 m² |
| N0-02 | Bureau | SN : Surface nette | SUP : Surface utile principale | 428 m² |
| N2-05 | Bureau | SN : Surface nette | SUP : Surface utile principale | 34 m² |
| N2-06 | Bureau | SN : Surface nette | SUP : Surface utile principale | 37 m² |
| N2-07 | Circulation | SN : Surface nette | SUP : Surface utile principale | 13 m² |
| N2-08 | Circulation | SN : Surface nette | SUP : Surface utile principale | 19 m² |
| SUP : Surface utile principale: 11 | | | | 5906 m² |
| SS-01 | Sous-sol | SN : Surface nette | SUS : Surface utile secondaire | 2275 m² |
| N2-01 | Balcon intérieure | SN : Surface nette | SUS : Surface utile secondaire | Non placée |
| SUS : Surface utile secondaire: 2 | | | | 2275 m² |

b

**Fig. 6** **a** Visitor model—rendered walkthrough. **b** Owner model—top: heated and non-heated areas (orange/green) —bottom: net areas detailed list. **c** Participatory model—structural elements. **d** Technical maintenance model—picture link to the original material from the model

This project responds to the essential need for methodological efficiency in acquiring and using digital data on existing built heritage. Public authorities and a large panel of construction professionals will find an approach scientifically approved and validated

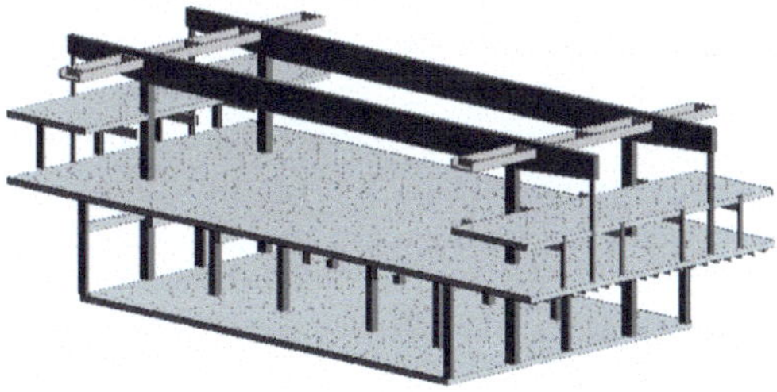

c

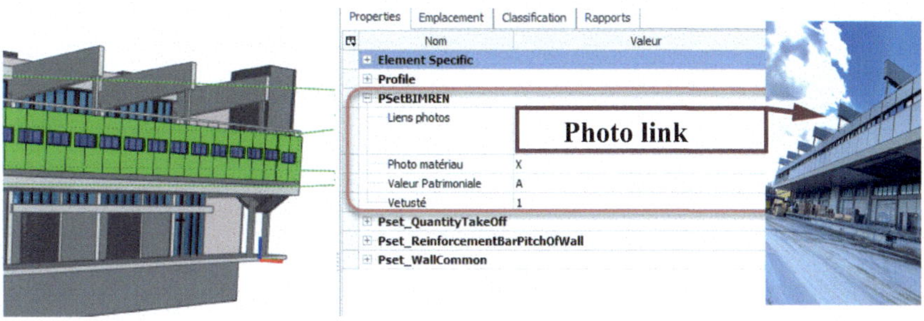


d

**Fig. 6** (*continued*)

by experts in the methodology developed. As the practice of digital data acquisition is still in the exploratory stage in Switzerland, no homogeneity has yet been accepted by all the players in the field. This research clarifies the procedure through an operational best practice reference framework that can be used throughout the life cycle. Ultimately the BIM model itself will be made sustainable.

## References


1. Mellado F, Wong PF, Amano K, Johnson C, Lou ECW (2020) Digitisation of existing buildings to support building assessment schemes: viability of automated sustainability-led design scan-to-BIM process. Archit Eng Des Manage 16:84–99. https://doi.org/10.1080/17452007.2019.1674126
2. Swiss Federal Office of Energy (2022) Gebäudepark 2050-Vision des BFE. https://pubdb.bfe.admin.ch/de/publication/download/8985. Last accessed 15 Nov 2022
3. Weber M, Schock B, Curschellas P, Jost A (2022) Building SMART homepage. https://bauen-digital.ch/dc/. Last accessed 21 Nov 2022
4. Alizadehsalehi S, Hadavi A, Huang JC (2020) From BIM to extended reality in AEC industry. Autom Constr 116: 103254. https://doi.org/10.1016/j.autcon.2020.103254
5. Rocha G, Mateus L, Fernández J, Ferreira V (2020) A scan-to-bim methodology applied to heritage buildings. Heritage 3:47–65. https://doi.org/10.3390/heritage3010004
6. Mark K, Ackermann S, Zeifang H, Kastner F (2021) NNBS Homepage. In: Standards et labels de la construction durable en Suisse. http://www.nnbs.ch. Last accessed 11 Nov 2022

7. Autodesk (2000) Revit. https://www.autodesk.com/. Last accessed 28 Nov 2022
8. AEC Systems Ltd (2021) Speckle. https://speckle.systems/features/connectors/. Last accessed 28 Nov 2022
9. Adán A, Quintana B, Prieto SA, Bosché F (2018) Scan-to-BIM for 'secondary' building components. Adv Eng Inf 37:119–138. https://doi.org/10.1016/j.aei.2018.05.001
10. Hooper E (2019) IFC (Industry Foundation Classes) properties in Autodesk Revit. https://www.bondbryandigital.co.uk/ifc-industry-foundation-classes-ifc-properties-in-autodesk-revit/. Last accessed 24 Nov 2022
11. Nemetschek (1985) Vectorworks. https://www.vectorworks.net/en-GB. Last accessed 28 Nov 2022
12. Datacomp (2022) BIM vision. https://bimvision.eu/. Last accessed 28 Nov 2022

# Spatial Conditions of Collectivity in the South-European Context

Mia Roth-Čerina(✉)

Faculty of Architecture, University of Zagreb, Zagreb, Croatia
mroth@arhitekt.hr

**Abstract.** The issues raised within the New European Bauhaus directly point to the question of the collective, both as a spatial as well as a social notion. In its physical manifestation, the spatial imprint of collectivity has a direct relation to the cultural and climactic context. While many implications of answering these issues can be shared regardless of geographical position, when talking about the physical framing of collectivity we can find common denominators among countries in southern Europe, in their traditions, way of life, relationship to open space, but also challenges facing urban expansion, heat islands, pollution, contemporary and active approaches to heritage etc. These were explored in one of the six conferences of the NEB goes South platform, an alliance of six South-European schools of architecture united in tailoring the New European Bauhaus policies to architectural education. The proposal investigates spatial conditions triggering appropriation by communities in both historic and contemporary South-European environments, as well as examples of embedding these topics within architectural education. The notion of togetherness and its various spatial expressions are illustrated in Croatian case studies, where built and designed spatial interventions are viewed as social acts, addressing inclusivity, and extending the notion of sustainability toward one that is societal, cultural, serving all species, bringing us into a shared productive ecology.

**Keywords:** Collectivity · Togetherness · Public space · South Europe

## 1 Introduction

Accomplishing goals of new European policies encompasses sustainable, inclusive and beautiful solutions as they were formulated in the New European Bauhaus initiative. However, looking at underlying meanings, strategies or tactics of their achievement, differentiation between geographic, climactic, cultural and historic contexts ask for a tailored approach. Prompted by the recognition of specificities in South-European contexts, six architectural schools from Portugal, Spain, France, Italy, Croatia, and Greece, came together to formulate the NEB goes South platform. "Countries in Southern Europe are dealing with, among other problems, increased signs of extreme climate events, rising temperatures, heat islands, desertification and drought. They also share rich histories and a cultural heritage which need careful protection while addressing these emergencies" [1]. Exploring possible answers to these challenges through a series of introductory


M. Blanco Lage et al. (eds.), *Towards a New European Bauhaus—Challenges in Design Education*,
https://doi.org/10.1007/978-3-031-40188-6_28

events and conferences [2], the platform focused on themes particularly relevant to relating current concerns with specificities of the southern contexts, mentalities, heritage, resources. These discussions, held in 2021 and 2022, can be found on the platform's YouTube channel and have informed a subsequent course of action.

The third NEB goes South conference was hosted by the Faculty of Architecture at the University of Zagreb in October 2021. Titled "Framing Collectivity", it explored the historic patterns of public and communal spaces; the specific conditions of warm climates making the interstitial space of the city a stage for public life; the effects of climate change on 'lived space' [3]; active reuse going beyond reprogramming to establish new identities, and contemporary expressions of togetherness in architecture and design education. This paper presents inquiries that informed the conference, as well as its conclusions aimed at future directions in examining collective spaces and the policies that condition it. The various expressions of collectivity are presented through case studies of architecture and educational content situated in Croatia.

## 2 Discussing Collectivity

The issues raised within the New European Bauhaus directly point to the question of the collective, both as a spatial as well as a social notion. In its physical manifestation, the spatial imprint of collectivity has a direct relation to the cultural and climactic context. The topics of togetherness and collectivity are acute and have become even more so during the past year. Though Hashim Sarkis proposed the topic of the Venice Biennale planned for 2020 [4] before the world went into lockdown, the overwhelming response and variety of approaches to the question of "How will we live together?" show this is a burning issue still aligned with Sarkis's original provocation: regarding our individual selves and longing for belonging be it physically or virtually; regarding our everydayness, places we live and work in; the common and public spaces we share in cities; the notions of accessibility and equality; the challenges that more-than-human collectives face in front of climate emergency. Answering these prompted an increasingly frequent use of the term 'togetherness' describing a condition of caring coexistence, enriching the notion of collectivity related to a predominantly human-oriented experience of being and acting together toward a state which relates to social, cultural, environmental, interspecies and planetary care. It also relates to Donna Haraway's notions of kinship, tentacular thinking, of making and being with others [5].

While many implications of addressing conditions of coming together can be shared regardless of geographical position, when talking about the physical framing of collectivity we can find common denominators among countries in southern Europe, in their traditions, way of life, relationship to open space and nature, but also challenges facing urban expansion, heat islands, pollution, contemporary and active approaches to heritage etc. Examining spatial conditions of framing and triggering collectivity, as well as the creation of space as a collective, inclusive endeavor, was a broad topic of the 'Framing Collectivity' conference, inviting architects, educators, researchers and students to approach the issue from various angles [6]. The talks investigated spatial conditions triggering appropriation by communities in both historic and contemporary South-European environments, as well as challenging the role of the architect and the contents of an architectural upbringing, posing the issue of design as a collective, inclusive process.

## 3 Spatial Expressions of Collectivity: Croatian Lessons

When talking about inclusivity and extending the notion of sustainability toward societal and cultural sustainability, we inevitably talk about the communities which make these goals possible, or inhabit the spaces designed to attain them. How can we frame them? What challenges does the contemporary way of life pose on the meaning of collectivity, of public space, of togetherness? An array of case studies is approached answering these questions from various facets, addressing possible meanings, histories and futures. Croatian examples were selected as a sample of shared South-European ambiances, based on how they represent the criteria of clearly representing a legible spatial pattern which coincides with behaviors of collectives, triggers their appropriation, or embodies expressions of nature and culture which communicate identities. It also looks at examples in which the architect acts as a public agent through the design itself, a medium communicating these inherited identities or facilitating appropriations in decisions which depart from solely rational or technologically sustainable solutions.

Spaces we share can stem from a myriad of functions or origins, and the forms they entail have become increasingly complex. Alongside past straightforward thresholds between the private and the public, the institutional and the open, the discussions and case-studies explored disappearing dichotomies of public–private, creator-consumer, extending the notion of commons or platforms of togetherness, or how we address collective identities in a broader sense. Notes on case-studies present illustrations of the various meanings of collectives in our shared spaces, how they were inherited, to what end they are reinterpreted today, how they relate to the notion of representing identities belonging to multispecies communities, or how symbolic meanings translate to reprogramming or reformulating inherited expressions of commons.

*Traditions.* Historically, spatial expressions of commons are self-evident and deeply embedded into South-European urban culture (Fig. 1). The specific conditions of warm climates made the interstitial space of the city the stage of public life, and this is legible even in minimal interventions such as the stone street bench or the communal loggia, and present in the spatial articulation of different scales: a house with its atria and eaves, public buildings with their colonnades, or narrow streets. Specific configurations intertwining these scales or public and private thresholds are a marking character of historic Southern towns. Mixed uses embody the essence of collective encounters and sharing spaces, such as the blurred border between the communal and the domestic where private property is relinquished to create a covered arcade—as Bernard Rudofsky described it: altruism turned architecture [7]. Investigated here are precisely these minimal traces of collectivity legible in historic patterns of public and communal spaces in Croatian traditional environments, but also typical for southern Europe: the built-in stone bench belonging to a neighborhood, the loggia used for communal discussions, covered from rain or sun, a self-explanatory space belonging to all.

*Architect as public agent.* Translating spatial interventions as social acts into now, how can we ensure this framing will not only meet the need, but instigate a need of coming together? The changing programs of social condensers affect the thresholds toward public space, and a recent example designed and built by the office X3M, show the elementary school of Pazdigrad (a neighborhood in the city of Split) placed in the middle of a typical heterogenous, unconsolidated environment where the school acts as a binder,

**Fig. 1** The stone bench. Left: Orebić, Pelješac peninsula. Right: Korčula, island of Korčula. *Photos* Author, 2022

introducing accessible public space in the in-between on different levels, thereby creating so many different public environments their surface exceeded the original plot size [8] (Fig. 2). These public spaces are connected to social spaces of the school, extending then into a double use even further, permeated by citizens regardless of when the school is open, enabling a new perspective of the surroundings. The approach communicates the agency of an architect when designing didactic environments in an unregulated post-transitional context, outgrowing the mere functional, aesthetic or technologically sustainable performance to embed new meaning for a neighborhood, adding a layer not asked for by the program but clearly representing the most prominent value of the intervention to a broader societal group.

**Fig. 2** X3M (Bošnjak, Buvinić, Furlan Zimmerman): the public spaces of the Pazdigrad school, Split. *Photos* Bosnić + Dorotić, 2017

*Shared identities.* Beyond the spatial imprint—or framing—of commons, collectivity can be examined as a question of identity. The visitor infrastructure in the Lonja wetlands in central Croatia establishes a dialogue between anthropological expressions and multispecies habitats, learning from existing patterns of nature and culture to use them to a different end. The visitor center samples rural morphologies, creating a new social focus, and tells the stories of the flora, fauna and cultural heritage. It repeats the scales of surrounding homesteads, placing an info-point and administration in one house, an exhibition and auditorium in the other. The two buildings frame a public square defined by deep trapezoid porticos and denivelations, initiating a new origin of rural communal life. The public space surrounded by these recessed covered spaces outlines an 'open room' and leads to a visitor center and a communal hall (Fig. 3). Inside, the spaces are dominated by sculptural installations connoting swarms, flocks, driftwood. The other part of the infrastructure, observatories within the nature park, are a new species in the landscape which relates to various species and forms. Going beyond the interpretation of an environment protected for its symbiotic equilibrium of humans, animals and nature, it raises the notion of interspecies collectives, of a respectful cohabitation, of using subliminal identities to raise awareness of a protected environment in danger. Although acting as an attractor, their purpose is to invite familiarizing oneself with a fragile equilibria we are a part of.

**Fig. 3** Roth&Čerina: the visitor infrastructure of the Lonja wetlands. Left: Osekovo visitor center. Right: observatory. *Photos* Marko Mihaljević, 2021

*Reframing*. Relating collective identities to existing spaces raises a particular approach to re-use. While South-European historic buildings and towns also share the increasingly complex issues of renewal which on one hand must meet current technological and energetic standards and on the other retain their original significance and language, they are faced with issues stemming from climate change, asking for intensifying green infrastructures or shade, mitigating urban heat islands etc. Furthermore, a specific issue in transitional contexts is the approach to the preservation and activation of modernist heritage (Fig. 4). Beyond the issue of advocating for their protection in the first place, a former political system they may represent faces a contested identity, and along with legislative knots burdening them, together they present an often-unsurmountable obstacle preventing their objective evaluation or subsequent renewal and re-use. They present a case study in themselves, a network of attractors to test reprogramming, continuously inviting engagement by artists, photographers, historians, a constant subject matter within education, yet remaining unused.

## 4 Collectivity in Architectural Education: Case Studies from the Zagreb Faculty of Architecture

Various aspects in approaching the notion of collectivity translate to the contemporary goals of architectural education: teaching agency, working collaboratively, and designing for—and with—society. They expand on the issues presented in the case-studies,

**Fig. 4** Left: Children Sanatorium by Rikard Marasović, Krvavica. Right: Hotel Pelegrin by David Finci, Kupari. *Photos* Author, 2020 and 2022

approaching them in the broadest sense, and relate not only to content but also the processes and places in which architectural education is set in. Several examples conducted through collaborative studios or extracurricular workshops at the Faculty of Architecture, University of Zagreb, overlap issues of transitional backdrops with specific South-European contexts. The post-socialist transition after the demise of the Yugoslav republic and subsequent war lead to a change in values and identities, where numerous spatial expressions of the former state's ambitions were left to ruin due to their embedded meaning, and the invisible infrastructures of the welfare state have yet to be replaced. Post-war migrations led to demographic declines and changes in social infrastructure. Alongside these, fates shared with other South-European countries grappling with consequences of touristification and climate change ask for tailored strategies aimed toward social sustainability. The presented case studies explore these topics from various approaches and goals. The examples explore the meaning and reframing of social infrastructure (such as schools) in rural communities; the programmatic shifts which could enable resilience of towns facing the eroding consequences of one-sided seasonal exploitation of historic environments; the capacity of social infrastructure to revitalize abandoned urban cores through reprogramming. Their lessons exceed their content, aimed at raising agency of the future architect.

*Learning spaces in transition.* A workshop done in collaboration of the Faculty of Architecture and the School of Design with UNICEF dealt with rural schools, exploring strategies of spatial rearticulation to reduce and modernize educational space of depopulated environments while creating scenarios to activate the community, bringing the students into the role of not only designers but almost social engineers. Through design they investigated spatial and programmatic scenarios to reformulate spaces whose reuse does not stem from their architectural quality, rather their presence in identity which could yet bring new meaning to communities. A three-level project task was formulated, which included a diagnostic one on how to improve the state of rural-school buildings carried out through research and immersive on-site collaborations, a designerly level which explored possible reformulations of space to adapt it to smaller class sizes while achieving additional communal space of gathering, and finally short and long-term strategies for community cooperation through reinterpretation of local traditions and resources [9].

*Sustainable Tourism.* Inclusivity and sustainability, two of New European Bauhaus' pillars, are related and can be observed from the aspect of environmental and social sustainability. Relating them to the realm of the public, a particular challenge of the South-European contexts is its touristification resulting in a manic-depressive trajectory of public life that takes place over the course of the seasons and increasingly erases the local everydayness. Politecnico di Torino and the University of Zagreb conducted a joint workshop in the small Istrian town of Motovun, an environment of intense film and food festivals with quiet, less populated intermediate periods. Departing from local patterns, but aware that irreversible population migrations were not to be ignored, the workshop explored speculative scenarios for a continuous inhabitation by another kind of dwellers, in parallel with considering reuse and reprogramming.

*Reclaiming the core.* A particular challenge in re-densifying cities is returning social infrastructure, or reclaiming the core with public programs of larger scales and contemporary requirements. Two studios carried out at the Faculty of Architecture, University of Zagreb, explored scenarios of alternative placement of educational programs to activate everyday life in the city center (Fig. 5). The Dispersed School Studio explored possibilities of reuse and proximity by dissecting the school's functional parts and creating an educational super-neighborhood making use of empty structures and in-between spaces. The School Forum Redux studio developed scenarios of grafting and adding generic educational spaces to historicist complexes, based on multiple scenarios and possible networks, following intense cooperation with a neighboring high-school struggling to achieve contemporary learning environments in their nineteenth century setting [10].

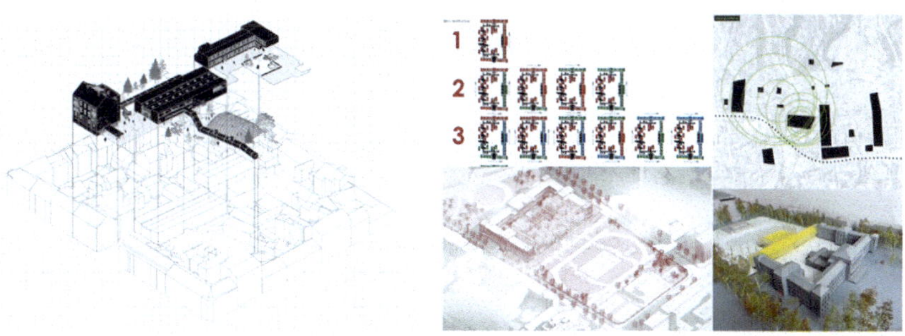


**Fig. 5** Left: The Dispersed School (students: D. Leko, M. Penava, I. Panza, mentors: V. Rister, M. Roth-Čerina), 2021, Zagreb. Right: The School Forum Redux (students: J. Fabijanac, D. Kolonić, F. Kovačević, mentors: V. Rister, M. Roth-Čerina, I. Martinis, J. Botušić Brebrić), 2022, Zagreb

## 5 Conclusion

The intent of the talks conducted on the topic of collectivity within the NEB goes South platform, as well as an array of examples serving to frame the notion of collectivity from the broadest spectrum of possible meanings, blends the foundations of achieving a spatial and social framework serving society. It entails the roles of the collectives acting in space as well as the spatial conditions triggering collectivity. Lessons found in spatial patterns of collectivity in the vernacular, identities legible in tacit collective memories, collaborative practices of space making, expanding the notion of the collective to more-than human entities, all entail a broad spectrum of meanings which lead to a common aspiration of togetherness.

Educational environments carry a particular responsibility in this regard, as we can look into ways of stronger bonds with the 'outside', raising students carrying agency, recognizing and interacting with the issues surrounding them. This asks for reframing and strengthening collectives within schools, blurring its boundaries, developing collaborative conditions, reaffirming the school as part of a cultural and societal collective.

Recognizing context-specific issues point to a shift in the focus and processes in which architectural education is carried out and dissolve the border between the school as an institution and its static object-oriented outcomes. 'Making with', as Donna Haraway describes it, rather than 'self- making', puts the notion of collectivity at the center, directing content and goals of education and practice, ultimately changing their value systems.

## References

1. NEB goes South (2021) Vision. Faculty of Architecture of the University of Porto; Valencia School of Architecture; Toulouse School of Architecture; Department of Architecture, Alma Mater Studiorum, University of Bologna; Faculty of Architecture, University of Zagreb; School of Architecture of the National Technical University of Athens. https://www.up.pt/neb-goes-south/vision/. Last accessed 26 June 2022
2. NEB goes South (2023) Events. https://www.neb-goes-south.pt/events. Last accessed 25 Mar 2023
3. Lefebvre H (1992) The production of space. Wiley-Blackwell, London
4. Sarkis H (2019) How will we live together? 17th Venice Biennale introductory statement, https://www.labiennale.org/en/architecture/2021/statement-hashim-sarkis. Last accessed 28 June 2022
5. Haraway D (2016) Staying with the trouble: making Kin in the Chthulucene. Duke University Press
6. Framing Collectivity. https://www.youtube.com/@nebgoessouth9904/playlists. Last accessed 28 June 2022
7. Rudofsky R (1964) Architecture without architects, an introduction to non-pedigreed architecture. Museum of Modern Art, New York
8. Glažar T (2018) A space for learning and creating togetherness (Žnjan-Pazdigrad Elementary School). Oris 111: 48–59 (Zagreb)
9. Fabrio I, Kapetanović Z, Rister V, Roth-Čerina M (2015) Architecture and design as tools of transforming the educational process in small communities. In: Vitale K (ed) Learning spaces in transition—typology for healthy learning environments. University of Zagreb, Faculty of Architecture, School of Design, UNICEF, Zagreb
10. Rister V, Roth-Čerina M (2022) Školski Forum Redux. University of Zagreb, Faculty of Architecture, Zagreb

# Multidisciplinarity in Action: Defining Collaborative Design

Federica Vannucchi[1(✉)] and Mia Roth-Čerina[2]

[1] School of Architecture, Pratt Rome Program, Pratt Institute, Brooklyn, NY, USA
fvannucc@pratt.edu
[2] Faculty of Architecture, University of Zagreb, Zagreb, Croatia
mroth@arhitekt.hr

**Abstract.** That architecture is a multidisciplinary practice is not novel. In western history, architects describe their profession as an assemblage of expertise which, one might think, can hardly be performed by just one individual. This is the case in Vitruvius' *The Ten Books on Architecture* where he stated that an architect has to be conversant in matters related to geometry, history, philosophy, music, medicine, jurisprudence, astronomy, and the theory of the heavens. This is repeated today in the Article 46 dedicated to "Training of Architects" included in the EU's Professional Qualifications Directive 2005/36/EC where an architect should only be knowledgeable of a variety of disciplines other than architecture, but that the profession of architecture is to exist "in society," mediating "between people and buildings, and between buildings and their environment," responding to "human needs and scale". What this being "in society" means and demands is at the core of architecture remaking today. This research focuses on the modes of collaboration between architects and users, drawing on the grounded theory research of the "Architecture's Afterlife" research project funded by the Erasmus + Program between 2019 and 2022.

**Keywords:** Collaborative practices · Communities · Inclusion · Architecture pedagogy · Architectural competence · Multidisciplinarity

## 1 Introduction

That architecture is a multidisciplinary practice is not novelty. In western history, architects describe their profession as an assemblage of expertise which, one might think, can hardly be performed by just one individual. This is the case in Vitruvius' *The Ten Books on Architecture* where he stated that an architect has to be conversant in matters related to geometry, history, philosophy, music, medicine, jurisprudence, astronomy, and the theory of the heavens [1]. This is repeated today in the Article 46 dedicated to "Training of Architects" included in the EU's Professional Qualifications Directive 2005/36/EC where an architect should understand fine arts, technology, human sciences, environmental issues, building regulations, just to name a few [2]. But an architect should not only to be versatile in a variety of different disciplines other than architecture, but—as the


M. Blanco Lage et al. (eds.), *Towards a New European Bauhaus—Challenges in Design Education*,
https://doi.org/10.1007/978-3-031-40188-6_29

Directive states as well—the profession of architecture is to exist "in society," mediating "between people and buildings, and between buildings and their environment," responding to "human needs and scale". What this being "in society" means and demands is at the core of architecture remaking today, yet, has taken on various meanings throughout the history of architecture practice and pedagogy.

If we were in a class of professional practice, we would simply answer that "society" is the client. But for architects (or society itself!) such a simplified answer is unsatisfactory. They have relentlessly questioned what defines the "society" that the profession is supposed to serve. Society has been renamed at any turn of architecture thinking, especially during modernity. The discussion of who is the user has also brought the problem of the relationship between user and space maker. This research focuses on the modes of collaboration between architects and users, or what the Directive calls "society". It does that by looking at the results of the Erasmus + program research project titled "Architecture's Afterlife: The multi-sector impact of an architectural qualification", a pan-European study awarded in 2019 [3] that seeks an answer for why 38% of European architecture graduates work exclusively or partially in creative and cultural professions other than architecture (Fig. 1). Through competencies mapping, the study identifies the multi-sector impact of an architecture degree and the extent to which competences taught to architecture students are needed in sectors other than architecture. Architecture's Afterlife reveals that, in their school years, architecture graduates are exploring collaborative, inclusive, interdisciplinary practices, while they also do evaluate the importance of personal competence and individual responsibilities in their practices. These skills are widely used in all kinds of expanded contemporary practice irrespective of how we define it, pointing to a shift of authorial definition as well as actively introducing a broader range of actors into the creative process.

## 2 The Notion of the Collective

### 2.1 Designing for Collectives

In history, "society" has been named in different ways depending on its role within the design and construction process. For instance: in the early twentieth century, architects understood society as the "mass" of people whose industrial progress demanded a process of standardization. In the 1950s CIAM discussions, society was "the greatest number" of individuals who might differ for habits and degrees of progress but were to ultimately conform to modern living through the application of modern architecture. Most recently, we no longer discuss society as a uniform entity but rather as composed by heterogenous parts. The term "communities" has become widely used to better describe people's cultural, racial, environmental specificity.

While questioning what is this "society" to serve, architects also have reflected on their own role within it. Today, the architectural history that we continue to tell, still complies with the architects' role that the avant-garde of the early twentieth century has defined; that is, the role of an intellectual guiding the masses of people. And yet, discussions on modes of collaborations between space makers and communities that they serve have been central to contemporary architecture discourse.

**Fig. 1** Percentages of architecture graduates in practice, combining practice with another field, working in a related sector and outside architecture

Parallel to the myth of an architect as the only creator, modern architecture has produced alternative models. The feminist experiments of the end of the nineteenth century and beginning of the twentieth century described by Dolores Hayden in her seminal text *The Grand Domestic Revolution* [4] are testimony of these. The 1960s have seen the definition of a participatory architecture. Among them, Giancarlo De Carlo's projects in Italy [5], and John Turner's "The Barriada Movement" [6] theorized a new collaboration between maker and user in the design and construction process.

That is to say that modern architecture has defined opposing roles for architects: on one side, architects are to be the only creator, on the other, the facilitator of a participatory process. Then—and here the main question of this article—how are these opposing roles taught in architecture schools in Europe? And do they translate to the architect existing "in society" today, as the Directives suggest?

### 2.2 Collective Practice

Since the early twentieth century, the teaching and practice of architecture have relied on the "avant-garde" model, that is, of a leading intellectual elite: the Fountainhead complex. As a consequence, architects do evaluate the importance of personal competencies and individual responsibilities in their practices. Yet, architecture graduates are today exploring different modes of practices, those that rely on collaborative, inclusive, and interdisciplinary exchanges. Space making is progressively turning into a shared process that actively involves all sources of actors relying on the various expertise needed by the construction process. But most importantly, space making demands the collaboration of the community for which it is destined. The demand to turn architecture into a consciously inclusive practice has brought to the fore questions regarding who is historically excluded and who is included in the process of making space.

The opening of the curriculum of the last few years to a global view of architecture practice has further questioned how architecture is, and has been, a shared experience, regardless of a certain persistence of naming a single author. Menna Agha, in her work on the Nubian House, explores the collaborative nature of construction, one that is superseded by women and performed by the men of the household [7]. While in the past collective work often has been treated as isolated experiments, enclosed in their own narrative, collectivity has changed and is changing architecture as both a field of knowledge and of practice.

## 3 Collectives and Collaboration as Discussed in the 'Architecture's Afterlife' Project

The 'Architecture's Afterlife's' project included an in-depth quantitative and qualitative study looking into the diverse trajectories of architecture alumni, aiming to identify the impact of an architecture degree within the context of Europe and the extent to which skills taught to architecture students are needed and used both in architectural practice and in other sectors. It sought to understand the skills gaps and mismatches between what is taught in architecture schools and what is needed by today's architecture practices as well as by other professions, industries and sectors, with the goal to identify opportunities for a multi-disciplinary and transdisciplinary curriculum that could more effectively serve student, labor market and societal needs.

Through its survey that attracted over 3600 responses and 48 in-depth interviews, it traces the various facets of the meaning of collective practice as well as serving collectives as described in contemporary practice. The Afterlife survey has highlighted that architecture graduates' most used skills are both 'personal competences' and 'cooperation competences' (Fig. 2a–c). The former include determination, work ethic endurance, handling criticism flexibility, constant learning and self-improvement, dealing with uncertainty and being able to function in conditions of uncertainty. The latter include working with clients, collaboration skills, and mediating skills. The in-depth interviews explore what kind of collaborative work is performed by the interviewees. During the school years, the most collaborative experience is certainly defined by the studio culture of architectural schools. In architectural practices, the communal aspect is measured as designing and collaborating within the working environment, as well as whether the work performed engages with communities, and whether it requires interdisciplinary expertise (thus a larger group of people) to be performed. At the same time, the interviews speak of the leading position and responsibility role they acquire in determining the formation of the built environment. Another aspect seems relevant here, which is often the self-learning process with which architecture students seem very familiar. This points to a contrasting

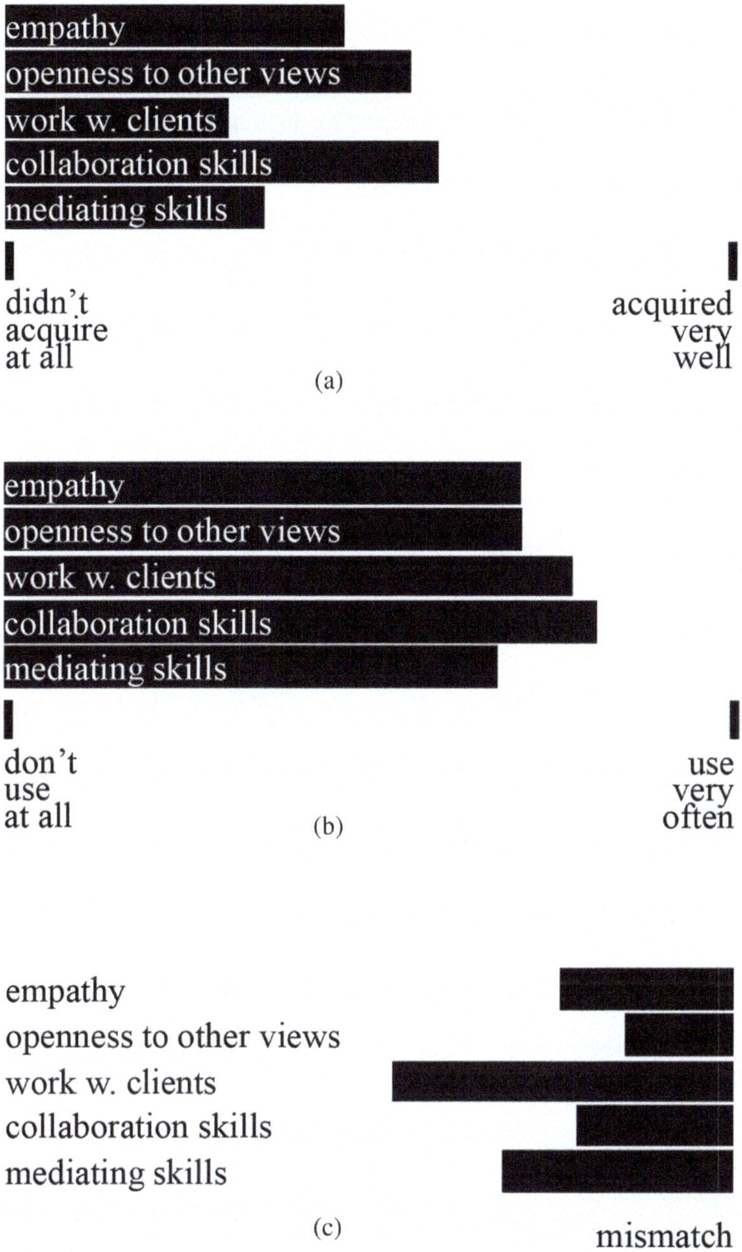


**Fig. 2** **a** Cooperation and diversity competences acquired during architectural education. **b** Cooperation and diversity competences used in current practice. **c** Mismatch between acquired and used

tendency: on one side, towards self-reliance; on the other, toward the necessity to learn by sharing with fellow students and colleagues.

The result of the quantitative analysis tells us that architecture graduates are exploring collaborative, inclusive, interdisciplinary practices. Yet architects do also evaluate the importance of personal competencies and individual responsibilities in their practices.

The Architecture's Afterlife Questionnaire has highlighted that the most used competences are both "personal competences" and "cooperation competences" but while the former ("personal competences") are adequately taught in school, the latter ("cooperation competences") are not. Instead, "diversity competences (empathy and openness to other views and ways of living)" are in between.

The in-depth interviews have further explored what kind of collaborative work is performed by the interviewees. Two aspects here are important: firstly, whether the work performed by the interviewees engages with communities; secondly, whether it requires interdisciplinary expertise, and a larger group of people to be performed. Both these aspects were often mentioned by the people interviewed, even if they are not currently engaged in architectural practice [8]. The majority of the interviewees described working collaboratively in some form.

The negotiation among people inevitably requires the use of various competences. This does not come as a surprise. In our western history, a fundamental idea of the architect's training is precisely the ability of mastering multiple fields of knowledge. Architecture's Afterlife public events and discussions have underlined the fluidity of the boundaries of architecture vis-à-vis the autonomy of the field. As professor at the California University Dana Cuff has suggested in one of these meetings [9], we are drawn to multidisciplinarity, and yet, "if architecture is not an autonomous operation, then what is it?" The question of architecture disciplinarity has persisted for the last seventy years (at least), and it is worth looking at why we are still pressed by this problem. While architecture schools comply with the idea that architects do need to learn how to design buildings, architects escape this definition by constantly reinventing their professional role. A possible answer was offered by Professor at the University of Brussels Hera van Sande who defined architecture as "a practice of change." Architects are actors of change both in reference to their own discipline as practitioners and to the world as makers. It is the very notion of architecture as a discipline of transformation that makes architecture education so versatile. However, this transformation has an ethical dimension. Following architect, activist and producer Felipe de Ferrari, an ethical transformation is an action that "reclaims what has been stolen by capitalism." Thus, the problem goes back to how architecture best serves societal needs.

The inherent interdisciplinarity of architecture is understood by looking at the answers of the in-depth interviews of Architecture's Afterlife. Particularly interesting are the answers given by graduates working in sectors different from architecture on how they perceive what they do in relation to their architecture training. When asked if they have ever left architecture, 75% answered that they do not think of themselves as having left architecture. Although working in fields different from architecture, they still understand their daily job as strictly related to their architectural training. When asked in what sector they are working in, the answers were unexpected: in architecture, in architecture combined with another field, in a sector related to architecture, in creative industries, in other sectors, and in interdisciplinary fields. These answers point both to the division of the notion of architect as a mindset, a habitus, and the architect as a professional qualification. It also subsequently showed that the skills of transversal cooperation were one of those used regardless of their profile or engagement in diverse fields of expanded practice.

## 4 Exemplary Practice as a Conclusion

What makes architecture graduates so open to other disciplines and to interdisciplinarity, does not simply rely on the skills that they have acquired but also on the working method that these skills have allowed them to form. The growing number of architecture graduates falling outside a traditional understanding of the profession, yet clearly employing skills attained in architectural education to achieve societal and cultural impact, point to the relevance of raising soft skills and behaviors uncovered and explored in both the qualitative and quantitative segment of the Architecture's Afterlife research.

In one of the interviews, the deputy mayor of a Croatian city stated: "I would say that the greatest benefit I got during my studies is an interdisciplinary approach, i.e. being open to all other experts and expertise, knowing how to learn from them and how to interconnect them. A city is an enormously complex mechanism." The Vice-director of an environmental association in Italy says that: "I think it is a good thing to be an architect. It's something that you can bring in practice, in the challenge that you have to face, it's an approach. […] An approach that always puts the problem or the target inside an environment, a landscape not only a natural landscape, but also an urban environment, and people." The director of a Belgian non-profit organization called Cultureghem said that "architecture gives you a toolbox which is invaluable. I'm almost like a translation machine between architects, engineers, university professors, […] and people, including homeless and disadvantaged people." For Cultureghem, founded in 2014 in Brussels, the "society" mentioned by the European Directives is the "community" to serve. In the very heart of the European Union, it looks at the disadvantages, such as migrants, single moms with their children, and the elderlies. Architecture becomes the means of communication and interaction. Most specifically, architecture negotiates how different people interact in space. In this, architecture not simply serves society, but makes this society form and grow. In fact, a growing number of meaningful, impactful practices today are collectives, challenging the notion of the single authorial figure. Subsequently, this raises the question of how we teach architecture for meaningful spatial agency, and how spatial conditions of creating for and with collectives are advanced through architectural education and practice. Finally, the notion of collaboration and the collective, in the many facets of its meaning, or 'society' in the broadest sense, is at the core of sustainable development as it is defined in global common goals: a prerequisite of our shared future.

## References

1. Vitruvius (1960) The ten books on architecture. Translated by Morris Hicky Morgan. Dover Publications, New York
2. Directive 2005/36/EC of the European Parliament and of the Council of 7 September 2005 on the recognition of professional qualifications. https://eur-lex.europa.eu/. Accessed 28 June 2022

3. Architecture's Afterlife: The Multi-sector impact of an architectural qualification (2022). Erasmus+ 2019–1-UK01-KA203–062062. Researchers: Dr. Harriet Harriss (Principal Investigator, Royal College of Art); Dr. Michela Barosio (Polytechnic of Turin); Dag Boutsen (KU Leuven); Dr. Johan De Walsche (University of Antwerp); Dr. Mia Roth Čerina (University of Zagreb); Dr. Carla Sentieri Omarrementeria (Polytechnic University of Valencia); Andrea Čeko (University of Zagreb); Dr. Haydée De Loof (Universiteit Antwerpen); Santiago Gomes (Polytechnic of Turin); Dr. Federica Vannucchi (Royal College of Art); Dr. Hanne Van Reusel (KU Leuven and University of Antwerp)
4. Hayden D (1981) The grand domestic revolution. MIT Press, Cambridge
5. De Carlo G (1972) An architecture of participation. Royal Australian Institute of Architects, London
6. Turner J (1977) Housing by people: towards autonomy in building environments. Pantheon Books, New York
7. Agha M (2022) The Nubian House: displacement, dispossession, and resilience. Making Home(s) in displacement: critical reflections on a spatial practice. Leuven University Press, Leuven, pp 327–346
8. "I'm almost like a translation machine between architects, engineers, university professors. And I'm smart enough to understand what they're saying, even without maybe getting all the nuances, but I'm equally comfortable talking to a homeless guy, having lunch with him who has come to eat cause he has no food or with the kids from the neighborhood who are bored as hell." (Director of an NGO, Belgium) "Communication and negotiation skills are the most important skills in my job. Each day I have a number of meetings with all kinds of experts, often on opposing sides. To understand their position and to mitigate them you have to be a sort of a diplomat, which isn't always easy." (Deputy Mayor of a major Croatian city) "My work is in teams. One of the most important items is the participation of communities in all the activities. That means that, in our teams, there are representatives of civil society. And so apart from technicians and politicians we work with, there is civil society in different forms." (National focal point for the UNESCO, Ministry of Culture of Italy)
9. "Archipelago of Practice", "Architecture's Afterlife" multiplier event 01, February 9th 2021, host: University of Antwerp, speakers: Dana Cuff, Hera van Sande, Pieterjan Ginckels, Sven Sorić, Grazia Trisciuoglio, Marco Brizzi, Evangelos Kotsioris, Menna Agha, Felipe de Ferrari, Iva Marčetić, Ana Dana Beroš, Davide Barreri and Ilaria Ariolfo, as well as project partners Harriet Harriss, Johan de Walsche, Dag Boutsen, Mia Roth-Čerina, Federica Vannucchi, Michela Barosio, Carla Sentieri. http://architectures-afterlife.com, last accessed 2022/06/28

# NEB and Critical Positions

# Introducing the BioDigital: Towards the Third Architectural Digital Turn

Rachel Armstrong(✉) and Rolf Hughes

KU Leuven, 9000 Ghent, Brussels, Belgium
rachel.armstrong@kuleuven.be

**Abstract.** We introduce the BioDigital platform as *the third Digital Turn*, enabled by the incorporation of "BioElectrical Systems" (BES) into urban contexts. Using metabolically generated electricity produced by microbial biofilms, BES are accessible by directly visualizing their electrical transactions on a screen (the BioDigital interface) and are characterized by three pedagogical and technological principles: (i) (up to 12 V) power generation, (ii) (microbial) information and (iii) biochemical transformation. Integrating life-flows between microbes (comprising the metabolic base of the biosphere), big (environmental) data governed by AI, electrochemistry and human inhabitation, this transactional platform is fundamentally environmental turning organic waste into electricity, cleaned water, bioremediation services, bodily monitoring, and biosynthesis. Introducing a new degree of "smart" resource circularity within architectural systems the BioDigital platform operates within the carrying capacity of any given site. For example, pioneering installations by Organica Inc., and Pee Power® have been scaled to meet community needs to treat the collected wastes of between 5000 and 30,000 people. Near future applications of the BioDigital include the laboratory prototype *Living Architecture*, which converts greywater/urine into bioelectricity, biomass and cleaned water; *999 years 13 sqm (the future belongs to ghosts)* a posthuman household gallery installation powered by microbes; and *Active Living Infrastructure: Controlled Environment (ALICE)*, a prototype microbial/human BioDigital interface powered by a microbial cyborg. As pedagogical systems, the BioDigital helps the architectural imagination envisage and implement new 12 V interfaces for its infrastructures starting within the home. Establishing limits to consumption and generating domestic resource from waste, innovation is promoted in new kinds of housework, inviting new protocols for inhabiting space, establishing new environmental rituals, making possible new types of living spaces and through social systems, activating a range of commons from which new microeconomies can arise.

**Keywords:** BioDigital · Microbes · Microbial fuel cell · Digital turn · Electrons

## 1 Introduction to the Digital Turn

Modern digital computers are powered by electricity, a form of energy carried by the flow of delocalized electrons along a metal wire. Their computations are regulated by components that govern the passage of packets of electrons (signal/information) through


M. Blanco Lage et al. (eds.), *Towards a New European Bauhaus—Challenges in Design Education*,
https://doi.org/10.1007/978-3-031-40188-6_30

logic gates (or logic circuits) based on a two-valued logic system—1/0, on/off, yes/no—that receive a "digital" input, process it, and change it into an output signal. For a computer to act, the output signal must inform another system that translates its calculations into physical actions such as digital manufacturing or instructions for robots.

The first *Digital Turn* is a milestone for architecture enabled by the advent of computer aided design and manufacturing (CAD/CAM), which shifted practice away from traditional media and towards digital media, with. Revolutionizing architectural construction in the 1990s, these new computer tools enabled the generation of novel "organic" structures with smooth and curving lines to parametrically mimic biological forms. By 2012, another new architectural movement began that further exploited further advances in computing techniques assisted by access to colossal volumes of information, or "big data," and its retrieval. This second *Digital Turn* enabled even more ambitious kinds of form-finding through simulations and heralded new types of artificial intelligence (AI) by converging parametric data models, robotics, artificial intelligence, machine intelligence, and big data, with physical systems such as digital manufacturing, robotics, and bio-manufacturing processes.

## 2 Introduction to the Digital Turn

Melika Aljukic observes that a third *Digital Turn* is emerging from this information-rich environment, where novel interpretations of geometric lineage are extending existing spatial skills and technologies to further alter the performance and behavior of architecture via Non-Standard architectural codification and production processes [1]. As part of this transition, the BioDigital platform refers to a specific disciplinary convergence between Information and Communication Technology (ICT) and biological systems (organic forms, biotechnology), which aims to use generative, complex systems that incorporate biological tactics into the production and performance of buildings so their impacts are aligned more closely with nature. In this essay, I argue that complexification via advanced computing techniques alone is not enough to characterize a Third Digital Turn and a fundamental shift towards material practice via a new thermoeconomics [2] is needed, where digital computing moves away from environmental impacts associated with industrialization and towards an ecological era of microeconomic production [3].

## 3 Introducing the BioDigital Realm

The BioDigital realm is more than a disciplinary synthesis but exists more specifically, as an actual communications interface provided by a range of technologies known as Bio-Electrical Systems (BES) [4]. Situated at the interface between organic biological and electronic digital systems, they incorporate microbes into their operating system, invoking an unconventional form of computing that results in the production of electricity, data, and chemical transformation, which microbes are most specialized in. Over eons, bacteria have made metabolism their specialty, at the expense of structural organization. Having developed remarkable abilities that keep their electron transfer chains going in many kinds of environment—even in the most inhospitable places—they have mastered the fine art of transforming their environment through a continual flow of electrons. The bacterial species *Shewanella* and *Geobacter* can harvest electrons from rocks and metals to make the universal energy storage molecule called ATP. This pared-down process is quite alien to all other life forms as they don't need a carbon source (e.g., sugar) for respiration. This means these microbes can thrive indefinitely by eating electrons from an electrode, use them as a source of energy, and then discard them to another electrode as a kind of simple "currency" [5]. Electron transfer is not a solipsistic activity, however, and requires a medium and community of actants that enable microbes to establish an electron marketplace using action-potential mechanisms, enabling communication to take place via waves of potassium-driven electrical activity. Communities of microbes can propagate these electrical signals at around 3 mm/h in tissue-like multi-species formations called biofilms.

## 4 Biofilms as Technology

In 1911, Michael Cressé Potter first brought the worlds of chemical electricity and organic biology together on a single technical platform using biofilms with microorganisms capable of electron transfer. The resultant "living" battery, or microbial fuel cell (MFC), produced several hundred millivolts of energy [6]. Acting as biocatalysts, the microbes converted the chemical energy of organic matter from waste streams into electrons that flowed into an external circuit to provide electrical power for as long as they can harvest energy from various environmental processes [7]. The biofilm thrives on different types of organic fuel that are abundant in 'waste' streams, to perform a range of metabolic tasks at room temperature such as cleaning wastewater, generating bioelectricity, and detoxifying pollutants. This highly mediated relationship established a power-sharing relationship across mechanical and natural bodies that is neither entirely biological, nor exclusively mechanical. The close exchanges between electronics and organisms results in a kind of holistic coupling that can be thought of as a *microbial cyborg*, where electrogenic biofilms metabolize organic 'waste' and excrete electrons that are captured by electrodes and introduced directly into a range of electronic devices—from LEDs to digital screens and capacitors. Sharing a 'language' or 'currency' of electrons, biological and digital systems are in close conversation with each other producing outcomes that exceed the modern understanding of electricity—specifically, the BioDigital interface can undergo actual physical change because of the continual, regulated flow of massless electrons between lively agents, without the need for interpretation by other manufacturing platforms [8]. Possessing an additional layer of capability than modern digital computers, electrons produced by biological processes operate via a range of different electron transfer systems known as redox reactions, which directly result in different kinds of chemical transformation. Characterized by a plethora of performative particles and ions, this irreducibly complex electro-chemical platform provides a material basis that embodies the fundamental life-force within vital matter. Generating a broad range of observable activities through the life-like transfer of electrons between different kinds of molecular systems, BioDigital actions include membrane modulation, change in protein structure, color change, charge, metabolism, and many more vitalizing actions, which express many of the qualities of living systems such as growth, movement, or sensitivity. While bioelectrical systems (BES) cannot compete with the sheer power of other electricity generating systems (renewables, fossil-fuels), their (material) circularity is unsurpassed operating within the limits of the carrying capacity of any given site and capping our consumption at source.

## 5 Towards the BioElectrical Household

The MFC is one example of bioelectrochemical systems (BES) that also include microbial electrolysis cells (MECs) and microbial desalination cells (MDCs) [9]. The MFC is, however, the only type of BES that produces harvestable amounts of electricity. While MFCs cannot produce enough electricity to support industrial manufacturing processes, they do have a real role to play within attaining the *circular home* by engaging vitalizing flows of electrons within our living spaces, instead of consuming fossil fuels. When linked in series, an array of MFC bioprocessors can reasonably produce useable amounts of bioelectricity and the potency of technology is steadily improving [10]. At a near-future date it would be reasonable to expect MFC arrays to provide a 12 V supply of electricity [11]. This would not be the same as a constant flow from a mains supply but periodic, using 'smart' digital tools (like artificial intelligence) to optimize electron storage and use, according to need. By integrating the smarter computing systems of the second Digital Turn into the BioDigital platform, households use less energy from the grid by obtaining it through their waste. Notions of a 12 V lifestyle are already being explored where low-voltage households use far less electricity than the norm (90 versus 900 kwh per month). Most savings can be made up simply by not using electricity to power major heating appliances (a water heater, stove, or space heater, for instance), which can be run off solar energy and passive solar heating, but the real benefit of a low-power domestic utilities system lies in the limits to consumption it sets that stimulates our creativity in how we perform housework. Specifically, the organic waste produced by human inhabitants (urine, feces, food waste, grey water) provides microbial resource (food) that is transformed downstream into a range of household goods (bioelectricity, biomass, bioremediation, high value compounds like vitamins), from which further waste is produced.

Since microbes cannot be seen with the naked eye and work on a different timescale than humans, ways of visualizing their activities are needed. Typically, microbial activity is deciphered using the tools of biochemistry. In human terms, these chemical signals are quite slow and are often associated with different forms of detectable electron transfer. By incorporating electrodes into the system, the bioelectricity produced by metabolic reactions can be directly observed as an indicator of overall (electrogenic) metabolic activity. The importance of bioelectricity as a household resource and system for translating between the human and microbial realms in this context, cannot be overstated. Generating a supply of low power 'energy,' the electrons generated by microbes in the process of metabolism also provide information about biofilm activity (e.g., thriving, under stress, stable), and bring about chemical change in the system as the result of the loss or gain of electrons associated with a molecule, which alters its physical properties. The BioDigital platform is therefore characterized by three unique pedagogical and technological principles: (i) producing low power electricity, (ii) generating (microbial/digital) information, and (iii) undergoing biochemical transformation.

The following case studies exemplify a range of BioDigital principles of practice where some apparatuses like the *Living Architecture* project [12] are exclusively realized in laboratory settings to successfully prove their scientific principles; others such as *999 years 13 sqm (the future belongs to ghosts)* [13] and the *Active Living Infrastructure Controlled Environment* (ALICE) apparatus, are explored as artistic research installations in various locations, which were successfully powered by microbes [14].

## 6 Living Architecture

The *Living Architecture project* is a 'living' combined utilities infrastructure that uses a series of bioprocessors, including MFCs, to turn liquid household waste, like urine and grey water, into valuable resources (electricity, biomass, water, reclaiming phosphate from washing-up liquids and removing poisonous gases from the air) that can be re-used in the household (Fig. 1). It not only cuts down on electricity and utilities bills, but also reduces the amount of untreated waste we put into the environment. The performance of the biofilms is optimized using an AI, that is powered by the MFCs, to operate simple mechanical controls that deliver feedstock where was needed within the arrays. Enabling the smarter use of electrons, multiple tasks are performed within the apparatus from generating power, to providing data and enabling metabolic transformations, which are compatible with the overall carrying capacity of the proposed household system. Through the creation of an accessible space for exchanging shared goods by the various household communities (humans, microbes), resources are used in a variety of ways. Founded on the (re)utilization of organic waste, microbial applications comprise a regenerative technological platform and infrastructure that coverts domestic waste streams into a kind of *fractionated composting system* to produce energy (bioelectricity, oils, biomass), useful substances, bioremediating processes (the removal of nitrous gasses) and biologically compatible materials that can be safely discharged. Valuing the contributions by all who carry out the work-of-life—the different microbial units that make up the *Living Architecture* system enables those that are not usually regarded as economically productive in a capitalist economy to take part in an ecological economy. Re-centering the site of value creation within the domestic sphere, our homes become wealth-generators. Inhabitants now have choices to make about how they use this ecological resource—perhaps they can reduce their own living costs but maybe too, they can donate some of their well-earned resource (formerly called 'waste') to help others.

While *Living Architecture* shows that a *circular economy of the household* is implementable, the economies of scale have not yet been established. Although more in situ installations and pilot studies are needed to collate the data on resource efficiency, it is likely more efficient for many households to contribute to a shared resource through their waste and what better site to process this than a garden—turning them, literally, into power plants. By scaling microbial operations in ways to serve a whole community which can be amplified through symbiotic exchange networks, we can provide access to resources that can be allocated at the scale of a local community. Working with Hungarian company *Organica Water* that designs urban wastewater gardens for municipal use: starting with human sludge [15], which is passed through a series of vats that break

**Fig. 1** Detail of the fully inoculated *Living Architecture* 'wall' and apparatus installed at the University of the West of England, Bristol, the Living Architecture project, 2019. Photography by Rolf Hughes © Rolf Hughes. An array of fifteen bioreactors within the 'living' wall that houses microbes in a large apparatus about the size of a bookcase

down the organic matter using the microbes on the roots of plants; our research group proposes to introduce arrays of MFCs into this system. When built, such an installation could use these natural microbial processes that typically take place in the soil to treat the human residues for up to 30,000 residents with no access to a formal sewage system [16]. The sludge generated by this process could provide nutrients for a garden space, as well as enough bioelectricity to power mobile phones, provide LED street lighting, power Wi-Fi transmitters, and activate screen displays that enable citizens to access, for example, websites to online council services. By creating an opportunity for citizens to obtain power, share the benefits of a public garden and find things in common with each other, this platform is a big stepping point towards activating the (human/microbial) commons, where the organic matter we currently call 'waste' becomes a shared and flexible resource.

## 7 999 years 13 sqm (The Future Belongs to Ghosts)

As the original version of *Living Architecture* could not be exposed directly to the public owing to the presence of genetically modified organisms, an alternative wild-type experience of a prototype bioprocessor wall system comprised only of MFCs was developed for the *Is This Tomorrow* exhibition at the Whitechapel Gallery, in collaboration with artist Cecile B Evans (Fig. 2) [13]. Entitled *999 years 13 sqm (the future belongs to ghosts)*, the installation was prefigured using the natural biofilms from the *Living Architecture* project to power an array of 15 MFCs, which was housed in a protected volume to the left of the space. The Perspex and steel construction acted as a screen on which a near future scenario was projected, which occupied the minimal housing space legally possible in London (13 sqm) being conferred with the longest possible lease (999 years). Occupied only by microbes, the only observable traces of humans were BioDigitally generated manifestations of the human past, present and future (ghosts), which were conjured by the microbial inhabitants from their wastes.

**Fig. 2** The installation 999 years 13 sqm (the future belongs to ghosts) is an arts installation embodying a posthuman apartment comprised of a bank of MFCs and digital screens. The installation is by Cecile B. Evans and Rachel Armstrong for the *Is This Tomorrow?* exhibition at the Whitechapel Gallery, London. Photograph by Rolf Hughes, 2019 © Rolf Hughes. Posthuman apartment inhabited by a bank of organic batteries on the left of the installation and the figure of a bird projected on mist to the right of the picture

The space itself was an artistic rendering of a posthuman apartment and was, therefore, unoccupied by human inhabitants, however, the system still symbiotically depended on 'us.' Natural biofilms within the MFC array were routinely fed with human waste that was introduced into the large black box at the top of the MFC array once a week. At the same time, the microbial excrements produced by this process (cleaned water and biomass) were collected in a black container at the bottom of the complex and removed from the apparatus. The products of this 'remote' symbiosis provided the bioelectricity to power the 'inner life' of the space, which was made visible by a screen-based system and a projection of a bird onto a wall of mist, a symbolic figure from the artist's films, continually rose in flight, only to fall again, cycling between life and death. Offering a vision of a lifestyle that comes after modernity, the BioDigital platform of the installation demonstrates a transactional interface between human and microbe enabling us to negotiate a new relationship with the natural realm—with the possibility of new ecological futures.

## 8 Active Living Infrastructure: Controlled Environment

While Living Architecture establishes a metabolic economy for transactions between humans and microbes, the *Active Living Infrastructure: Controlled Environment* (ALICE) prototype (2019–2021), valorizes the BioDigital platform by establishing the principles for a real-time 'language' for communicating with microbes. Using electrons produced by the anaerobic biofilm of 15 MFCs as 'data,' the prototype explored how a direct, real-time link between bacterial metabolism could be made using electronic systems that can interpret and visualize this data. Electrical activity from the biofilm was a source for both power and data, which was translated by software into animations that conveyed the overall status of the biofilm in relatable terms (Fig. 3). Audiences could, therefore, respond to the microbial behavior, not by looking at unpleasant 'slime' (the natural 'face' of microbial colonies), but by interacting with appealing forms on a familiar screen-based interface. Since BES are *living*, possessing a force and agency of their own, they require our appropriate care and attention if they are to engage with us in a productive, symbiotic manner [17]. The outputs visualized by the BioDigital interface are an averaging of the overall biofilm activity. Adaptation and evolution are important for the survival of the species and/or microbial communities so there is also much competition within microbial communities for nutrients and space. While the animations provide an aestheticized view of these relationships, the microbial environment not always a *peaceful* collective where all beings flourish. No matter how carefully the microbial commons is engaged through our acts of care and BioDigital negotiations, if convivial outcomes are desired then, like Victor Frankenstein's creature, our socialized engagement with them is obligatory and invites us to develop an ethics of care towards these 'living' apparatuses [18]. Possessing a very particular kind of environmental intelligence, we can learn a lot from bacterial data that reveals a great deal about the character

**Fig. 3** "Mobes," from the ALICE website (http://alice-interface.eu) showing dynamic, interactive, graphical representations of microbes, courtesy the ALICE consortium, 2021. © ALICE consortium, screen shot from website. Digital animation patterns in pale pink, purple, yellow, and blue that are reminiscent of butterflies, and which are overlaid with geometric symbols that represent different parameters, which influence microbial activity inside the ALICE apparatus

of a place. By generating a relatable communications interface participants can play with resident microbes through data and performance in an exploratory exchange, as if they were a pot plant, or even a pet. This world of "Mobes," a characterful term coined for the data-based representations of microbes, offers a simple, probiotic approach to interspecies communication within the highly situated realm of microbes, in a relatable manner that could even become part of our everyday routines. Being in conversation with, rather than exploiting microbes, means we may start to learn-along-with them through their ability to generate clear and direct signals and data that relate to shared concerns, like transforming waste streams into household resources based on new value systems that invite different kinds of (house)work and domestic routines for our living spaces.

ALICE exists as a permanent online exhibit. It can be accessed under the section Bio-Digital Interface by clicking the Launch Artwork button, where visitors are taken to an animated set of 'Mobes.' On selecting different options from drop-down menus, the environmental parameters (temperature, pH) and performance (power output) of a real-world microbial community can be interrogated that inhabits a permanent MFC array installed in a scientific laboratory. Depending on how the visitor reads the health of the microbes, they can respond to the 'Mobes' by feeding them using a remotely operated valve system, or by speeding up their metabolisms by activating an LED to warm them gently. The graphical symbols provide a language where factual propositions (environmental parameters) are represented, and further truths can be inferred to directly, or by means of a calculating system, which can be overridden by human intervention. In ALICE's case these calculations are performed by the system software, but in the case of *Living Architecture*, an artificial intelligence powered by the MFCs observes the system and adjusts the inputs to optimize its performance accordingly.

ALICE also exists as a real-world installation that premiered at the Digital Design Weekend, V&A, London, UK, as part of the London Design Festival from 24 to 26 September 2021 [19], and was installed at the Electromagnetic field Festival, from 2 to 5 June 2022 (Fig. 4) [20]. Inviting meaningful human/microbial transactions (if I give you food then will you give me data, chemistry, and power?) (Fig. 5), ALICE demonstrates the potential for BES to become creature-like. Collectively, these *living* microbial installations suggest that various kind of BioDigital interfaces between humans and microbes can establish readable and even relatable transactional platforms that transform household wastes into new resources, like natural soils. The combination of microbial 'flesh,' traditional electronics, artificial and microbial intelligence confer the work with the status of microbial cyborg, a relatable entity that does not ooze, stink, or repulse. In all its situated expressions, ALICE is becoming the human-relatable face of the microbial realm that invites new relationships, interactions and inter-species understanding. Centering our experiences on household economics as a way of unleashing situated ecopolitical decisions, ALICE helps in making ecopolitical decisions, changing how we use resources and organizing our living spaces so that our environmental impacts change from being merely consumptive, to life-promoting.

**Fig. 4** The ALICE installation, a transparent orb powered by microbes that animate LEDs and iPads. The work was installed at the Electromagnetic Field festival, Eastnor, June 2022, embodying the bio-digital platform through the integration of microbial and artificial intelligences with biological and technical bodies. Courtesy of the ALICE consortium: Ioannis Ieropoulos, Julie Freeman, and Rachel Armstrong. Photograph by Rachel Armstrong. © ALICE consortium. Clear orb in the centre of the photograph standing on a royal blue carpet. The orb is lit from the inside by arrays of LEDs and iPads powered by microbes. A child points to the orb while his parents watch him carefully interact with it

## 9 The "Power Plant:" Activating the Microbial Commons

The BioDigital platform establishes the tipping point that enables the third Digital Turn where digital technologies no longer consume fossil fuels but work in concert with nature, generating a whole range of biomolecules that are integral to a domestic circular economy. By incorporating the advanced computing platforms of the second Digital Turn and coupling them with forms of biosynthesis microbial technologies can free us from the static infrastructures of modern cities and the utilities they provide. At a time of climate crisis and the displacement of peoples everywhere—such as during times of severe flooding—having access to clean water, shelter, power, and sanitation will literally save lives. By incorporating these life-bearing BioDigital infrastructures into our homes and living spaces the architect's role is to incorporate microbial infrastructures within our living spaces, so waste streams can be upcycled as microbial goods, which are returned into the home and biosphere thereby increasing the overall life-bearing potential of a site. Thus, inhabitants are not indebted to the commercial marketplace, but through their

**Fig. 5** ALICE's microbial fuel cell arrays powering LEDs and iPads for the electromagnetic field festival, Eastnor, June 2022, courtesy the ALICE consortium, 2021. Photograph by Rachel Armstrong. © ALICE consortium. Printed white "living bricks" connected to tangles of clear piping and electronic wiring inside the clear ALICE orb lighting up yellow and pink LEDs

very activities of living can contribute valuable resource to the well-being of a place, even when they are not productively included in the capitalist work system.

Perhaps the most radical step suggested by these microbial technologies is to challenge our basic assumptions about the baselines of what we need to live comfortably and healthily. Imagine the reduced impact of human development if every home that is now connected to a 230 V grid, could operate comfortably on a 12 V battery supply. While this would require innovation in some of the things that we do every day that we solve by consuming a lot of energy, like washing machines and fridges, these same tasks could be done differently, such as using advanced new materials to help with refrigeration and finding alternatives to mechanical agitation like ultrasound to carry out this housework. The hardest part to altering our impacts is changing our thinking, our habits, and our concepts of what a "good life" entail. Whether we like it, or not, the rules for living on this planet have changed. The good news is that the BioDigital platform can help us make the necessary adjustments to work within the carrying capacity of our lands, draw on our collective creativity and help us find much, much, ways of better working alongside nature. The bad news is that nobody has demonstrated at scale that this is possible in Western communities and so, this requires the participation of adventurous and bold pioneers. Despite the present challenges, the BioDigital platform is set to establish the third Digital Turn and change the present stalemate in making a paradigm shift towards an ecological era of human development, where ultimately, the human species can become a fundamentally bioremediating and enlivening force within the biosphere.

## References


1. Aljukic M (2020) The third digital turn of non-standard architecture (thesis). University of Sydney, Sydney
2. Garrett TJ, Grasselli M, Keen S (2020) Past world economic production constrains current energy demands: persistent scaling with implications for economic growth and climate change mitigation. PLoS ONE 15(8):e0237672
3. Hughes R, Armstrong R (2021) The art of experiment: post-pandemic knowledge practices for 21st century architecture and design. Routledge, London
4. PHOENIX Homepage. https://www.researchgate.net/project/PHOENIX-Project-EU-COST-Action
5. Brahic C (2014) Meet the electric life forms that live on pure energy. New Sci. https://institutions.newscientist.com/article/dn25894-meet-the-electric-life-forms-that-live-on-pure-energy/
6. Potter MC (1911) Electrical effects accompanying the decomposition of organic compounds. Proceed Royal Soc B 571(84):260–276
7. Sharma V, Kundu PP (2010) Biocatalysts in microbial fuel cells. Enzyme Microb Technol 47:179–188
8. McFarland BJ (2016) A world from dust: how the periodic table shaped life. Oxford University Press, New York, p 25
9. Greenman J, Gajda I, You J, Mendis BA, Obata O, Pasternak G, Ieropoulos I (2021) Microbial fuel cells and their electrified biofilms. Biofilm 3:100057
10. Younggy K, Hatzell MC, Hutchinson AJ, Logan BE (2011) Capturing power at higher voltages from arrays of microbial fuel cells without voltage reversal. Energy Environ Sci 4:4662–4667
11. Walter XA, Forbes S, Greenman J, Ieropoulos I (2016) From single MFC to cascade configuration: the relationship between size, hydraulic retention time and power density. Sustain Energy Technol Assess 14:74–79
12. Armstrong R, Ferracina S, Caldwell G, Ieropoulos I, Rimbu G, Adamatzky A, Phillips N, De Lucrezia D, Imhof B, Hanczyc MM, Nogales J, Garcia J (2017) Living architecture (LIAR): metabolically engineered building units. In: Hebel DE, Heisel F (eds) Cultivated building materials: industrialized natural resources for architecture and construction. Birkhauser, Berlin, pp 170–177
13. Bevan R (2019) Is this tomorrow? Review: installations show a troubled mood of the future, evening standard. https://www.standard.co.uk/culture/is-this-tomorrow-review-installations-show-a-troubled-mood-of-the-future-a4066551.html
14. ALICE Homepage. https://alice-interface.eu
15. Organica Water Homepage. https://www.organicawater.com
16. Armstrong R (2022) Safe as houses: the more-than-human home. Lund Humphries, London
17. Bellacasa MP (2017) Matters of care: speculative ethics in more than human worlds. University of Minnesota Press, Minneapolis, p 5
18. Latour B (2012). Love your monsters: why we must care for our technologies as we do our children. In: Breakthrough Institute. https://thebreakthrough.org/journal/issue-2/love-your-monsters
19. Barto M (2021) Digital design weekend. https://www.vam.ac.uk/blog/design-and-society/digital-design-weekend-2021
20. Electromagnetic Field Camp (2022) Pee is powerful! From artwork to new world infrastructures with ALICE. https://www.emfcamp.org/schedule/2022/307-pee-is-powerful-from-artwork-to-new-world-infrastructures

# Towards a New European Landscape: Countryside as Inhabited Ecosystem Model

Adriano Dessì(✉)

Dipartimento Di Ingegneria Civile, Ambientale e Architettura – Università di Cagliari, Cagliari, Italy
adriano.dessi@unica.it

**Abstract.** The New European Bauhaus doesn't propose just a new way of interpreting health and welfare into a new sustainable productive paradigm but may well determine a deep re-establishment of our way of conceiving and inhabiting the European area. All along the Twentieth Century, cities have been the greatest human aspiration places but also consumption models of natural resources until the degree of "being human on earth" [1] reached a critical point. Meanwhile countryside has been "forgotten" although it is the place where human knowledge grows certainly more than in the cities [2]. The new Bauhaus deal, over the very pandemic period, has enabled to reflect on the manifold continental crises consequences but has also increased that human space re-establishing role of "design" starting from models radically different from the past. To this effect, the paper proposes two perspectives: Tracing into the multiple forms of historical European countryside—from the "inhabited" one [3] of the major continental plains to peripheral polyculture areas, from Mediterranean gardens and agricultural cities to multifunctional farmhouses—some paradigms which capture these new deep-changing energies. Interpreting European countryside as a "coevolution" field with cities and as a theatre of techno-ecological development enables to reformulate the human combination of producing/inhabiting into SDGs. Both perspectives will be illustrated through some modern and contemporary projects which were outstanding for their innovation with the aim of constructing a first state of art of countryside design into a new sustainable continental landscape vision.

**Keywords:** Countryside · Rural landscape · Agricolture · Natural resources · Coevolution

## 1 The Time of Rural Landscape

### 1.1 The Long-Term of Landscape

The European Countryside has been modeled on the really hight biodiversity of human and other life-beings relationships, and that is true for every cultures which have built it along centuries. If we consider Countryside as a specific—but really large and significant—landscape form, we could interpretate the historical evolution of Europe space like a big transitional composition of Landscape forms. In addition, as Carlo Tosco stated on


M. Blanco Lage et al. (eds.), *Towards a New European Bauhaus—Challenges in Design Education*,
https://doi.org/10.1007/978-3-031-40188-6_31

its "Il Paesaggio come Storia" -Landscape as History- [4], we are able to recognize on Landscape, instead its own modifications, a certain continuity of human habitat, a continuity which is really strong to discover on the city that is a weak and fragile organism, easily destroyable by natural events or wars and—at the same time—easily refundable, in other or in the same places—and with other or the same people.

The landscape, in opposition of that, and specifically the rural landscape, is an a to-regenerative and resilient organism in which we could notice big daily transformations but in which the "big picture" is always clear and deeply linked to the preview conditions and forms. In fact, regarding the landscape we must think about three conditions of the same idea of human beings.

The first one is that every landscape form—and everything we could do by transforming it—is more related to the landscape evolution processes than just to the human one. In an agricultural system or in a riverside park, the dynamics of plantations or the velocity of river flows are surely the main issues of design, and we are allowed to think just specific areas or routes for humans—like manutention paths or very delimited production areas.

The second one is related to the time of landscape forms evolution which is evidently longer than the human's life cycle one.

The third is related to the "palimpsest concept" of Landscape, in respect of which we are allowed to understand it and manipulate it just if we deeply know the number of layers and strata that have composed it in the past.

## 1.2 The Agriculture as a Co-evolution Ecosystem in European Landscape

One of the most experimental—and not only necessary—activities on landscape, which is traditionally true on the European one, is agriculture. The origin of agriculture can be traced back to the first Middle Eastern nomadic societies, where the continuous migration between the arid areas of the desert and the fertile areas of the river valleys allowed the continuous exchange of human and plant habitats.

But it was in Roman culture that the agricultural patterns permeate the city open-spaces in different ways like patios, courtyards, orchards, gardens, canals, fountains, empty fields—a process which is well-described by John Dixon Hunt with the Latin sentence "rus in urbe" [5]. But also in previews civilizations, agriculture has been the main tool on transforming the continental space and the very shared unique code on constructing landscape in the Mediterranean. Furthermore, since the origins, agriculture has been the field where more co-evolutionary processes have been incorporated and constitutive of the same productive practices. This has been true since the first eastern Mediterranean civilizations and the fertile crescent in which, as early as 9000 BC, human "disturbance" actions through localized fires in the evergreen forests—relatively poor in species—created "micro-crags" in which the first herbaceous plants of alimentary genus such as oats, wheat and barley, typical of the hills but unsuitable for the expansion of the forest. In this system between human action and the expansion of vegetation, not only rich in food plants but characterized by a great biodiversity, a synergy has been established, a process of plant-man co-evolution which will be essential for the development of Mediterranean culture. Therefore, since agricultural practice is intrinsically linked to the concept of "coevolution", constituting the main historical factor in the rooting of

Mediterranean landscape societies, considering it within a new "system of values" in the transformation of space through the project means directing the project itself towards a horizon of balance between man and earth and between man and other living beings.

With respect to these two themes declined in the "man-soil" and "man-living" relationship, it can be said that agricultural landscape—and this is historically true, but even more so today—represents exactly the necessary meeting between utilitarian and symbolic practices of the primary productive activities of man with the "self poietic" and "ecological" regenerative dynamics of a specific way of organizing the terrestrial space. As Augustin Berque states "the landscape is always something both ecological and symbolic, it is always eco-symbolic" [1].

## 2 City and Countryside in Europe

### 2.1 About Some Historical Forms

Throughout the Mediterranean area, there are historical links between the city and the forms of agriculture that surrounds it. In the Arab settlement, the countryside—as well as being a source of income for the city (*medina*) in which the horticultural spaces were already used for relaxing and free time—allowed the creation of a succession of microclimates, particularly suited to provide houses resistance during top temperatures. In cities by the sea, when the logic of agricultural self-sustainability was secondary to the mercantile one, systems of urban gardens were located within the walls but in the peripheral parts, so that they were protected but close to the ports for the products sale and maintained through collective irrigation systems (dams or large tanks) or individual (mostly domestic wells) [3]. Thanks to different plant densities, these spaces were those of waters detention and where humidity was functional to the general control of urban climate, places of extreme biological richness and therefore carriers of the system of urban food supply. It is especially in these structural forms' aspects of the Mediterranean agricultural landscape which, with different declinations, we can identify and unravel those internal qualities based on balance and historical union between city and territory that make up a large designing potential for urban development. The multifunctional character of the Mediterranean landscape is based on this system, derived by Roman *hortus conclusus*: gardens for food, for society, therapeutic, for loisir. Due to proximity between the city and these enclosed gardens, which were sophisticated water cycle and daily and seasonal temperature changes monitoring mechanisms, and thanks to the extensive capillarity of that system, it has been possible to observe the developing of different forms of enclosed gardens—the Île de France *murs à pêches*, the sub-Saharan and mid-East palm-orchards and also the olive gardens of the Peloponnesus and Andalusian citrus-groves and Iberian *huertas*, the *bocages* of Normandy and also the terraced gardens of the Douro valley, have survived in time and represent, at the moment, a very achieved model through which is possible to rethink the European space regeneration (Fig. 1).

### 2.2 About Some Practices

In some European countries like France and Spain, the agriculture peri urban fabric derives by merging the ability of the *bocages* and ancient canals (*acequias*) to protect

**Fig. 1** Murs à pêches in Montreuil, northern Paris (montreuil.fr) and, on the right, aerial photo of Montreuil taken in 1942 (geoportail.fr). Image copyright: Revue Sur-Mesure (En ligne, n. 2, 2017); aerialphoto copyright (free caption): geoportail.gouvr.fr

and irrigate fields and orchards and, at the same time, to mark the paths of crossing and monitoring territory, are today the basis of many strategies for sustainable landscape—and especially for the city—development. It happens, of course, for a certain renewed attention of municipalities, but also of communities, for agricultural policies—in the past often focused just on the ability to produce income from agricultural territories—but mostly for a growing awareness on the importance of the qualitative housing factors expressed by these territories and for the urban will of benefiting ecologies, symbols and quality of living spaces of rural world. In its "Campagnes Urbaines", Pierre Donadieu highlights the case of Plouzanè, a small suburban coastal village in Breton, where the will of inhabitants, often non-local and therefore devoid of a historical memory of places, to reconstitute an "urban *bocage*", in total continuity with the rural one, doesn't give just a strong sense of naturalness to the public spaces, but creates a real ecological network that allows inland environmental corridors to communicate with Atlantic coastal ecosystems: « (…) by rebuilding slopes and hedges in the city, they have restored the lost *bocage* to find a harmony with the original one. (…) A hedged farm ideal, but very real, which protects against the wind and favors wildlife, but especially that the inhabitants themselves have planted on Breton soil» [3]. In southern France as like as in the Mediterranean Spain—but also in some areas of Bavaria, northern Italy and the Western Balkan, there are ever-increasing spaces instead residual margin between urban and the natural primary elements (rivers, hills, forests), in which it is possible to observe participatory processes of "organized agriculture", made up of amateurs which are allotted small equal territories. The multitude of techniques and people cultures exploit the diversity of these soils transforming it into the diversity of agriculture and production.

In Carcassonne, on the interstitial peri urban fringes between built blocks and river, peripheral public spaces communicate directly with a fabric of vegetable gardens organized in equal lots and with common management furniture, which guarantees the protection of the edge and the riparian vegetation, the biological enrichment of an otherwise unproductive soils and the establishment of alternative spaces related to the residence determining a sort of "urban loisir" in some "rururban" territories.

The agriculture into open urban space seems an interesting proposition—and probably also an aesthetic culture manifestation—of new forms of living together and developing fertile community relations, and this might have consequences in our traditional way of conceiving urban forms and public spaces. The really diffuse appreciation of the vegetable gardens presupposes a shift in sensibility from the visual, mental and design sphere of recent landscape architecture to the physiological, relational and empirical one more related to the rural habitat, «where getting your hands dirty, but also tasting, eating, toiling, etc. implies an opening to the contamination between different realms of the sensible world» [6] (Fig. 2).

**Fig. 2** Michel desvigne paysagiste "cité nature", Arras, France, 2001–2005, (Lotus 149, "Lotus in the fields", 2012). Copyright: Editoriale Lotus International (Lotus n. 149, 2012)

# 3 New Perspectives in Landscape Architecture

## 3.1 The Idea of "Weak Urbanization" in Agronica

At the beginning of 90's, the searching for a new model of urban development that could start from an idea of coexistence and continuity with the existing structures of agricultural landscape, has been the basis of interesting design explorations. One of them has been certainly the research of Andrea Branzi for heavily urbanized agricultural area in suburban of Eindhoven and, mostly, in the northern territory of Milan.

The Agronica project in fact, more than a real masterplan, consists of schemes that represent different ways of land use which refer to agricultural structures, even if not more productively, and who are able to incorporate and organize, through a weak infrastructure, all those activities urban plants that can be located in a peri urban area. As Branzi said himself: «Agronica develops a model of "weak urbanization", which consists of a system of diffuse furniture and facilities that guarantee the survival of agricultural and natural landscape, in presence of advanced urban services but no longer all-encompassing» [7]. The concept behind these models is to be found precisely in the structural role that agriculture can have in the generative process of new urban areas: diversification, cyclical renewal, adaptability, naturalness, productivity, territorialization, are the new paradigms that the contemporary city needs to transpose from the agricultural world so that it can see a sustainable alternative to pan-urban models that marked the modern era (Fig. 3).

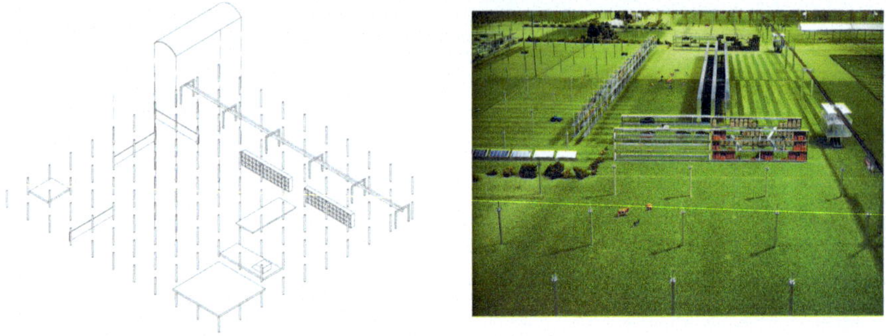

**Fig. 3** Axonometric study (author) and model image of Andrea Branzi's "Agronica". Author's drawing (left) and image copyright (right): Editoriale Domus (Domus n. 800, 1998)

### 3.2 A New "Rurban Paradigm" in Valencia

Similarly, but with different goals and programs for the new urban districts, in the Valencian peri-urban area, the Sociopolis project explores, through an articulated masterplan, the possibility that the layouts of the Iberian *huertas*, historic system of closed vegetable gardens, could redefine a neighborhood housing in which agricultural diversities constitute the functional program and the main feature of public and common spaces at urban-scale. In this vision, the fruit and vegetable markets will be the new places of socialization, the public spaces and sports equipment are merged with large orchards, the terraces and public spaces into residential areas become vegetable gardens for leisure. Among the first works of urbanization for the neighborhood construction, as well as roads, pedestrian walkways, car parks and technological networks, the district has been provided by irrigation channels for the maintenance of vegetable public gardens becoming a connector system of facilities. In the Valencian project, the awareness of a new "rururban" character of the city boundaries, therefore pushes to a new idea of city itself, which it could rediscover in its old rural structures and in its internal agricultural diversities, an evolutionary code—not only formal, but also social and ecological. Vicente

Guallart, the masterplan coordinator, in this sense, states: «the challenge of building a new neighborhood on the edge between the city and the *huertas* of Valencia allowed us to explore this new, open, hybrid condition and dynamics of the territory and to propose a new model for the construction of urban margins» [8] (Fig. 4).

**Fig. 4** Modern "acequias" and first "farmers" on Sociopolis terrain vague. Images copyright: Vicente Guallart Office

### 3.3 The "Pre-Landscaping" Strategy

In some other experiences the countryside is seen as a paradigm of "urban regeneration" in which the idea that the city growth and implementation could be occasion of larger landscape construction and not just specific solutions and functionalistic improvements of urban programs. The Masterplan by Dominique Perrault for the development of a new district in the industrial and disused area of Unimetal, in Caen, can be framed in this family. This experience is interesting, from the point of view of agricultural landscape that design the city, because Perrault introduces an urban redevelopment strategy based on the so-called "pre-landscaping" method, as he himself defines it, not based on a deterministic design but on tracing on the ground a clear formal rule that could be able to direct the future development of urban programs. The structure proposed is the historical square planting mesh which becomes a gradient of circulation, land use, permeability and productivity of the surfaces. Each quadrant includes the planting layouts of tree crops, herbaceous clearings, or pieces of controlled nature and the crops inside will form the first courtyards of the future residential areas. Exactly in the rule of planting layout lies this way of colonizing action but also this new idea of urban design: the concepts of 'hectare type', of 'optimum distance', of 'crop diversity', of 'natural radiation', of 'system rationalization' and 'controlled ventilation', are built in a 'temporary park' which will become a new urban organism; the planting grid is interpreted as an 'operating morphology' which links these ecological needs to the city (Fig. 5).

**Fig. 5** Dominique Perrault's model design for ex-steel area of Unimetal in Caen (Arquitectura Viva—Monografías n. 91 "Pragmatisme and Landscapes", pp. 76–81). Copyright: Arquitectura Viva

### 3.4 A New Concept of Ground Sedimentation

The project for a new residential district in Montreuil addresses another great question of metropolitan diffusion in the territory, the topic of residual landscape. In many peripheral fringes, this form of residual agriculture is trapped within large hollow blocks. In Montreuil they are made up of long and narrow lots, historically divided by the constant presence of the so-called *murs à pêche*, walls about three meters high and with a considerable thickness, fences to protect the crops and, at the same time, local ventilations ducts during the hottest seasons. They are built with large thermal masses capable of absorbing and retaining heat during the day and releasing it slowly during the night, in which internal crops, mainly orchards and vegetable gardens, underwent gelation processes. In 2002, the *murs à pêche* of Montreuil have been interpretated by BNR Architectes for 36 social housing design: they were not only the object of recovery, but they constituted the main thorns for reconnecting the urban territorial plots and a new settlement way for a low-charge residential program. BNR reused the traces and "sediments" of the agricultural walls as elements of the human-habitat dimension: the social housing blocks are instead set up, through the walls, as multi-family dwellings and the walls, never defining real separations, have the ability to define gardens in front of houses and private backyard. These blocks are derived on the murs à pêche fabric, used and extended on the ground floors of the apartments, defining their interior spaces and—at the same time—exploiting their thermo-hygrometric capacities. The continuity between the historical system and the contemporary does not end with formal or structural persistence: the change of intended use and the new model of social housing is here made consistent with the historical practice of cultivating, of maintaining, of treating one's own garden and participating in the collective care of shared gardens (Fig. 6).

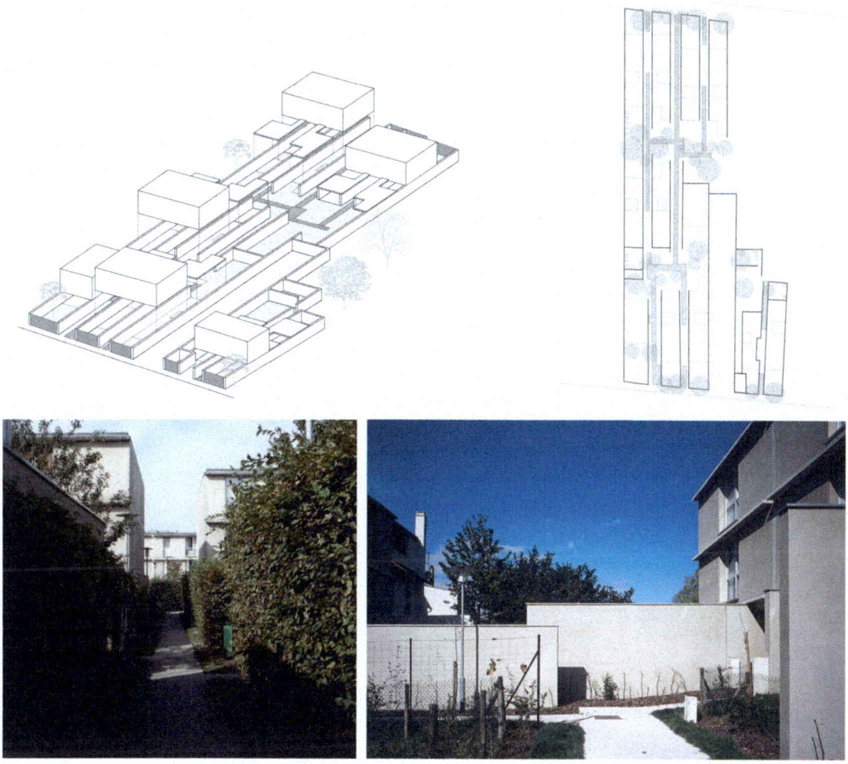

**Fig. 6** Plan, axonometric view (author) and pictures of gardens in BNR's Montreuil housing. Author's drawing (top) and image copyright (downside): DC by AMC architecture

## 4 European City Should Be Still Formed by Countryside?

The paper tries to provide evidence to one of future scenarios of New European Bauhaus, the recovery of the historic relationship between the countryside and the city in the European Landscape. João Ferreira Nunes stated that the dissolution of city—and the loss of its characteristic shape—has started when it has broken the links with agriculture. It is not just about spatial or productive dissolution, but much more cultural, which concerns to the irreversible loss of a sustainable way of relating to natural resources and of establishing co-evolutionary modalities with other living beings.

The countryside, in fact, is the field for development of complex and diversified habitats, unlike the city. Up to modern times, the "medium" size of European city has ensured a certain balance of this relationship. Afterwards, the relationship between inhabited city and productive countryside, city and nature, city and open spaces has considerably altered, causing deep imbalances in the ecological structure of the city in its territory [9]. Canals, gardens, enclosures and hedges, which from countryside penetrated the city, provided water, shadow, heat balance and major food; at the present, they can be revisited in a most sustainable and resilient way for the city.

The projects presented show different ways of possible reconnection between city and countryside, recognizing and selecting some historical practices and projecting them into a process of regeneration of the contemporary European city even in different conditions: the suburbs, the new districts, the abandoned post-industrial spaces, the residual lots in the consolidated urban fabrics.

In all of them, we can trace two levels of possible reconnection "city-countryside":

- A "structural" one, in which the lines of force of countryside—ecological, hydraulic, agricoltural—provide continuity between urban and rural fabrics;
- A "minimal" one, in which some elements belonging to countryside—canals, gardens, fences, abandoned infrastructures and farms—could become devices for the ecological reconnection between city and territory, but also an opportunity to experiment new forms of housing and bring a deeper dimension of the relationship between human living and natural spaces.

In this design culture it is possible to focus a perspective of countryside not like the "background" of the city but like an operational space—an universe of sustainable ecologies, productions and techniques—which is still capable of giving it shape.

## References

1. Berque A (1996) Etre humains sur la terre. Gallimard, Paris
2. Koolhaas R AMO (2020) Countryside, a report, Taschen, Cologne
3. Donadieu P (1998) Campaignes urbaines. Actes Sud, Paris
4. Tosco C (2009) Il Paesaggio Storico, Laterza, Bari
5. Dixon-Hunt J (1993) Nel concetto delle tre nature/In the concept of the three natures. Casabella, Il disegno degli spazi aperti/The design of open spaces, pp 597–598
6. MacLean A (2012) Orchard landscape. Lotus 139(2):36–46
7. Branzi A (2006) Modernità debole e diffusa. In: Skira (ed) Milano
8. Guallart V (2004) Sociopolis. Project for a city of the future. Actar, Barcelona
9. Dessì A (2019) Le città della campagna. Il paesaggio rurale nel progetto urbano. Franco Angeli, Milano

# Designing with the Landscape

Riva Lava[1](✉) and Maria Frantzi[2]

[1] National Technical University of Athens, Athens, Greece
rlava@arch.ntua.gr
[2] DESS Amenagement Rural, Lyon, France

**Abstract.** This paper argues that, while European legislative frameworks, assessment methodologies and systems of visual resources are invaluable tools for designing with the landscape and its contemporary connotations, it is difficult to 'unlock' the cultural properties of landscape without a creative choice of values made by the architect who designs with it. Architecture that evolves with the landscape—both natural and cultural—is a call to re-examine today's realities within the framework of climate change, limited resources and an unprecedented flow of migration around the globe. Landscape is investigated in its multitude and intrinsic nature, as imaginary locus, mindscape and landform. As part of the New European Bauhaus discussions which were conducted across Europe, landscape emerges as a significant driver for architectural aesthetics and life supporting designs, while it offers the context for reconsidering the true needs of existence. Furthermore, landscape is an important constituent of place making along the European South, where well-being and a sense of freedom stem from the moderate climate and the prolonged life outdoors. Natural environment, local materials and landscape are also discussed as the treasures of the Southern European cultural continuum where inherent values of heritage are preserved. Case studies of Greek architects A. Couvelas, B. Babalou-Noukaki and A. Noukakis are juxtaposed to landscape assessment methodologies in order illustrate the discourse presented here.

**Keywords:** Landscape · Architecture · Asset · Visual resource · Diatopia · Interpretation

## 1 Assessing Landscape

### 1.1 A Complex Continuum

The European Landscape Convention, also known as the Florence Declaration on Heritage and Landscape as Human Values (2014), states that landscape—urban or rural—is directly linked to harmonious development (2.1.a). It also suggests that the 'artificial segregation between conservation and innovation' should be abandoned in light of the challenges of climate change, risk management, biodiversity conservation and human well-being (2.2.c). It becomes clear that the protection of our environs cannot rely on quantitative methods only, as we are spiritually and emotionally invested in the landscapes we live in—as much as we depend on their material faucets to survive. A growing consensus in Europe has led to in-depth landscape assessment studies across the


M. Blanco Lage et al. (eds.), *Towards a New European Bauhaus—Challenges in Design Education*,
https://doi.org/10.1007/978-3-031-40188-6_32

European South, among them three case studies on in Valencia, Galicia and Catalonia [1], which call for public participation within the framework of the convention. A comparative study across Europe, in Catalonia (Spain), France, Italy, Switzerland, the Netherlands and the United Kingdom, confirms that governments and cultures vary considerably when they cater to the protection of European landscapes [2]. Nevertheless, the legal recognition of landscape implies rights and responsibilities on the part of all institutions and citizens towards their living environment [3]

As part of the New European Bauhaus discussions which were conducted across Europe, landscape emerges as a significant driver for architectural aesthetics and life supporting designs, while it offers the context for reconsidering the true needs of existence. This article discusses our collective, democratic commitment towards our shared heritage, which is the foundation of the existing EU legislation and methodology, as juxtaposed to architectural authorship—a different kind of responsibility towards landscape.

In order to protect our environment, a qualitative approach is sought in order for it to be regenerated, assimilated or solely maintain its qualities (Pearce and Turner [4].

Sustainability is the way to ensure a renewable resource base so that the use of resources does not exceed their capacity to regenerate. It has been found that the depletion of the environment and its natural resources can harmfully affect the living standard. Hence it is important that the timeless character (which will be analyzed further as *diatopia*), the identity or genius loci of the surroundings where a settlement is built is preserved and that the bearing capacity of its ecosystem—both natural and manmade—is not overloaded. It is not possible to asses the bearing capacity of a cultural or environmental good by economic means only; in order to balance the assessment of values and qualitative parameters one must bring in intangible values, which are difficult to measure.

Such values are preserved by a holistic—and at the same time—site specific study of the inherent character of the continuum of a landscape (natural or/and constructed), its current physiognomy and the desired development in terms of sustainability and resilience against social and environmental threats. Such a study would take into account (i) a bottom up spatial approach, from edifice to landscape (ii) the implementation of projects based on true facts and not 'virtual' data (iii) a continuous survey of data and metadata so that all figures can be combined and assessed at the same time.

In the cases of cities and settlements, where there is no documentation in place, the following steps, in the order suggested, can facilitate the preservation of the landscape continuum:

- The compilation of the historical, spatial and typological analysis of the entity
- The establishment of GIS data frames as a basis for all following steps
- The inclusion of the existing heritage, both cultural and natural, in the developmental design as described in the Amsterdam Declaration so that the emerging surplus in societal, cultural, environmental and financial value is monitored.
- The definition of boundaries for settlements, networks of public spaces and land uses in order to ensure a balanced outcome for the local community and the place
- Linking spatial design with the framework for combatting climate change

Within this context, landscape functions as the unifying element of all the properties of a geographic region [5], bringing together social groups and their political choices. The protection, management and design of landscape cannot be approached separately, these processes are agents of a solid act of sustainable and balanced development. Categories of natural features, tangible and intangible heritage, infrastructure and services as well as spiritual, social and institutional properties are all interrelated by the term 'territorial capital'. The territorial capital of an area brings under one roof its financial, social and environmental components. Such components span "from geographical features of place to characteristics which are difficult to pinpoint, such as quality of life, relation structures, local mores, traditions and more" [6].

### 1.2 Visual Recourses

In order to utilize landscape as an asset there must be a pertinent framework defining its usefulness within its particular environment and in relation to a desired outcome. Indexes facilitate the transformation of such desired outcomes into normative rules. The research on landscape suggests that quality of life is intertwined with the visual environment of man [7]. The subsequences of economic growth for the quality of landscape led us to understand how important it is to include the visual dimension of landscape into the process of design.

Making visual resources part of the sustainability agenda ensures to a degree that what we (want to) see, including our collective memories of place, can be preserved and protected. Sustainable development professes the protection of visual resources as a dynamic process, and not as something decorative, through which the sources of information and inspiration can be saved. The utilization of spatial visual resources as a set of objectives in scenario analysis for management of networks of cultural identity, contributes to their preservation [8].

A visual resource can pertain to measurable features such as topography, water, flora, human presence, structures etc, but can also express the way such features interact. Hence the visual character of a landscape rests on the way its spatial elements are combined and synthesized. The description and evaluation of visual resources is based on how we decipher the visual organization of the environment which is studied, where the consequences of ecological, social, financial and cultural elements are all inherent traits of the landscape. The methodology for the implementation of visual resources focuses on the visual and geometrical orders and systems within the landscape examined.

In the framework of this landscape research and the evaluation of environmental "visual resources" as part of the present approach, landscape is viewed as an areal entity, where landform and landcover imply a distinct visual pattern, as well as a living organism, where the functional relations of its elements are also considered. Land cover comprises water, vegetation, and man-made development, including cities [9].

As architecture and built heritage are part of such visual entities, the built environment description and analysis are likely to be essential parts of the context of sustainable development. A system for portraying the "image" of a traditional settlement, which incorporates the decoding of the architectural patterns and identifies the architectural resources composing it, is considered one of the possible applications of the visual resource tool for the protection of landscape and heritage [10].

Systems for decoding visual resources, pertain—among other components—to what is called "decoding subjectivity": value judgements and responses are motivated by subjective attributes of the person/entity, as each person is subjected to a set of values (e.g. cultural, social, etc.).

Grasping the "subjective code" in the perceived message of visual resources is crucial. The development of a system for decoding the subjective element of perception when confronting a landscape is key for the process of architectural design. The value formation process which is incorporated in the mental structure employed by an architect is the foundation for the next steps when designing with the landscape.

## 2 Designing with the Landscape

### 2.1 *Diatopia* and the Question of Time

Landscape assessment may or may not hint towards its cultural context. Mitchell [11] speaks of landscape as a natural setting which is mediated by civilization, bringing into play the frame and what is framed, the real place and its idol. Integrating new architecture into the landscape largely implies that a new system is being placed into an existing one, after having comprehended the dynamics of the latter and while searching for new spatial dynamics. Nature and human works produce new concepts, symbols, values and meanings which transform spaces into landscapes. Through conscious and unconscious processes, we scan the system of a landscape forming a field of energy which delivers new stimuli to be read and perceived. A landscape can transform into a state of mind, fueling new concepts and ideas.

Landscape as the backdrop of the design process can be assessed and analyzed—as we saw—within a certain normative framework. Yet, such a framework cannot gear the design towards the cultural challenges to be confronted; in order to do so, the architect is called to choose and prioritize among the tangible and intangible values of the landscape. Such choices will give form to his or her design concept.

As landscape is the merger of place and time, an approach which can encompass time proves to be a significant driver for design. Such a contemporary approach is what has been coined as *diatopia* [12], p. 58 or inter-locality [13, 14], a canvas on which our contemporary spatial culture can best be synthesized. Diatopia absorbs what has been and what is, bringing the traces and anthropological marks into play.

With her 'House of the Winds' [11] Greek Architect Agnes Couvelas employs the *diatopic* method seeking to harmonize construction and traditional materials with landscape, in that case, the unique landscape on the island of Santorini. Santorini bears distinct traces of volcanic, aeolic and human activity so intensely that its landscape has also been described as 'an archive of the earth' [15], p. 18. More particularly, the remainders of fortresses and the natural composition of eroded rocks on the Vlychada shore of Santorini (Fig. 1) have been origins of design for Couvelas. The House of the Winds in the southern tip of Santorini takes on the debate of how to achieve balance when building in naturally and culturally unique environments, such as this volcanic corner of the Mediterranean.

A lone building, the house 'sits' heavily, firmly rooted in the ground so as not to be blown away by the wind, while another part of the building appears 'detached', as if it

**Fig. 1** The shore of Vlychada, Santorini

has come from afar, like the enormous boulders hurled by the volcano. Together with the neighbouring columns of the veranda they recall the form of the cliffs on the seashore that stand precariously, their base undermined by the sea [16] Fig. 2. The openings are designed as windbreakers, according to the Bernoulli laws, protecting the indoors as a natural windshield Fig. 3.

Couvelas made a decisive shift by adopting both value systems—cultural and natural—into her architectural design with the House of Winds. The house's facades are informed as much by architectural evolution on the isle, as by forces of nature, like the wind and the patterns of erosion. Her design is informed by and also informs the cultural continuum, while using raw, natural elements like the wind to shield the interiors. Couvelas (re)turns to nature: the cliffs of Vlychada, a natural heritage site at the southern part of the island, lends a feeling of 'suspension' to her architecture [15, 17].

The discourse on *diatopia* or inter-locality takes on a new meaning if the natural environment, the local materials and the landscape are discussed as the treasures of the Southern European and Mediterranean cultural continuum, as masterfully described by Braudel [18], where inherent values of heritage and anonymous architecture are preserved. In Greece, the concept of landscape is no longer specified as solely a "beautiful landscape", as was the case until recently. Landscapes to be preserved according to Greek law 1650/86, as well as the Greek law 3937/11 for the 'preservation of biodiversity', now relate closer to the European framework and the idea that a landscape carries 'man made features' as well as cultural value.

**Fig. 2** "Dug-out" mass on the rocks of Vlychada in both the cliffs and the house

**Fig. 3** Wind deflectors at the northeast edge of the upper storey

## 2.2 Interpretation: The Question of Ancient and Contemporary

Greek architects Bouki Babalou-Noukaki and Antonis Noukakis, in their project for an open air sculpture exhibition space on the ancient site of Filopappos engage with a historical site of magnitude. The intervention is located the wider area of Filopappou Hill, including the Hills of the Nymphs and Muses at the foot of the Acropolis of Athens. At the south part of the Muses Hill there is an abandoned quarry. This area, even though it is located inside the archaeological site, it doesn't contain archaeological findings. The landscape has been altered through the extensive quarrying in the nineteenth and early twentieth centuries.

With the intention to capture and attribute the particular physiognomy of the place, the project investigates the relationship between landscape and cityscape; it sets out to restore the relationship between ancient and contemporary city; it features the historicity of place and its transformations through time. The Filopappou Hill Open Air Museum focuses on interpreting locus within the framework of historic time (Fig. 4).

**Fig. 4** The Filopappou hill open air museum is based on the relation between space and time

The project confronts the area of the hills as a whole, taking into account the significance of its boundaries. The central idea of the design is the highlighting and the reconstruction of the "δια Κοίλης" (*Koile*) road as the most important ancient route of the site and as a functional and organizing axis of the whole area. The regeneration around the Long Walls gate is also proposed in order to make a direct connection to the neighborhood and the city, to create an alternate access to the site and to propose the restoration of the ancient route (Figs. 5 and 6).

Landscape is investigated in its multitude and intrinsic nature, as imaginary locus, mindscape and landform. For architects, the interpretation of landscape is geared towards new prospects and designs. There is no idle interpretation of landscape for architecture; revisiting a historic site and 'seeing through' its landscape, as well as constructing a narrative of one's own, are legitimate aspects of architectural authorship.

## 3 Conclusions

When assessing a landscape, rational approaches regarding the meaning of landscape are usually based on the classification of similar traits where the attributes of landscapes are analyzed. Likewise, methodologies may employ spatial entities such as spaces, subspaces, sites, microsites etc. Nevertheless, there is always the interpretative intention behind the formation of landscapes as an investigation into the natural cause behind the form which has resulted. GIS data and planning standards consist the foundation for all holistic, socially adept approaches aiming at societal and territorial cohesion as well as environmental protection. The Florence Declaration provides for geospatial data and terms of geographical and institutional definition.

**Fig. 5** Plan showing the evolution of Athens since 338-86 BC. *Source* Travlos [19], p. 92 and in red the Filopappos ring road, in blue the ancient wall, in yellow the Koile road

**Fig. 6** Athens today with the same diagram as Fig. 5

Interpretation of landscape may pertain to the perception of space as a delimited area of sensory—not geographical—boundaries. Furthermore, there is the correlation between ecological polymorphy and the visual complexity [20] of a landscape without the necessary overlapping the two. The visual quality of a landscape is often linked to its function and based on an ecological interpretation. The tangible and intangible values contained in landscapes form the potential for designing with the landscape. When we design with the landscape, more often than not the anthropological—societal, historical, poetic—interpretation of the formation of the landscape remains latent and it is up to the architect to make the cultural choice that unlocks the potential of a landscape.

Based on EU preservation legislation, methodologies of landscape assessment and the systematic compilation of visual resources undoubtedly offer a sound foundation to architectural design, but, they cannot per se lead to it; both examples employed in this paper, the House of the Winds and the Filopappou Hill Open Air Museum, heavily rely on the architects' perception of what is important about the landscape and how the cultural continuum can be restored. In the examples showcased, the power of nature and the power of history are called forth as vital parts of the design, while the way architects designed *with* the landscape resulted to genuine architectural authorship.

## References


1. Marine N (2022) Landscape assessment methods derived from the European landscape convention: comparison of three Spanish cases. Earth 3:522–536
2. De Montis A (2014) Impacts of the European landscape convention on national planning systems: a comparative investigation of six case studies. Landscape Urban Plan 124:53–65
3. Déjeant-Pons M (2019) The environmental, cultural, social and economic dimensions of the landscape: a look at current experiences on the threshold of the 20th anniversary of the adoption of the convention. Econ Cult 3:357–370
4. Pearce D, Turner RK (1990) Economics of natural resources and the environment. Johns Hopkins University Press, Baltimore
5. Clement G (2002) Eloge des Vagabondes, Arbres Et Fleurs A La Conquête Du Monde. NiL, Paris
6. Davoudi S, Evans N, Governa F, Santangelo M (2008) Territorial governance in the making: approaches, methodologies, practices. Boletin de la Asociación Española de Geografía 46:33–52
7. Cicerchia A (2019) Landscape in the well-being measures. Economia della Cultura, Società editrice il Mulino 3:333–348
8. Frantzi M, Seferlis A (2000) The adventure of the architectural heritage management in Arachova of central Greece: effects and prospects. In: Proceedings of the 5th international congress on restoration of architectural heritage, CICOP, Firenze
9. Jones and Jones (1977) Aesthetics and resource management for highways. U.S. Dept. of Transportation, Washington DC
10. Giaoutzi M, Frantzi M (1992) Ecologically sustainable development and visual resources. In: Paper presented at the 32nd congress of RSA, Brussels
11. Mitchell WJT (1994) Picture theory: essays on verbal and visual representation. The University of Chicago Press, Chicago
12. Fatouros D (2007) The insistence of architecture. Kastaniotis, Athens (Φατούρος Δ (2007) Η Επιμονή της αρχιτεκτονικής. Εκδόσεις Καστανιώτη, Αθήνα)
13. Couvelas A (2008) Lecture at the Hellenic institute of architecture. Athens
14. Couvelas A (2008) Lecture at the KAMRA TAL-PERITI. Valetta, Malta
15. Lava R (2017) Architecture in the middle of [now] here: how locus overwrites style. In: Paper in archi-medes international conference, place and locality versus modernism. Examples of emerging new paradigms in Architectural Design NTUA, Athens
16. Couvelas A (1997) House at Santorini. Period Domus 789:32–36

17. Couvelas A (2016) The house of the winds. Shape IKE, Athens
18. Braudel F (1995) The Mediterranean and the Mediterranean world in the age of Philip II. University of California Press, Berkeley
19. Travlos I (1993) The evolution of Athens (Η πολεοδομική εξέλιξις των Αθηνών). Editions KAPON, Athens
20. Lima M (2011) Visual complexity: mapping patterns of information. Princeton Architectural Press, Princeton

# Integrative Architectural and Strategic Planning: Application of Quadruple Helix Methods in Croatian Context

Rene Lisac(✉) and Kristina Careva

Faculty of Architecture, University of Zagreb, Kačićeva 26, 10000 Zagreb, Croatia
rene.lisac@arhitekt.hr

**Abstract.** In the last ten years on the Faculty of Architecture at the University of Zagreb integrative methodologies for architectural and strategic planning and programming developments in the public domain have been developed and put into practice. Stakeholders from different sectors (civil, government, academic and business) have various perspectives on what public interest is and how it should be achieved. In complex strategic decision-making Sustainable Development Goals (SDG) play an important role as criteria in evaluating certain action, solution or policy and its benefits to the environment, the social realm, or economics. Intensive participatory and interdisciplinary Quadruple Helix (Quad Helix, QH) processes are proven to ensure that innovative practice is in line with public interest, sustainable development goals, and, most of all, is harmonized between different sector stakeholders. Parallel and intertwined with architectural and strategic planning activities, designing and managing an integrative QH process is a task by itself, with specific challenges to overcome: detecting the challenge holder as well as key stakeholders to be fully inclusive but efficient; involve stakeholders continuously from the beginning, balance involvement depending on their role importance, and provide relevant feedback; moderate discussions dynamically so participants can equally express opinions and achieve consensus; process participative data to be relevant for planning; design the QH process individually for each task. Three recent projects, developed within the Faculty of Architecture that follow above mentioned planning principles, have been selected and analyzed to conclude on the challenges of contemporary integrative planning.

**Keywords:** Integrative methods · Quadruple helix · City acupuncture · Strategic planning · Architectural and urban planning · Smart city · Fab city

## 1 Introduction

Transition from the twentieth century economic city to the sustainable city for tomorrow is a necessary global change, rapidly taking place from the 2nd decade of the twenty-first century. New paradigms like Smart City or Fab City emerged, having more or less compatible fundamental motivations and approaches, like new pathways to achieve sustainability goals. In the process of this transition questions arise: do new implementations planned benefit the citizen needs and aspirations, or do they reflect sustainability in a wholesome manner?


M. Blanco Lage et al. (eds.), *Towards a New European Bauhaus—Challenges in Design Education*,
https://doi.org/10.1007/978-3-031-40188-6_33

### 1.1 Smart City and Fab City Paradigms

The issues related to smart cities became the subject of scientific studies in the 1990s. They reflected urban development associated with the use of modern technologies, innovation, and globalization [1]. In his 2007 lectures, William J. Mitchell, Professor of Architecture and Media Arts and Sciences at the MIT, defined "intelligent cities" as the areas that function based on the combination of increasingly effective digital telecommunication networks (compared to nerves), widespread intelligence (compared to brains), sensors and markers (compared to sensory organs) and software (compared to the knowledge and cognitive competences) [2]. This new "technological organ system" for the city does not necessary imply that it contains sound purpose how it should be used, criteria which changes in urban functioning should it bring or vision of the goals to achieve. These statements are firmly advocated by Rem Koolhaas in his lecture Are Smart Cities Condemned to Be Stupid? [3]. Therefore, in the first place we can observe it merely as a tool, and before it becomes a new paradigm it must thoughtfully acquire criteria from sustainable development, environmental and especially social principles.

Fab City concept, that emerged around 2011 between City of Barcelona and MIT Centre for Bits and Atoms, aims to bring the changes in the cities by reforming its productivity aspect. Aiming at locally productive, globally connected and self-sufficient cities, in its Whitepaper by Tomas Diez it brings tools as follows: Advanced Manufacturing Ecosystem, Distributed Energy Production, Cryptocurrencies for a New Value Chain, Food Production and Urban Permaculture, Educating for the Future, Building the Spiral Economy, Collaboration between Governments and the Civil Society [4]. Unlike most Smart City visions, we can say that Fab city has more thoughtfully integrated sustainability criteria, but still a specific chosen set of principles that support the new vision how cities should produce and connect.

### 1.2 Involving Social Structures in Planning

Primary focus in contemporary urban transitions on ecological issues in the city started to shift towards and include social aspects of sustainability. The ecological problems of pollution, depletion of energy and other resources, waste management, recycling, increased traffic, etc. [5] are being complemented with new dimensions, the ideas of social justice, equality, cohesion, diversity, cultural values, human rights, social capital, access to employment, health care and various services, economic stability and growth, civic participation, identity, safety, social networks, sense of identity, etc. [6, 7]. All major theories of planning over the last 30 years have placed emphasis on the importance of involving a wide range of social actors in spatial planning processes [8]. Weather the task is to understand and involve complex dimensions of city social structures and social capital in planning as a important part of social sustainability, or to integrally develop and implement urban inventions from Smart and Fab city paradigms, public participation is the way to go. Therefore, it is not a surprise that public participation and integral planning that include stakeholders form all sectors have become a inevitable method in sustainable planning.

In the last 10 years the Faculty of Architecture at the University of Zagreb has been developing and practicing integrative methodologies for innovative planning and programming developments in the public domain. Stakeholders from different sectors (civil, government, academic and business) have various perspectives on what public interest is and how it should be achieved. In complex strategic decision-making SDGs play an important role as criteria in evaluating certain action, solution or policy and its benefits to the environment, the social realm or economics. Intensive participation and interdisciplinary QH processes are proven to ensure that innovative practice is in line with public interest, sustainable development goals, and, most of all, is harmonized between different sector stakeholders.

Parallel and intertwined with architectural and planning activities, designing, and managing an integrative QH process is a task by itself, with specific challenges to overcome. We have selected and analyzed three recent projects developed within the Faculty of Architecture that follow above mentioned planning principles, to conclude on the challenges of contemporary integrative planning.

## 2 Quadruple Helix in Architectural and Strategic Planning

The Quadruple Helix (Quad Helix, QH) model was initially developed and can be defined as "an innovation cooperation model or innovation environment in which individual users, commercial ventures, universities, and public authorities cooperate in order to produce innovations" [9] (Fig. 1). The core thesis is that innovation is the outcome of an interactive process between the four stakeholder groups—university, industry, government, and civil society—and each group contributes according to its institutional function in society [10]. QH Model has proven its value in various innovation processes, but why would it be valuable in architectural and strategic planning?

Firstly, creative and innovative processes are so to say ventures into the unknown territory that offers multiple scenarios, depending on the stakeholders involved. Overlaying objective facts, their subjective realities confront, interact, and tend to create new, common reality. As wide stakeholders are, the result becomes les subjective, so to say it "objectifies". QH vortex model without doubt handles better than straight forward disciplinary or non-participative processes. Urban ecosystems are structured exactly the same, over more or less objective layers of Physical space and Infrastructure opens subjective realms of Collaboration Ecosystems and Innovative Solutions [12]. In 2010, Carayannis and Campbell, based on the Triple Helix and Quadruple Helix, proposed the Quintuple Helix model [13]. Knowledge, which is supposed to lead to more sustainable development, is a driving force for progress in this model. The fifth element of the model in question is the natural environment [14]. Still, for the urban issues it is more accurate to refer to the QH model, mainly because they prefer a subjective approach to issues related to sustainable urban development. The environment does not generate knowledge and cannot be treated as subjective in this sense [15].

Secondly, decision making processes in urban ecosystems can greatly benefit form including all stakeholders. Although it is not necessarily creative or innovation process, implementing QH can bring improved decisions with better involvement, acceptance of changes and so on. twentieth century characterizes extreme growth in knowledge, specialization of profession and sciences, development of complex and sophisticated systems how our society works. Dispersion of knowledge together with separation between

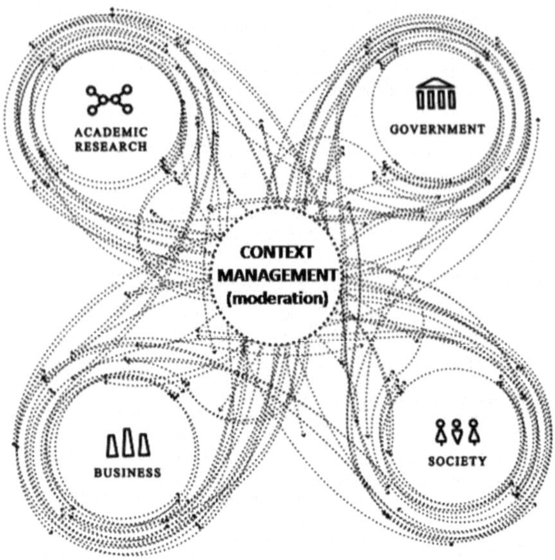


**Fig. 1** QH model with context management agent—moderation of process by authors, based on Schütz et al. [11]

different stakeholders lead to numerous disadvantages for planning and implementation in urban areas. Natural complementary reaction to this are new trends of interdisciplinary approach and participative methods that tend to balance omnipresent dispersion with new cohesion. An optimal model of urban management and creation of a healthy Smart Sustainable City is based on knowledge properly disseminated and distributed among all stakeholders of QH process [15].

This is especially present in former socialist countries in SE Europe. Transition from socialist system to democratic, based on market economy, brought general change in relations between QH stakeholders (Fig. 2). Academic and Government sector held power for the benefit of civil and business, but after 90's power shifted between government and selected layer of business, not necessarily considering other sectors. Often associated with corruption, it brought lack of confidence in government by all sectors, general dissatisfaction as well as barrenness of plans. Strengthening relations between Academic and Civil sector has proven to be the only way to balance unhealthy axis government—business. Introducing cohesive inclusive methods in planning on Faculty of architecture in Zagreb follows these trends in Croatian context.

## 3 Integrative Planning at Faculty of Architecture in Zagreb

Among first efforts in introducing integrative methods that exceed boundaries of architectural and urban planning profession on Faculty of architecture, an initiative called City Acupuncture was developed between the faculty and local NGO "Zagreb Society of Architects" in 2009. Initiative aimed on small interventions in public space to improve quality of city life: spatial, aesthetic, and functional improvements to empower

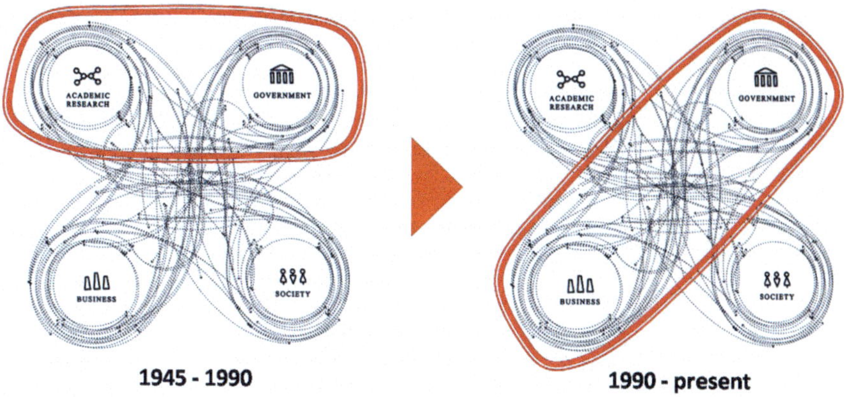


**Fig. 2** Knowledge and power distribution in Croatian urban ecosystems. Transition from socialist (1945–1990) to market economy democratic society (1990-present) by authors, based on Schütz et al. [11]

city social capital like social cohesion, democratization of public space, involvement of citizens and so on. Locations in specific city quarter were chosen and interventions designed through interdisciplinary and participative workshops [16]. Almost 20 workshops, including elective course on the faculty Participatory Design of Space, included more than 250 young professionals and students from different fields, resulted in around 150 proposals, more than 20 realizations of interventions and five methodology transfers to other initiatives and cities, sharing inspiration and experience in integrative planning. Being focused on small and simple interventions it was possible experiment the potential of participation and interdisciplinary methods put into practice, in their fullest or purest form which turns to be much more difficult for bigger scale projects. Four such projects presented in this work were inspired to follow wholesome methods from City Acupuncture and confront it against "business as usual": how cities are planned by professions, governed by municipality, commercialized by developers, and perceived by citizens. And thus, bringing to light challenges in introducing integrative planning methods.

### 3.1 Zagreb for Me Project

Being directly inspired by City Acupuncture, Zagreb for Me project from 2015 to 2017 brought the methods on a bigger scale, regeneration of large public spaces like squares, parks, promenades came into focus, on the whole city of Zagreb level. Since there is no commercial interest involved, Triple Helix process (excluding business sector) was custom designed to choose locations and program the integrative regeneration process (Fig. 3). Project was recognized by extreme introduction of integrative methods throughout all 3 sectors [17]. Public participation was ensured through vast interviews on several public locations, as well as focus groups with relevant NGO's. In terms of academic, architectural, and urban planning profession, as well as sociology and anthropology were involved through urban planning analysis and moderation of process. Discussions on several city levels from local municipalities of quarters to the top city planning institutions included government sector. Unfortunately, due to project complex and advanced nature, together with lack of persistence in the city institutions, it did not pass into implementation phase.

**Fig. 3** Overlapping top-down urban planning and bottom-up participation methods in detecting and programming city locations for regeneration (Zagreb For Me project)

## 3.2 Brdovec New Central Public Zone

Brdovec is a small rural county near Zagreb, developing quickly into a small satellite city. The need for developing New Central Public and residential zone was recognized, but to precisely define the need for urban spaces and facilities, as well as specific housing typologies, an integrative master plan for the area was designed on the Faculty of Architecture (Fig. 4). QH process was designed to target key issues, wide survey among citizens was combined with focused discussions in the county municipality, developer representatives were involved in housing typologies issues. The inclusive process resulted in precise and diverse program of central facilities and spaces to fit the needs and future identity of the county between urban and rural, mix of four different housing typologies provide diverse living options to cover that span. Master plan proposed general spatial disposition, to be forwarded as a foundation for standard urban planning procedures.

## 3.3 Petrinja Central Public Zone Reconstruction

In last days of 2020, strong 6.4 magnitude earthquake struck Petrinja, a city 60 km SE of Zagreb, and left vast destruction in the area as well as the centre of the city. To bring

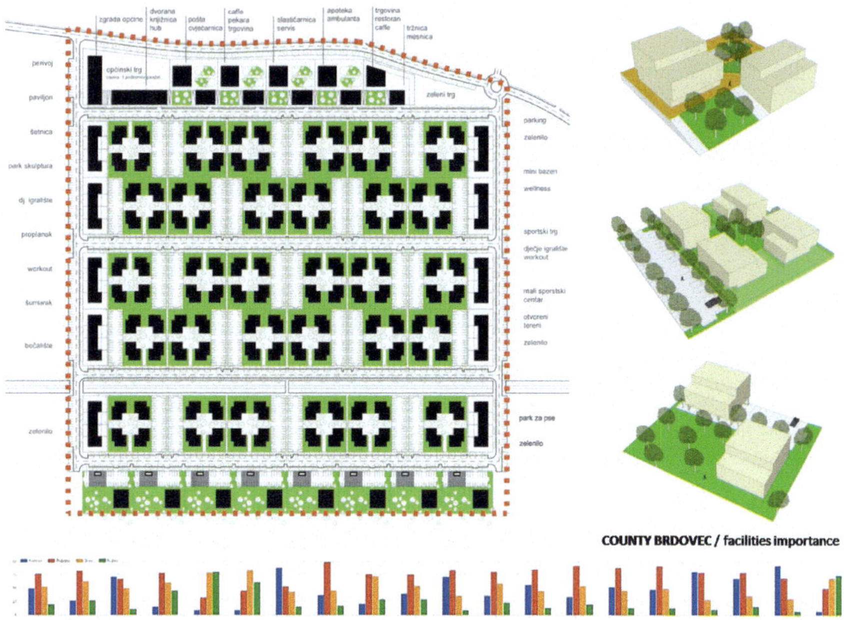


**Fig. 4** Survey results and master plan for Brdovec New Central Public Zone

new optimism and preserve life in the city, reconstruction of central public zone around Main Square proved to be of high importance. Since the existing urban plans were not prepared for such destructive change and introduction of new contemporary identity, quick reaction following standard procedures was not possible. To avoid spontaneous and unplanned interventions that could in the long run result in poor urban planning, triple helix process was designed that includes wide survey, focus groups with city municipality and local NGO initiatives, as well as interviews with local urban planning experts (Fig. 5).

## 4 Conclusion: Challenges of Contemporary Integrative Planning

Experience from selected project has shown potential as well as challenges that follow integrative QH processes. Parallel and intertwined with architectural and planning activities, designing, and managing an integrative QH process represents a task by itself, involving all stakeholders together brings integrity but also means balancing their different perspectives and interests. While government and academia tend to observe issues top down, business, and civil sectors nurture bottom-up perspective. QH processes, similar as interdisciplinary environment, are not only creative, wholesome, and cohesive, but also educative and evolutionary. In these processes each sector has its natural weaknesses to overcome: academia with rigid concepts that are sometimes not applicable; government with political interests; business being focused on primary financial results; and civil sector with private, individual interests. Between all this, crucial decision-making criteria to not to stray from public interest can be principles of sustainable development:

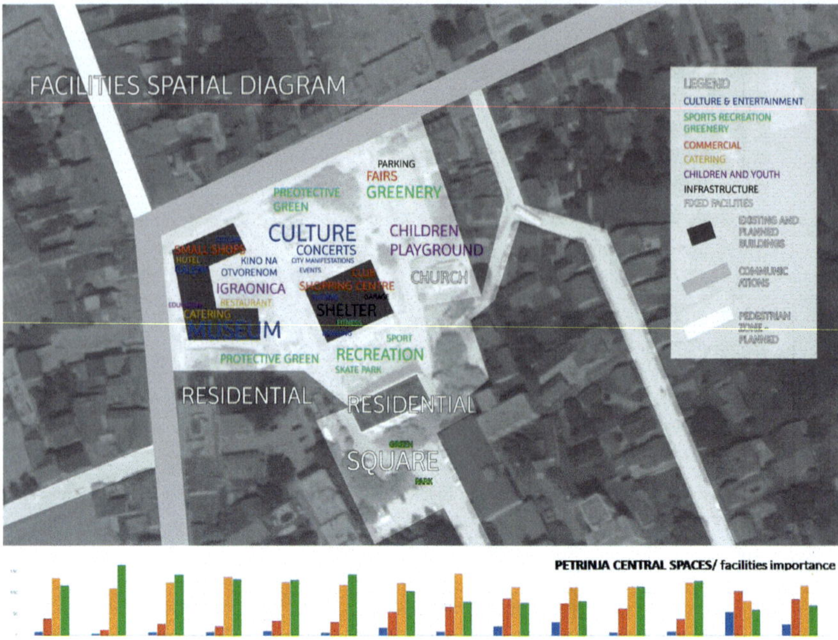


**Fig. 5** Survey results and facilities spatial diagram for Petrinja central zone regeneration

If it benefits firstly the environment, then the society and in the end it brings profit—it can be considered public interest (Fig. 6).

Aside from balancing stakeholders' interests, it is necessary to detect the challenge holder (main stakeholder) as well as other key stakeholders, to be fully and continuously integrative from the very beginning of the process but stay efficient. Further, it is important to involve stakeholders continuously from the beginning, balance their involvement depending on their role importance, and provide relevant feedback. The role of moderator agent in the process is to moderate discussions dynamically so participants can equally express opinions and achieve consensus, as well as to ensure all issues are precisely addressed. Special challenge is to process participative data to be relevant for planning, but not to lose the original intent as well as the feeling for the end user.

What the experience has precisely shown that without custom designing the QH process individually for each task, it is not possible to ensure it can be carried out or the results implemented. This flexibility is necessary for methods that tend to slowly change the paradigm of how we approach city planning, to improve rigid conventional procedures.

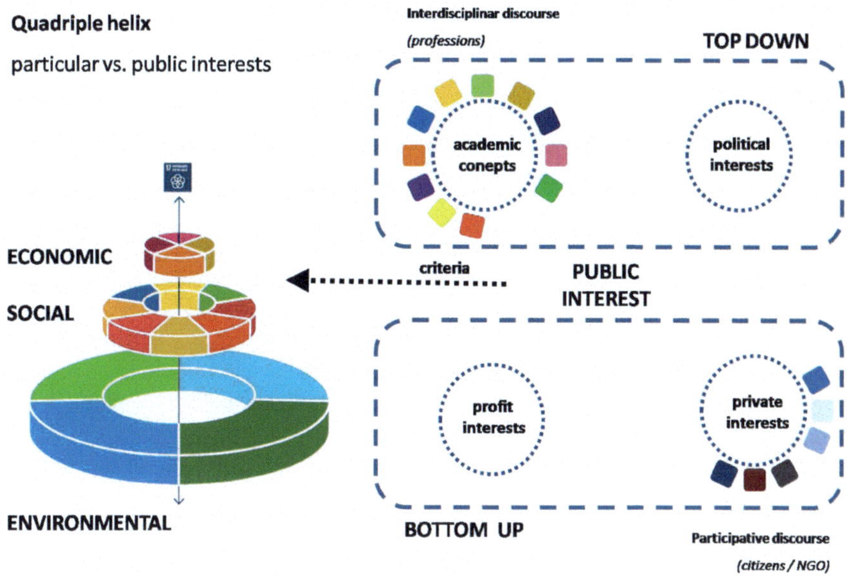


**Fig. 6** Weaknesses between sectors and sustainable development criteria for defining public interest

## References


1. Gibson DV, Kozmetsky G, Smilor RW (eds) (1992) The technopolis phenomenon: smart cities, fast systems, global networks. Rowman and Littlefield Publishers, New York
2. Mitchell WJ (2007) Intelligent cities. UOC Papers, p 5. http://www.uoc.edu/uocpapers/5/dt/eng/mitchell.pdf. Accessed 05 Dec 2022
3. Koolhaas R (2014) Are smart cities condemned to be stupid? https://www.archdaily.com/576480/rem-koolhaas-asks-are-smart-cities-condemned-to-be-stupid. Accessed 05 Dec 2022
4. Diez Ladera T (2013) The new production ecosystem. Personal, distributed, open fabrication. CCCB LAB RSS, Barcelona
5. Bromley R, Tallon A, Thomas C (2005) City centre regeneration through residential development: contributing to sustainability. Urban Stud 42(13):2407–2429. https://doi.org/10.1080/00420980500379537
6. ODPM (Office of the Deputy Prime Minister) (2003) Sustainable com163 munities: building for the future. ODPM, London
7. Colantonio A, Dixon T (2011) Urban regeneration and social sustainability: best practice from European cities. Wiley-Blackwell, Oxford. https://doi.org/10.1002/9781444329445
8. Mišetić A, Krnić R (2012) Actors in the planning and development of Varaždin: the contribution of social sustainability research in the urban context. Društvena istraživanja: časopis za opća društvena pitanja, Vol. 22 No. 1. Institut društvenih znanosti IVO PILAR, Zagreb
9. Arnkil R, Järvensivu A, Koski P, Piirainen T (2010) Exploring quadruple helix, tampereen yliopistopaino Oy Juvenes Print: Tampere CLIQ, p 70. http://urn.fi/urn:isbn:978-951-44-8209-0. Accessed 05 Dec 2022
10. Cavallini S, Soldi R, Friedl J, Volpe M (2016) Using the quadruple helix approach to accelerate the transfer of research and innovation results to regional growth Eur. Union Publ. https://doi.org/10.2863/408040

11. Schütz F, Heidingsfelder M, Schraudner M (2019) Co-shaping the future in quadruple helix innovation systems: uncovering public preferences toward participatory research and innovation. She Ji J Des Econ Innov 5(2):128–146. https://doi.org/10.1016/j.sheji.2019.04.002
12. Taratori R, Rodriguez-Fiscal P, Pacho M, Koutra S, Pareja-Eastaway M, Thomas D (2021) Unveiling the evolution of innovation ecosystems: an analysis of triple, quadruple, and quintuple helix model innovation systems in European case studies. Sustainability 13(14):7582. https://doi.org/10.3390/su13147582
13. Carayannis EG, Campbell DFJ (2010) Triple helix, quadruple helix and quintuple helix and how do knowledge, innovation and the environment relate to each other? A proposed framework for a trans-disciplinary analysis of sustainable development and social ecology. Int J Soc Ecol Sustain Develop 1(1):41–69. https://doi.org/10.4018/jsesd.2010010105
14. Carayannis EG, Barth TD, Campbell DFJ (2012) The quintuple helix innovation model: global warming as a challenge and driver for innovation. J Innov Entrep 1:2. https://doi.org/10.1186/2192-5372-1-2
15. Kuzior A, Kuzior P (2020) The quadruple helix model as a smart city design principle. Virtual Econ 3:1
16. Careva K (ed) (2014) Improving the city life quality through small and precise interventions in urban structures. Zagreb Society of Architects, Zagreb
17. Careva K, Lisac R, Pletenac T, Vukić J (2017) Akupunktura grada kao participativni alat za revitalizaciju javnog prostora. Prostor 25(2):190–199

# Towards a Future Past: The New and the European in the Bauhaus

Sophie Mak-Schram(✉)

Zeppelin University, Friedrichshafen, Germany
sophie.mak-schram@zu.de

**Abstract.** What does it mean to 'rescue' a modernist project like the Bauhaus to envision a future? The pedagogical emphasis of the Bauhaus, which sat at the core of its approach to society and the role aesthetic and design practices could play in this, was one premised on learning through making, interdisciplinary practices and design as an enabler, not just mirror, of socio-cultural aspirations and ambitions. What is striking over a century later is how this alternative model of education continues to be espoused as the future. Recent publications, such as *Bauhaus Imaginista* (2019), trace the impact of the Bauhaus' socio-cultural approach and aesthetic across the world, emphasising its enduring international legacy. In the realm of inclusion, however, recent texts signal the limitations of the Bauhaus' vision. Artists such as Anni Albers were sidelined into departments 'appropriate' for their gender, and the school was supported and bore close relation to the nation-state via its funding and staffing. This paper will reflect on the ways that the 'old' Bauhaus has become used as a symbol, and what critical reflections and distance the New European Bauhaus might need to take from its modernist referent. This consideration will focus on the issues of gender inclusion within the Bauhaus' otherwise radical educational structure.

**Keywords:** Bauhaus · New European Bauhaus · Inclusion · Internationalism · Pedagogy · Gender

## 1 Introduction: More Than a Name

What does it mean to 'rescue' a modernist project like the Bauhaus to envision a future? The New European Bauhaus, as an initiative by the EU aiming to "connect" the European Green Deal to "living spaces and experiences", is curiously named in relation to the architectural and art historical reference it makes. The Bauhaus is an art school founded by Walter Gropius in 1919 in Weimar, Germany, that holds a central position in art history of the twentieth century for its formative influence on pedagogies in and of the arts. This paper will reflect on which elements of the 'old' Bauhaus are referred to in the New European Bauhaus, and how the use of the Bauhaus as symbol and referent might create connotations that contradict or shift the intentions of future-oriented pan-European policies.

---

The original version of this chapter was previously published without open access. A correction to this chapter is available at https://doi.org/10.1007/978-3-031-40188-6_40


M. Blanco Lage et al. (eds.), *Towards a New European Bauhaus—Challenges in Design Education*,
https://doi.org/10.1007/978-3-031-40188-6_34

The Bauhaus is not just a name of an art school that existed between 1919 and 1933, but also an agglomeration of associations. Both its pedagogical practices and its positioning of art have subsequentially been influential and shaped ideologies of alternative art education and the role of art in society above and beyond those directly involved in educating and being educated within its walls. The Bauhaus was distinct from previous art schools in terms of how it actively worked to put art in relation to industry and commerce, and how it moved away from the distinction between craft and art. Instead, the various Workshops combined technique and experimentation with a focus on outputs for society. Pedagogically, the specific structure of the Master teaching learners was not in itself novel, but as a "school *by* practitioners, *for* practitioners" [1] the Bauhaus modelled a form of education where practice led for the end of function, relevance and industry rather than for art's sake alone. This model became internationally relevant. Josef Albers' colour studies, for instance, and his rigorous methods of developing new colour combinations via mathematical method, were pedagogical practices that influenced educational practices as much as produced licensable, marketable things. After the Bauhaus closed in 1933, Albers went straight on to the newly founded art college, Black Mountain College (North Carolina, USA), where he both continued to teach and expand on the Bauhaus ideals, and supported his students in the devising of patterns and prints for commercial textiles and similar industry-proximate production.

The question that guides this paper is whether it is the 'old' Bauhaus' specific practices, or its eminence and subsequent associations produced through historicization, that the New European Bauhaus makes reference to. Specifically, this paper is interested in how the continued interest in the Bauhaus is suggestive of the assumptions made around its idealism and their continued relevance not just in the present but also for the future. What is striking over a century since the Bauhaus' founding, is how this alternative model of education continues to be espoused as, as much as for, the future. Why are we still looking to a historic model to set the horizon of a future? And, what do we overlook or gloss over when doing so?

Recent publications, such as *Bauhaus Imaginista* (2019), trace the impact of the Bauhaus' socio-cultural approach and aesthetic across the world, emphasising its enduring international legacy. In the realm of inclusion, however, recent texts signal the limitations of the Bauhaus' vision. Artists such as Anni Albers were sidelined into departments 'appropriate' for their gender, and the school was supported and bore close relation to the nation-state via its funding and staffing. This paper will reflect on the ways that 'old' Bauhaus has become used as a symbol, and what critical reflections and distance New Bauhaus might need to take from its modernist referent. This consideration will focus on the issues of gender inclusion faced within its otherwise radical educational structure.

## 2 The New European Bauhaus

In a March 2021 policy briefing, the European Parliament said the following:

> Both the Bauhaus and the New European Bauhaus share a concern for aesthetics and inclusiveness. At a time of economic scarcity, Gropius sought to spread democracy by educating students to create designs for everyone—at once industrial and ideally beautiful, useful and economical. At a time of saturated markets,

the New European Bauhaus initiative is applying this inspiration to a modified set of societal, environmental and political issues [2].

There are several conflations in this statement. Gropius and the Bauhaus become associated with democracy, the turn of the twentieth century is aligned to the present day via a strange equivocation of economic scarcity and saturated markets, and the Bauhaus becomes an inspiration for societal, environmental and political issues. A specific, nationally grounded and state-funded art school becomes a potent symbol for ways of being together with art in society, and hereby transformed into an EU-wide source for resolving current social and ecological concerns through policy. There appears to be no reflection on the simple fact that the Bauhaus functioned as a school rather than as a movement, with all the disciplinary as much as temporal constraints that this comes with. Schools function very differently to policy, even if we were to boldly imagine policy as an expanded education.

The New European Bauhaus is related to the European Green Deal, which aims to "transform the EU into a modern, resource-efficient and competitive economy" [3]. Within this, the New European Bauhaus seeks to foster creativity with an end, something that is also reflected in the lead quote on the New European Bauhaus' site, where Ursula Von der Leyen, president of the European Commission, is cited as saying "If the European Green Deal has a soul, then it is the New European Bauhaus which has led to an explosion of creativity across our Union" [4]. The New European Bauhaus is, vitally: (1) A policy imperative; (2) A pan-EU policy and; (3) Focused on sustainability in where and how European inhabitants live.

The tensions that arise between this vision and framing of the New European Bauhaus and its namesake, arise from the question of international or interlocal impact and the assumption of inclusiveness that underlies the policy ambitions. The Bauhaus' interest in the international is at odds with current sustainability practices that might now operate on a more local level, yet also the Bauhaus' specific relation to the nation-state of Germany also renders it curious as a model for a pan-EU policy. The assumption of inclusiveness as finding itself a model source in the Bauhaus is also something that recent scholarship complicates, as I will expand on below.

## 3 Inclusion in the Old Bauhaus

The recent celebration of the Bauhaus' centenary of founding in 2019, led to renewed scholarship. This scholarship included some revitalised discussions around the Bauhaus' gendered division and sidelining of women. The pedagogical approach of the Bauhaus and its emphasis on Workshops was one in which students learned in a hands-on manner. Everything they made during as much as outside of these Workshops both belonged to the Bauhaus and became part of its associated aesthetic. There was a strong focus on practical application and the bringing together of art and craft, with a focus on what we might now call access or inclusion as an outcome of this (art) education.

Their designs for everyday life, as William Smock rightly points out in his otherwise overly casual book [5], have however since been reabsorbed into the cultural sphere. One can think here of museum chairs and generally the modernist aesthetic, which despite

Gropius' adamancy against a Bauhaus style, do aesthetically date pieces of art and design as from the early twentieth century and are now associated with a certain geopolitical site (the Euro-American early to mid-century) and level of socio-economic access. The inclusive approach of the Bauhaus, in the sense of being 'democratic' (as the European Parliament might have it), is tripped up by the elitism of the modernist aesthetic which is no longer comfortably affordable or aesthetically accessible for all. The Bauhaus has become synonymous with modernist aesthetic to such a degree that high end interior and vintage sites tag and group other manufacturers, designers and anything with tubular steel, a lack of decorative quality and an emphasis on function under the school's name as a category. With vintage Bauhaus and Bauhaus-related furniture now easily priced at €2000–€4000 if not more, this is a far cry away from the democratic emphasis of the school's drive to design [6].

Further, and vitally, there is a question of who this 'all' was that the aesthetic sought to include. How can an 'all' be appropriate for the global context? Is there a collective 'all' within Europe for whom buildings and cities can be designed? The able-bodied, often male, likely northern European majority white default is often assumed. The Bauhaus, and the context within which its pedagogies emerged, is strongly rooted in these identities. Whilst the Bauhaus positioned itself as working with(in) society, this society was specifically that of urban German in the period between two World Wars. The Bauhaus actively worked and negotiated within this German societal context, given its dependency on state funding and need to politically position itself: this eventually led to its move from Weimar to Dessau, for instance. Notes on the student body's diversity indicate that in the first years in Weimar, the student body make-up included 17–33% students of foreign origin and 25–50% female students [7]. Whilst these percentages seem healthy, two further questions easily arise: they may have been educating a somewhat diverse population in the German context, but who is the 'they' who was doing so, and how did school structures restrict the inclusion of these diverse identities (albeit limited to nation-state and binary gender identifiers here) by way of processes, rules and forms of education? A brief glance over the lists of faculty over the years signposts to who the 'they' educating and thereby setting the content and values of what was learnt were: the majority of the faculty was German, white and male. This is not to suggest that they should not have been teaching, but rather to signal a question around what the reprisal or revisiting of this vision does when one now—as with the New European Bauhaus—seeks to make it applicable and empowering for a wider subset of identities and socio-political needs.

Recent texts signal the limitations of the Bauhaus' vision in relation to this 'all' in the realm of gender in particular. Artists such as Anni Albers were sidelined into departments 'appropriate' for their gender. Despite the Bauhaus consisting of 1/3 women—a total of 462 women over the course of its history—they were almost all placed into the Weaving Workshop regardless of many's ambitions to become painters or architects [8]. Women were limited to choosing between weaving, pottery and bookbinding as the Workshops they could attend, whilst men could freely choose from all of them. In 1924, this meant that male students could choose from 12 Workshops, whilst carpentry, metal, sculpture, glass and stage, amongst others, were off bounds for women [9]. The pottery and bookbinding Workshops eventually closed, leaving female students only a single choice: Weaving.

The Weaving Workshop has since become a rich site of study and has been influential in its pedagogies, particularly in relation to its mathematical rigour, creative use of material for functional textiles and its drawing on non-Eurocentric traditions and sources of weaving. There is an irony in the fact that one of the recent centenary celebration's higher-profile exhibitions was that of Anni Albers and the Weaving Workshop's work (at Tate Modern in 2019), given that these elements of the Bauhaus were not actively or fully supported at the time. Internal notes between Gropius and the male Masters, reveal a concern with the enrolment numbers of women. Their concern was that the perceived overrepresentation of women shifted the balance between art and craft, as women were understood to be producing craft rather than art [10]. Even if this was one of the few institutions at the time that even permitted women to attend, any contemporary reference to the Bauhaus now needs to contend with the other associations—and even allegations—involved in their approach to gender.

This concern about art and craft later shifted for the Masters when the Bauhaus' interests changed to emphasise industry. Then the problem became the idea that women's overrepresentation was moving the school towards applied art rather than art for industry. In either framing of this impact of women, two things are revealed: the idea of art and craft as being brought together or renegotiated is undermined by this defensive concern against craft entering or being overrepresented in the Bauhaus, and women were perceived to have more influence and presence than they numerically did. The inclusive act of allowing women to apply to the Bauhaus, which was historically unusual, was thus not followed through with an active embedding of female agency, leadership or even choice across the wider school structure.

## 4 Conclusion: Towards a Future Present

What does it mean, then, when we look to the Bauhaus, as an alternative model of education, and espouse it as the future for policy horizons across the EU? Are we seeking to strive towards state-funded art and design education that unifies the arts and focuses them towards industry and mass production? Or even, dare we read into the Bauhaus' impact for the community of staff and students involved, as the EU dreaming of a post-capitalist society full of artists? Probably not. There are limitations to the applicability of a historic school in interwar Germany to contemporary needs within the EU context. Ambiguities sustained in the history of the Bauhaus include the unclear international intent of the school (to what end, and in what relation to the school's situated reality of Weimar and Dessau?) and the methods through which inclusion was gestured towards but not embedded into practice. These are ambiguities present in much modernist art practices and alternative pedagogies. International networks and the vision of a society unfettered by tradition's accoutrements that could be facilitated through the 'neutral' medium of art was one shared by movements such as De Stijl and later projects such as Black Mountain College. Like the Bauhaus, however, they tended to struggle to actualise ideals of access and were hamstrung by the societal norms they were responding to. From the contemporary in which conversations around decoloniality, Diversity, Equity and Inclusion, the climate crisis and accessibility across class, ability and gender, are prevalent, the reference of the Bauhaus needs nuancing if it is to bear relevance on current policy.

The implications of bringing this historical model forward without specification, risks misleading and divergent interpretations. We might not need a Bauhaus, whether European or New, for new policy and lived experiences in European cities. The constraints, even if only briefly outlined here, of the Bauhaus' approach to inclusion in terms of who was in charge of the kinds of education offered and the differential levels of access permitted within the school, signal a need for a contemporary update. Certainly, many key Bauhaus teachers are worth citing in the realm of arts and education. To list a few: Itten's pedagogy, Gropius' prototyping and the rigorous creativity of Albers and Stolzl, amongst others, might concretely feed into further curricula, as they did with Black Mountain College. This could subsequently lead to policy that might foster more schools and practices like this, where art and aesthetics are not separated or imagined as distinct from daily life. However, the scale of these projects, which were embedded into local contexts and often, as with the Bauhaus, networked and responsive to these contexts, is often small and specific. Any referent would need to carefully untangle the vested interests and consider scalability: the Bauhaus may have sought to play a role in industry and design, but it did not seek to become a wholesale solution for living and urban design.

The usage of a modernist referent for a future-oriented policy focused on living together, risks overlooking a number of concerns: the recent reprisals and reflections on the actualities of inclusion in the Bauhaus itself, the specificities of it being a school and thereby a structured learning site in which pedagogies were practiced in relation to particular educational needs, and the ways in which its symbolism is an accruement of associations and historical eminence, all of which risk further changing over time. More widely, the use of a historic school to set a horizon for future policy raises simpler questions about the century that has passed between then and now, and the conflation of progress with something that has since been safely enshrined within history as 'innovative'. There is certainly something new about the Bauhaus, but to move towards *a* new and an emphatically European one feels at odds with both the policy and its referent.

## References


1. Von Osten M, Watson G (2019) Bauhaus imaginista. Thames and Hudson, London
2. European Parliament (2021) Education and the New European Bauhaus. https://www.europarl.europa.eu/RegData/etudes/ATAG/2021/689363/EPRS_ATA(2021)689363_EN.pdf
3. A European Green Deal homepage. https://ec.europa.eu/info/strategy/priorities-2019-2024/european-green-deal_en. Accessed 31 Aug 2022
4. New European Bauhaus homepage. https://new-european-bauhaus.europa.eu/index_en. Accessed 31 Aug 2022
5. Smock W (2004) The Bauhaus ideal: then and now, 2nd edn. Academy Chicago Publishers, Chicago
6. See online interior sites with an emphasis on vintage and antique furniture, *1stdibs* and *(smow)*, for instance. One might also think here of the pop culture story around Kourtney Kardashian and Pierre Jeanneret's Committee chair, which was originally designed for Le Corbusier's

design of the Indian city of Chandigarh. This chair was intended for general everyday use but has since been so coveted in the revival of a modernist aesthetic that India has had to put a ban on the export and sale of any remaining chairs in Chandigarh without the permission of the Ministry of Culture. Tellingly, both *(smow)* and *1stdibs* categorise modernist design like Jeanneret's chair, as well as work by Poul Henningsen (1894–1967) who was known for Danish modern design, under the tag Bauhaus. See for more on Karadshian and the chair: Riddiford, Margie, 'The captivating story behind the design world's adored Chandigarh chair', *Denizen* (2021/05/14). https://www.thedenizen.co.nz/design/pierre-jeanneret-chair-cassina/. Accessed 03 April 2023
7. Bauhaus Weimar. Bauhaus Kooperation. https://www.bauhauskooperation.com/knowledge/the-bauhaus/phases/bauhaus-weimar/. Accessed 03 April 2023
8. Otto E, Rössler P (2019) Bauhaus bodies. Bloomsbury, London
9. Gropius W, Moholy-Nagy L (1924) Bauhaubücher 7. Albert Länger Verlag, Munich
10. Weltge-Wortmann S (1998) Bauhaus textiles: women artists and the weaving workshop. Thames and Hudson, London

# A New Aesthetic for a New Bauhaus

Garreth Miley(✉)

South East Technological University, Cork Road, Waterford X91 K0EK, Ireland
garry.miley@setu.ie

**Abstract.** In addition to its demand for a more sustainable architecture, the New European Bauhaus calls for the increased inclusivity of disadvantaged groups in both the architectural design process as well as in the outcome of this process. It also calls for an improvement of the aesthetic quality of the architectural experience. In response to the NEB call, this paper explores the possibility of the development of a new type of architectural aesthetic which takes its cue directly from the need to make architecture more inclusive. Starting with an examination of the term 'aesthetic' as it applies to the creation of an architectural work and building on the ideas of Peter Blundell Jones in relation to the expression of ritual in architectural form this paper suggests that, viewed through the lens of the concept of ritual, a deep examination of the programme may lead to architectural solutions which provide a more meaningful response to the rituals of everyday life. The paper further explores the possibility that as communities around the world become ever more diverse and as the interactions of these new communities lead to new forms of ritual, an architecturally sensitive response to these new rituals offers the possibility for a fresh approach to the idea of designing for inclusivity.

**Keywords:** New European Bauhaus · Ritual · Inclusivity · Community · Aesthetic · Peter Blundell Jones

The New European Bauhaus has three stated objectives with regard to architecture: that it should be sustainable (which the NEB takes to mean that it should be biodiverse, zero-polluting, contributing to the circular economy and in line with accepted climate goals); that it should be aesthetic (by which the NEB suggests it should offer an enhanced quality of spatial experience by virtue of style); and that it should be inclusive (implying that works of architecture should place increased value on diversity, accessibility and affordability).

The first goal—that architecture should be sustainable—is for the most part related to issues concerning environmental planning and the techniques of construction. These topics have been much discussed in recent years in the academic press as well as in the popular media and this paper will not attempt to advance the discussion any further.

The focus of this paper concerns the second and third of the NEB objectives dealing with aesthetic satisfaction and inclusivity and, in particular, whether it might be possible to advance these themes in a simultaneous and related way. In short, the question posed is: can an architectural aesthetic be developed which specifically takes its cue from an idea about inclusion?


M. Blanco Lage et al. (eds.), *Towards a New European Bauhaus—Challenges in Design Education*,
https://doi.org/10.1007/978-3-031-40188-6_35

On the face of it, such a proposal might not seem very likely; however, on further reflection a possible line of enquiry does emerge. To approach this investigation in an explicable way we must first reflect on the idea of what constitutes an aesthetic. While the word aesthetic these days is sometimes seen as synonymous with 'style' or 'architectural language', in philosophy and cultural theory the term has more typically been taken to mean 'the nature of beauty' and not a mere manifestation of it. Some of the best known philosophers in history—Plato [1], Aristotle [2] and Kant [3] amongst them—have attempted to explain the nature of beauty in various ways. While all of their contributions offer valuable insights into the topic it is fair to say that philosophers, in their analysis of the nature of beauty, tend to treat the term 'beauty' in a very broad and generalized way. The result is that architects, particularly after the arrival of Modernism, tend to approach the aesthetic element of their work without reference to a robust set of guiding ideas.

Nevertheless, there are some concepts which recur in architecture which, while not offering a precise definition, occupy a position which we might say is adjacent to a definition of the term aesthetic. These concepts contain enough signification to give us a clear sense of what the word might want to mean. For example, there is a line of argument first developed in Aristotle's Metaphysics [2] and later taken up in Alberti's De Re Aedificatoria [4] which advances the notion that beauty in building relates to 'the harmony of all parts in relation to one another' and that this harmony has its foundation in a rigorous understanding of geometry.

Louis Kahn also relies on harmony and geometry in creating beauty in his work but his greater contribution to the idea of aesthetics relates to his approach to materiality. In a famous quote [5] he stated that the brick, by its nature, aspires to form an arch. This simple formulation elevated the idea of the brick—something rectangular and having mass—to an abstract aesthetic concept: the brick arranged in relation to other bricks. Mies van der Rohe, in his unbuilt Country House of 1923, also elevated the notion of the humble brick to the level of an aesthetic concept. In this project Mies demonstrated that, besides wanting to form an arch, the brick is also comfortable in making a garden wall.

In addition to Aristotelian ideas about harmony and Kahn's sensibility to materiality, we find aesthetic concepts within the notions of the immanent and the transcendent in nature, purity of physical form, the dispersal of light, the tectonics of construction, and so on.

There is another source of aesthetic possibility which is often overlooked but which is very clearly illustrated in an interesting book by Blundell Jones [6]. It concerns the idea of the relationship between architecture and human ritual behaviour.[1] The position advanced by Blundell Jones is a useful starting point for the development of the thesis at the core of this paper.

When architecture critics attempt to describe the power or effect of a piece of architecture they do so almost exclusively in relation to the effect of the work in question

[1] The study of ritual typically takes place in the domains of sociology and anthropology where there has been considerable debate around the very definition of the term. For the purposes of this paper, Leach's [7] generally held position that ritual comprises a wide spectrum of human activity, from the sacred to the profane, is the preferred position.

on the individual. I'll quote the critic Roger Scruton as an example of how this type of criticism works although any number of other writers might just as easily have been cited. In this quotation from 'The Aesthetics of Architecture' [8], Scruton describes the effect of Chartres Cathedral on the visitor:

> The experience of Chartres is the apprehension of a divine light penetrating all things, all matter made permeable to Soul, of a universal harmony which transforms every stone from its material roughness into a minute symbol of the intellectual love of God. But to understand that meaning – understand it fully, in all its elaboration – nothing short of a visit to Chartres will suffice ... it is the experience which remains the important thing.

In reading this excerpt I picture myself in the role of the visitor that Scruton is referring to. It's late on an autumn afternoon and the cathedral is quite empty. I'm sitting in a pew quietly reflecting on the extraordinary light as it falls on the stone's smooth surface and this deep contemplation of the quality of the architecture makes me think of the nature of God.

But for all that I appear to be having a very profound moment of reflection, at some level I am aware of the fact that I'm not appreciating the architectural potential of the cathedral to its fullest. This is because the true architectural significance of one of the great French cathedrals is only really revealed to me when I happen to be present on the occasion of an important ceremonial function, a function where a procession of religious in extraordinary garb is taking place, when the choir is singing in Latin and the air is thick with burning incense.

When I visit a great French cathedral on a day of such religious significance, I experience its architecture in two ways simultaneously. At one level I am the visitor that Scruton describes, consciously observing what's going on all around me, watching the shadow slowly advance along the shaft of a column. However, at another level I'm experiencing the structure as though my own sense of myself is subsumed into the collective consciousness of the congregation. The choir, the procession, the light, the stone, the sound of my own voice, the incense, and so on, all become one thing. In this elaborate ritual, both the building and the individual (in this case, me) are only ingredients in something which is much more significant and, in some way, this makes the experience of its architecture all the more intense.

Taking this reading to its furthest conclusion we could go so far as to say, as Greenhough [9] has previously suggested, that the full architectural experience of St. Peter's in Rome is only ever really available when a Pope has died and another is in the process of being selected. On all other occasions St. Peter's is an object, albeit one from which I can derive considerable pleasure in the way Scruton suggests, waiting to be called upon to fulfil its architectural potential.

The example chosen to illustrate the point is at the extreme end of the ritual spectrum. We might call it Ritual with an upper case R. This type of Ritual is most certainly not exclusive to Roman Catholicism. Those from other religious cultures will find equivalents in their own traditions—the ablutions at the Blue Mosque in Istanbul before Friday prayers is another acute example. Neither is this type of Ritual exclusive to organised religion. Soccer fans experience something similar on the terraces of the stadium on a

Saturday afternoon. We may also feel this type of Ritual on a visit to the theatre or when a momentous vote is taking in the national parliament (provided of course that we're in the privileged position of being present).

But there are also rituals which come with a lower case r. Building on a scenario proposed by Douglas [10], we imagine a group of friends who decide to meet one Friday evening for dinner. In arranging their rendez vous, it is unlikely that these friends have in mind a meal of pre-made sandwiches to be consumed standing in the middle of the street in a howling gale. It is far more likely that they'll agree to meet at a popular restaurant. This restaurant may well have a reputation for the quality of its food but equally important to its selection will be its ambience, which is to say, the ritual architectural experience that it offers—the maitre d' making a fuss over the group as he meets them at the door; the crispness of the table linen; the glint of the cutlery; the drama of seeing-and-been-seen, and so on. It may well happen that the restaurant was designed by a well known architect with a reputation for working in a tectonic style and for whom the quality of the connection of the beam to the column is a matter of immense personal pride. But in the overall 'architecture-ness' of the dining experience, this particular detail isn't that significant. The architecture of the restaurant is to be found in the ritual of dining and this ritual is expressed in all the elements that go into the making of it.

There are even more routine ritual architectural experiences available to us than a date at the restaurant. There is the trip to the hair barber, the visit to the dentist, the picking up of the children from the crèche, the browsing through the new releases in the favourite bookstore, and so on. In fact, it could be argued that in every event of our daily lives some element of ritual is at play and the successful resolution of this ritual is dependent on the appropriateness of the architectural setting.

Viewed in this light, the work of architecture might be seen in a different way to the way we normally see it. Rather than a virtuoso masterwork by an auteur-architect or a marketing tool for a private developer, the work might be envisaged as a facilitator of the rituals of our daily routines. The development of the brief, in this reading, now equates to a deep understanding of the ritual involved. The architect's task is to display a sensitivity to what is crucial to the nature of this ritual. If the architect's proposal permits the true and proper functioning of the ritual everyone is satisfied.

So far, I've attempted to make a case for how ritual may inform an aesthetic aspiration in architecture. But the task which I set for myself at the beginning of this paper was to see if there's a way this aesthetic might allow for a greater sense of social inclusion. I'll attempt to address this now.

Speaking without the benefit of statistical analysis, I think it's fair to say that the idea of inclusivity in architecture typically plays out in the realms of the technical and the political where underprivileged or overlooked minority groups negotiate improved circumstances. These improvements result in the provision of special or additional facilities.

But, bearing in mind our previous discussion on the idea of ritual, there might be another way that this design problem can be approached.

In its broadest, most liberal and, perhaps, its most architectural definition ritual is the way in which the group structures the interaction of its constituents in the performance of a necessary routine. In former times when population groups were more homogenous, the rituals which developed reflected the traditional culture of the homogenous group. But in the contemporary world population groups are becoming ever more diverse and this diversity can only be expected to increase in future years. This allows for the possibility of the evolution of new and heretofore unimagined structured social interactions, some of them a blend of the disparate choreographies inherited from the various groups which go to make up the new community and some of them emerging from entirely new forms of social engagement. These new and/or blended forms of ritual will require particular architectural responses to permit their satisfactory outcomes.

Therefore an architectural approach which strives to address inclusion in a meaningful way might take as it's starting point the careful examination of the interactions of new and evolving social groups with a view to proposing their realization in architectural form. This will require that the programme be seen as something other than the singular vision set down by a privileged individual or group. It will also require that the architect reconsider their role in the process of proposing the spaces which form the backdrop to our daily routines.

## References

1. Plato (2003) The symposium, Revised edn. Penguin Classics, London
2. Aristotle (1998) The metaphysics, Revised edn. Penguin Classics, London
3. Kant I (2007) The critique of judgement, Revised. Oxford University Press, Oxford
4. Alberti L (1988) De Re Aedificatoria: on the art of building in ten books, Revised. MIT Press, Cambridge Massachusetts
5. Hall W (2015) Brick, 1st edn. Phaidon, New York
6. Blundell Jones P (2016) Architecture and ritual: how buildings shape society, 1st edn. Bloomsbury, London
7. Leach E (1970) Genesis as myth and other essays, 1st edn. Jonathan Cape Ltd., London
8. Scruton R (2013) The aesthetics of architecture, 1st edn. Princeton University Press, Princeton
9. Greenhough H (1947) Form and function, 1st edn. University of California Press, Berkeley
10. Douglas M (1972) Deciphering a meal. Daedalus 101(1):61–81

# New and Old Bauhaus: What Is Happening to Modern Architecture?

Josep Muntañola Thornberg[1(✉)], Magda Saura[1], Carmen Escoda Pastor[1], Regina Garcia[2], and Bürkle[1]

[1] Universitat Politecnica de Catalunya, Barcelona, Spain
{jose.muntanola,magdalena.saura,carmen.escoda}@upc.edu
[2] Universitat Internacional de Catalunya, Barcelona, Spain
regina.grc@uic.es

**Abstract.** There is a key point in architectural education that should be clarified to improve the legitimation of architects and urban planners in the digital and global world of today: The relationships between experience and knowledge in architectural and urban design theories and practices. Our proposal refers to the historical seminar in MOMA in new York in 1948 about a similar topic, since it is the definition of modernity related to the old and new Bauhaus experiences that is involved in architectural education today. Considering the impact of new disciplines in education, such as neurology, the cognitive studies and the socio-cultural historical studies, we want to study how this link between experience and knowledge can disappear, and with it, the possibility of a truthful architectural education when historical analyses falsify the existential social conditions in which knowledge emerges in concrete educational processes. This is extremely relevant today, when artificial intelligence wants to produce knowledge from very limited scientific laws and without considering the artistic and cultural dimensions of architecture and planning.

**Keywords:** Architectural education · Architectural cognition · Critical theory

## 1 Introduction

The lack of discernment of why modern architecture is good or bad and the reasons of it needs to be carefully analyzed. This fact is extremely complex and involves different disciplines besides a good knowledge on the architectural design theories and practices of today. This text wants only to be a first step asking for further research.

To start, consider architecture because of the classical triad between design, building and dwelling. No architecture is possible without design, building and dwelling, however the interrelations between these three fundamental architectural dimensions change permanently and produce very different human historical situations.

From a phenomenological and hermeneutic point of view the diagram I analyses the differences between intersubjective communication thanks to verbal languages, that we call "interlocution", and the intersubjective communication thanks to architecture that


M. Blanco Lage et al. (eds.), *Towards a New European Bauhaus—Challenges in Design Education*,
https://doi.org/10.1007/978-3-031-40188-6_36

we call "interlocation", between human bodies and places from the beginning of the humanity until today.

The role of space in "interlocution" is metaphorical. Books do not change its meaning when they change of place but, in "interlocation", it has a direct impact upon each different intersubjectivity process of communication, where time is a metaphorical substitute of the metaphorical role of space in verbal communication. In fact, as the center of Fig. 1 shows [1].

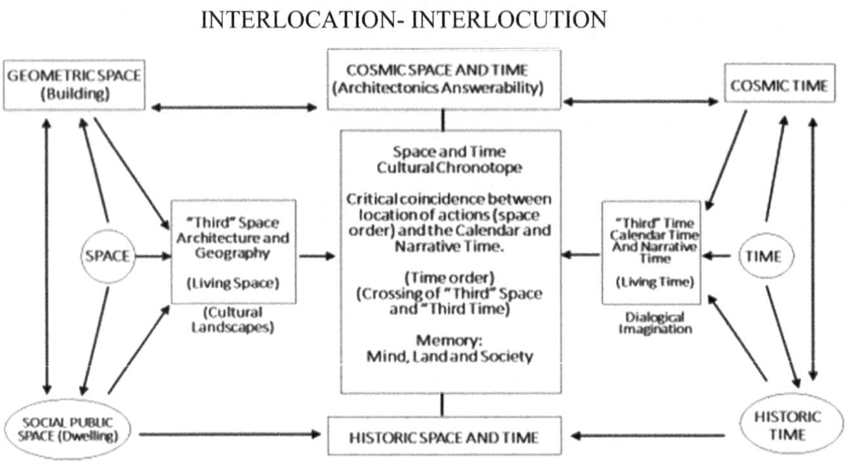


**Fig. 1** Phenomenology of space and time according to Paul Ricoeur

The real day by day intersubjectivity communication articulates both in a permanent action of the human life, as the last neurological findings analyze [2].

Just look to Fig. 2 where Karl Friston analyses the relations between subjects and environments, as a double chain in the brains by crossing a bottom-up process with a top bottom one, where human knowledge is, nor a totally autonomous process, neither a process totally determined by its environmental influences. The crossing of these two epigenetic chains, that is, of these two ways of knowing, is the fundamental condition of our human mind. This crossing point correlates interlocution with interlocation, and consequently, articulates the role of writers with the role of architects, and should be the main point of the discussion on the lack of discernment in our students of architecture today.

## 2 A Historical Excursion to Architectural Histories and Stories

In the title of this text, we refer to one of the most important facts in the modern architectural history in 1948.

Immersed into a hard battle for the best discourse about what is modern architecture, a seminar in the MOMA in New York confronted Lewis Mumford, a lieder art historian at that moment with twenty years of significant articles in The New Yorker [3] and the Old

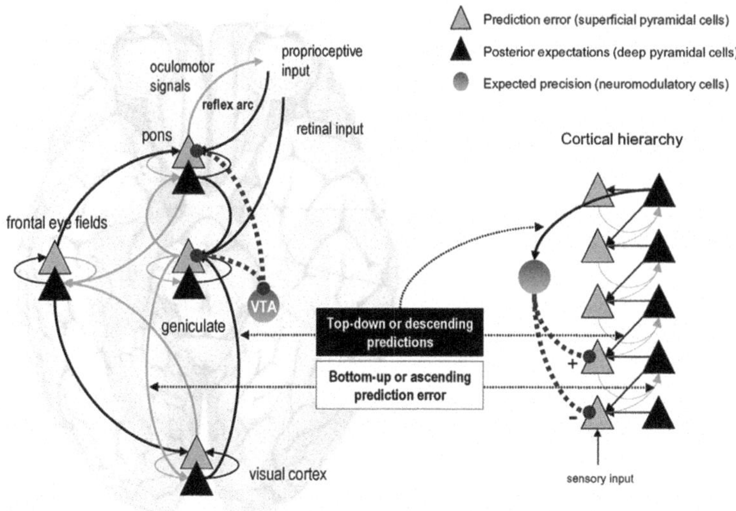


**Fig. 2** Neurological fundamental feedback between senses and simulation according to Karl Friston

Bauhaus architects Walter Gropius and Mies van Der Rohe, who were building the new American and modern skyscrapers of the city and that have been criticized by Lewis Mumford articles for years. This seminar and exhibit from this year represented the triumph of the true modernity in architecture, with the ideas of the Old Bauhaus and was organized by H. R. Hitchcock and Ph. Johnson important critics at the time in New York, and the failure of Mumford, who was invited to coordinate the meeting, to defend his ideas. He was totally discredited and, as explained to professor Josep Muntañola fifteen years later [4], he decided to consider architecture as a little and residual element in the anthropological books that he wrote from that moment on. His ideas were considered by Hitchcock as "toxic contaminations" that could destroy the emerging modernity in USA, since modernity had nothing to do with vernacular facts or other traditions, and they were only produced by Mumford's national chauvinistic head [5, 6]. Sennett needed seventy-five years to claim that the Mumford position in relation to urban planning was wrongly forgotten without mention this seminar, but the social meaning of architecture was clearly involved on his comments on the book [7].

The reactions by Tzonis and Tafuri have often been ignored [8, 9]. Still today, these positions about the social meaning of modern architecture are present in our schools of architecture, in a way that we have denominated as and antidemocratic positions, where only one opinion is accepted and all the others should be violently eliminated out of the school [10]. In this way, architectural criticism is weak and there are not open and free debates. Then, architects have poor creativity immersed in ethical and political confusions, since they like, and they have good experiences, in architectures that they need to reject because of these anti-democratic positions coming from totally different cultural and political powers. The confrontation between ethics and politics, without debates, is killing creativity instead of empowering it. As Aristotle indicated, the first victim of this lack of authenticity is the subject itself that immediately is losing freedom, singularity, and intelligence [1, 10–13].

It is also, a pity that, the role of Klee, Kandinsky, Schlemmer and Moholy-Nagy, in the old Bauhaus, have been ignored and distorted, only the role of architects is underlined … A lot of studies today identify the Old Bauhaus with the architects' positions, when Klee, Kandinsky and Schlemmer, that were not architects, did a lot of basic courses of design to architects in the Bauhaus. A fact that today could be considered as a destructive move against the purity of the disciplinary value of architecture itself, in a lot of schools of architecture, where only architects can teach, in a complete contradiction with the real situation in art and science today based upon an interdisciplinary and inter-artistic performance everywhere.

Even the significant last book by Giedion is today misinterpreted as a continuity of previous books, when it was, in fact, a total change with a social and anthropological point of view, in relation to the "architectural historical transitions," where technology was no more the only origin of the historical changes in architecture. Giedion really pointed to the chronotopic social and physical structures as the key causes of historical transformations [10, 14–17].

## 3 The Children Reference

Picasso said that who cannot paint as children do, they would not be really good painters.

In Figs. 3, 4 and 5 we summarize how children build architecture. The last diagram in Fig. 5, it is a representation of a city by a two-year-old girl is a good example. When she was asked to explain, in verbal "interlocutive" terms. The "interlocation" in the drawing, she said immediately: These are buildings and balconies. She forecasted twenty years before the covid effect in the cities when housing units without balconies, or with poor views, decreased prices dramatically. The main differences between schools are based upon the dialogical dimensions of the curriculums in each school as Fig. 3 indicated [18, 19]. What is important here is the social kernel of architecture as a specific kind of artistic human action where functions are essential and cannot be eliminated.

Figure 6 indicates the emergence of the concept of "chronotope" in the human mind as an articulation of physical, social, and mental dimensions of the human being, Bakhtin explained how this concept, fundamental in his dialogical social theory of artistic communication, is the structural narrative knot both, in verbal and in architectural intersubjective communication [20]. In Figs. 7 and 8 the role of the chronotope "narrative knots" is represented in an abstract way by the architect Enric Miralles in a design proposal for an urban frame in old Barcelona after it was destroyed. Finally, it was later rebuilt in a totally different way [21].

## 4 How to Teach Criticism Today

The professor Salama explained which the conditions are to have design workshops able to develop a good power of criticism in architectural education today, in a recent and excellent book [22].

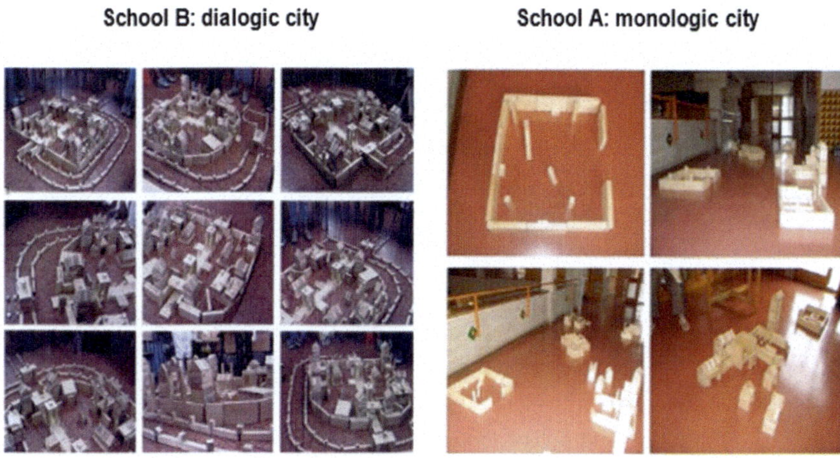


**Fig. 3** Children construction of cities: monological and dialogical

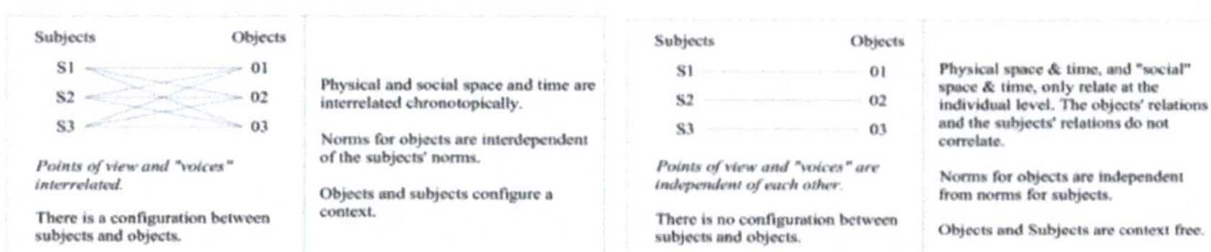


**Fig. 4** Monological and dialogical chronotopic structures on children architecture

**Fig. 5** Representation of a city by a two-year-old girl

Therefore, to understand the verbal arguments of each student in relation to a building, they should be related to the subjects very different interlocative experiences so they can

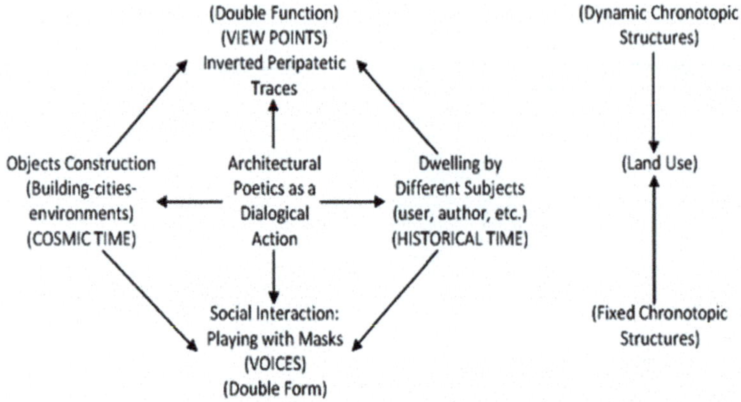


**Fig. 6** The structure of the chronotope

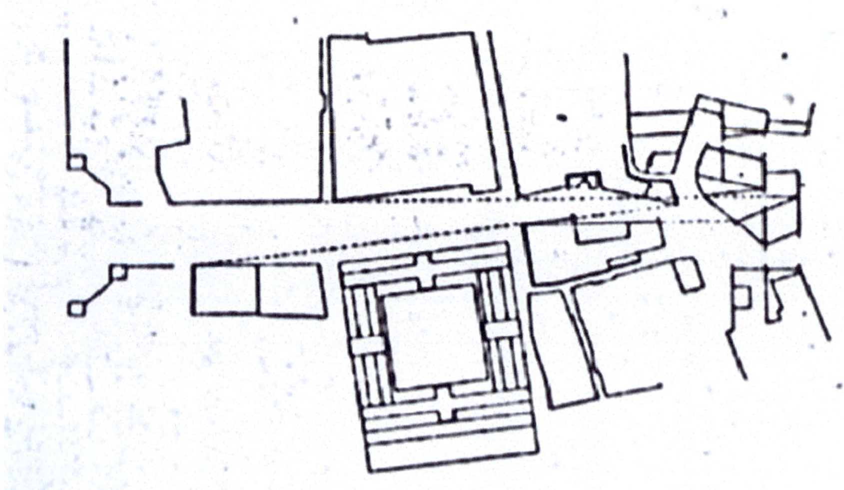

**Fig. 7** An innovative urban plan by Enric Miralles as a chronotopic design process with real subjects and real points of view

agree verbally but not in relation to the aesthetic dimensions of the same building. Also, to live in a building is related to the historical experience and to the historical knowledge of related buildings of the users so, again, to live in the same place can be interpretated in very different ways. In a design workshop all these different meanings are mixed as Salama described, and the professor cannot deal with students as if they had the same experiences and knowledges. Similar verbal intentions can correlate with extreme different experiences and even if they are alike, these differences should be considered in a pedagogical evaluation. The discussion in the seminar in New York are no significant because Mumford is criticized, this is usual in a good democratic country, was the total intolerance and the identification of modernity with only one position without real

**Fig. 8** Proposal of a new urban plan in Barcelona by Enric Miralles

dialogue, that start a wrong attitude of a monological truth, without taking care of the variety of positions that the modern avant-garde trends involved.

So, in Fig. 9 we represent the complexity of the traces that impact in our brains, making some basic distinctions between the three ways our mind is affected by architecture in our students. It is the philosopher Paul Ricoeur who developed in the last years of his life several books and texts defining these different traces that affect our memory [23–26].

Good design education should consider these three different traces, since they are at the origin of very different health pathologies. These three kinds of traces have been summarized in the following way Ricoeur itself [27].

The first kind of traces, the cortical traces, to develop they need social relationships and social experiences, fundamental in the in the first years of age, or at the beginning of the basic design courses. To develop these traces, students need the have social contrast of their ideas and feelings, and social support. They need social interactions and debates where they can relate their designs with the other designs.

The second kind of traces the historical traces, emerge from psycho-social experiences with constructions and their environments, and with experience in historical physical references. Today, in Barcelona, doctors warn of a fifty per cent increase in addictions to the screens, with a heavy impact in the mental health of children and teenagers. This is because without physical and psychosocial correlative experiences, by living only in virtual representations, they cannot have friends and healthy sexual relations.

The third kind of traces, the document considered as a trace, cannot be produced because of the total lack of mental freedom. Similarly, if you reject your own feelings in order to be a member of one prestigious group, suddenly you lose the critical power. Good critics go away of architecture when they are discredited in an aggressive way.

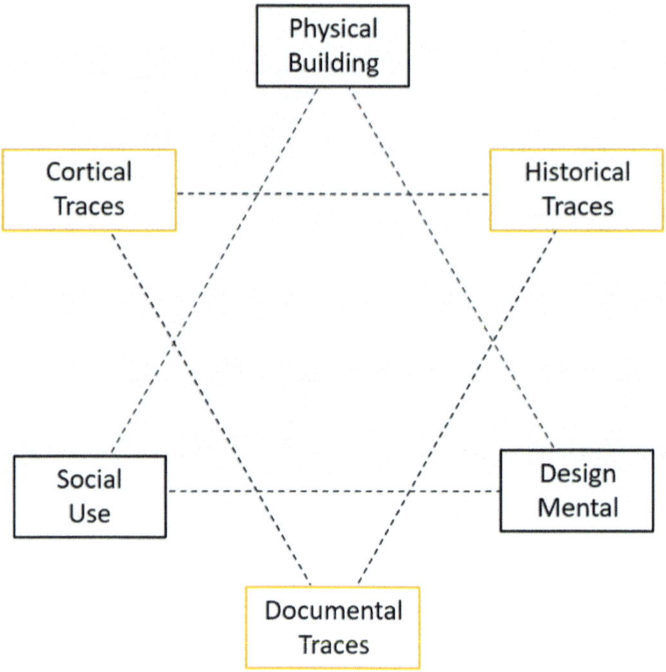


**Fig. 9** The formalization of the memory according to Paul Ricoeur

What is relevant here is that the three kinds of traces are intertwined and if one does not work the others are affected. We have described these processes in other texts, however the best scientific prove has been the long term studies about the health of children after thirty years of living in non-healthy urban environments, made in the Columbia University some years ago [28]. The pathologies multiply their powers and the deterioration of children health in unhealthy environments is spectacular. Also the therapeutic healing of these pathological wounds is difficult and very expensive, because in order to put again the three traces of Fig. 9 in good shape you need a long therapeutic process with teams of experts, because you need doctors, biologist, psychologist and psychoanalyst altogether. A very expensive work indeed. This is the case in Barcelona today, a very busy and modern city with a high level of social interaction in the schools between different social backgrounds, but with a progressive growth of mental pathologies when these three pathological processes, that we have described above, instead of working together start to fight each other, and then the subject life and the social life do not intercommunicate, and, a lack of needed traces, or an excess of some traces, affects finally mental health.

## 5 Going to the Beginning from the End

Then, at the end it is a matter of equilibrium. Like in architecture itself, a lack of equilibrium between the physical, social, and mental human dimensions, that is, in architecture, between building, use and design, destroys both the environment and cultures. In education, as each subject is in fact a singular and unique equilibrium, nobody escapes depression when this equilibrium is damaged, and a lot of times we discover the danger too late. Nobody can live without an equilibrated intertwining among the cortical, the historical and the mental documental traces the lack of one kind of traces affects the other two kinds and both subjects and cultures die.

Good criticism can analyze building as an innovative and new equilibrium between the three traces and between design, building and dwelling, but never considering only one of them. A difficult task indeed, but a fundamental human task too that architects must undertake …

## References


1. Muntañola J (2022) Lotman and Bakhtin in the architectural design theories of today. In: Human being: image and essence. Humanitarian aspects, vol 1, no 49. Semiosphere: special aspects. To the 100th anniversary of Yuri Lotman. Inion Ran, Moscow, pp 7–27. https://doi.org/10.31249/chel/2022.01.01I.30
2. Friston K et al (2000) Principles of neural sciences. McGraw Hill Companies, New York
3. Mumford L (1947) The skyline. The New Yorker, 6 Oct. New York
4. Muntañola J (1965) Un home a New York Una visita a Lewis Mumford. In: Serra d'Or Marc 1965. Ediciones Monestir de Monserrat, Barcelona
5. Fenske G (1997) Lewis Mumford, Henry Russell Hitchcock and the bay region style. In: Pollak M (ed) The education of the architect. MIT Press
6. Fenske G (2011) What is happening to modern architecture. In: Muntañola J (ed) Architecture and virtuality. Arquitectonics, vol 21–22. Iniciativa Digital Politécnica, Barcelona
7. Sennett R (2019) Building and dwelling. Penguin Books, London
8. Tzonis A, Lefaivre L (1990) Why critical regionalism today? Archit Urban 236:22–33. Editions du Seuil, Paris
9. Tafuri M (1980) La Sfera e il labirinto: vanguardia e architettura da Piranesi agli anni. Edicions UPC, Barcelona
10. Muntañola J (ed) (2016) Arquitectura y Modernidad: ¿suicidio o reactivación? In: Arquitectonics: mind, land, and society, vol 29. Universitat Politécnica de Catalunya, Barcelona
11. Muntañola J (ed) (2022) Artificial intelligence for architects & designers. In: Arquitectonics, vol 33. UPC Editions, Barcelona
12. Muntañola J (ed) (2021) Building the new sense of place. In: Arquitectonics: mind, land, and society, vol 32. Universitat Politécnica de Catalunya, Barcelona
13. Muntañola J (ed) (2022) Artificial intelligence and architectural design: an introduction. In: Arquitectonics, vol 34. Iniciativa Digital Politécnica, Barcelona
14. Muntañola J (1975) Prólogo. In: Giedion S (ed) Arquitectura y los Fenómenos de transición. Gustavo Gili, Barcelona, pp 3–21
15. Giedion S (1963) Space time and architecture: the growth of a new tradition. Harvard University Press, Cambridge
16. Giedion S (1971) Architecture and the phenomena of transition: the three space conceptions in architecture. Harvard University Press, Cambridge

17. Giedion S (1975) Arquitectura y los fenómenos de transición: las tres edades del espacio en arquitectura. Editorial Gustavo Gili, Barcelona
18. Muntañola J (1980) Developmental architecture cognition in children's conception of place. In: Broadbent G, Bunt R, Llorens T (eds) Meaning and behavior in the built environment. Wiley, London; Editorial Gustavo Gili, Barcelona (1974)
19. Muntañola J (ed) (2012) Arquitectura e investigación. In: Arquitectonics, vol 24. Iniciativa Digital Politécnica. UPC, Barcelona
20. Fenske Bakhtin M (1985) Speech genres and other late essays. University of Texas Press, Austin
21. Muntañola J (2004) Arquitectura 2000 (proyectos, territorios y culturas). In: Arquitectonics, series número 11
22. Salama A (2016) Spatial design education. Routledge Press
23. Ricoeur P (2000) La Memoire L'Historie et L'Oublie. Editions de Seuil, Paris
24. Ricoeur P (2001) Discours Honoris Causae Barcelona University Ramon Llull. Comp Rev Catalana Filos 3(2):87–99
25. Ricoeur P (2005) Parcours de la Reconnaissance. Gallimard, Paris
26. Ricoeur P (2016) Architecture et Narrativite. Études Ricœuriennes [Ricœur Stud] 7(2):20–30. ISSN 2156-7808 (online). https://doi.org/10.5195/errs.2016.377
27. Ricoeur P (2021) Conference of the Doctoral honorary degree at The Raimond Llull University Barcelona 2000
28. Shonkoff JP, Phillips D (2000) From neurons to neighborhoods. National Academy of Sciences, Washington

# Gendered Bodies. Feminist Approaches to Spatial Design

Arianna Scaioli(✉)

DAStU, Politecnico di Milano, Milan, Italy
ariannaluisa.scaioli@polimi.it

**Abstract.** Looking at the Old Bauhaus of Gropius, the topic of gender equality was already present; there would be no distinction between men and women, showing a progressive vision. It was, however, equality only on paper, with little validation in practice. The New European Bauhaus reframes past ambitions, encompassing the new SDGs and assuming a crucial role in guiding spatial transformations bridging the intersection between spatial and social/gender justice. The contribution proposes to recast the discipline of architecture by its encounter with feminist methodologies, investigating through a research-by-design approach the spatialization of gender equality. Accordingly, it assumes a gendered perspective—where architecture is called upon to (re)construct women's condition, movement and visibility in space. By recognizing the relevance of design experimentation that modifies the 'hardware and the software' namely the deep formal urban structure and the ethical-cultural dimension, it embraces the concepts of "care" (Tronto) "complexity and multiplicity" (Deleuze and Guattari) and the "right to the city" (Lefebvre). It questions the traditional narrative of the male standard, understanding whose perspectives are valued, how they are represented in the built environment and which are the uses. This approach re-weaves the relationship between gender, body, and architecture, spatializing them in the materiality of space, and unveiling the networks of power and how people carve out space, which is negotiated and appropriated. The paper presents a series of case studies, discussing themes, tools and methods of a feminist approach in architectural design, fostering gender equality in space.

**Keywords:** Feminist critique · Body-space-city · Gender and architecture · Gender equality · New European Bauhaus

## 1 First Section

> A social transformation, to be truly revolutionary in character, must manifest a creative capacity in its effects on daily life, on language and on space. [1: p. 54]

Observing the built environment from a gendered perspective opens up new horizons of interpretation of urban dynamics and modes of inhabiting space, unveiling the power relations encompassed in our cities. Built space has mainly been shaped by a traditionally male-dominated vision of cities—that produces and reproduces space division according


M. Blanco Lage et al. (eds.), *Towards a New European Bauhaus—Challenges in Design Education*,
https://doi.org/10.1007/978-3-031-40188-6_37

to gender roles rather than meeting the needs of the whole population [2]. Positioned within a feminist critique of architecture, this contribution becomes an exploration, a terrain of discussion—of (be)coming together—embodying the openness of feminist spatial futures [3].

Feminism, as a broad vision for the rights of all bodies and voices [4], offers a critical framework to explore differences, rethink the "microphysics of everyday life" [5], counteract hegemonic power relations both in space (by replacing 'power over' with 'power for') and in the design process. Through participatory research, co-design, and co-creation, it values solidarity and connectivity, recasting the role of the architect from genius to enabler, breaking down hierarchies. Feminist practices engaging inclusive design, ethics, and politics give voice to often ignored groups by looking at how gender relations—which are socially, culturally, spatially, and temporally constructed—shape space. Operating at different levels, from daily life, with an attention to the body's experience in space, at the level of collectivity and multiplicity, generating spatial complexity, and at the theoretical level, it creates the possibility of practicing otherwise counteracting the status quo [3, 6–8]. However, the discussion about the spatialization—or translation—of gender equality into spaces and architectures is somewhat complex and in need of further explorations where architects, scholars, communities, and institutions are called upon to set up a dialogue jointly.

Looking at the Old Bauhaus, the topic of gender equality was already present. Gropius' inaugural speech in 1919 showed a progressive vision: "No distinction between fair and strong sex. Absolute equality, but also the same obligations." It was, however, equality only on paper, with little validation in practice [9, 10]. The New European Bauhaus (NEB) reframes past ambitions, encompassing the SDGs and assuming a crucial role in guiding spatial transformations bridging the intersection between spatial and social/gender justice [11]. In this framework, gender-sensitive design and women empowerment are brought to the forefront of the research agenda, generating a set of theoretical values that can inform the design process and outcome from a creative and interdisciplinary perspective.

The contribution proposes to recast the discipline of architecture by its encounter with feminist methodologies, investigating the spatialization of gender equality. By bringing into the discussion a feminist critique, dwelling at the intersection/interaction of theory and practice, it allows for stretching the disciplinary boundaries grounding architecture as a practice of resistance and advocacy for women, explored through its materiality. By presenting two case studies, one by the feminist collective Equal Saree and the other by muf, it contributes to a discussion on themes, tools, and methods of a feminist approach in architecture, fostering gender equality. The first, Equal Saree, is a feminist architecture collective based in Barcelona. It was founded by Helena Cardona Tamayo, Julia Goula Mejón, and Dafne Saldaña Blasco. It features a feminist critique in all its projects, promoting gender equality through spatial design. muf is a collaborative practice of art and architecture established in London in 1994 by Liza Fior, Katherine Clarke, and Juliet Bidgood, informing a feminist critique of spatial practice for over 30 years.

## 2 Engendering Cities. Going Beyond the Man-Made City

Our cities are patriarchy written in stone, brick, glass and concrete. [12: p. 88]

One of the most relevant concerns of feminist and gendered critiques of architecture has been questioning bodies' experience in space as a mediator of spatial experience, revealing how they are "psychically, socially, sexually, and representationally produced" [8: p. 104], and how they inscribe and project themselves holding a generative power to inform spatial transformations. This attitude breaks down the consolidated male-dominated vision of cities based on the notion of a neutral user—usually a universal man—i.e., Caucasian, able-bodied, working male, contributing to shape throughout centuries our built environment. This androcentric city makes itself explicit on several levels: the symbolic one (street names, male standards ...), the visual (sexist advertisement ...), and the physical (use of space, safety, accessibility, comfort ...), reiterating gender roles in space [13].

Counter-designing the androcentric city through a feminist critique builds on the recognition that gender norms impact materially and symbolically space [14], enabling or preventing different (gendered) bodies from using, inhabiting, or modifying it [15, 16] defining "appropriate" spaces for women. Recognizing that (by both habitual use and by metaphor) "spaces and places are not only themselves gendered but, in their being so, also reflect and affect the ways in which gender is understood" [15: p. 177] reveals how women's segregation reduces their access to resources and knowledge, and how the built environment legitimizes and normalizes gender discrimination. Women construct personal paths, encompassing different itineraries and rhythms in their everyday lives, carving out space from their direct experience with the physical environment. However, building on the notion of gender as a relational category [17] allows us to recompose the dualism between the androcentric city and the women-centered one, not considering the two of them as separate spheres but recognizing how the notion of masculinity and femininity were constructed in relation to each other. This approach reframes the question of gender equality in space and feminist methodologies shifting from a 'women-only issue' to an 'everyone issue,' where recognizing gender differences would not mean denying the 'other' but deconstructing and recomposing our cities in relation to the 'one' with the 'other.'

Mainstreaming gender "offers an opportunity to re-examine socio-environmental relations in an urban milieu" [18: p. 218] to create spaces where women "can participate as equals, introducing their innovation and knowledge" [18: p. 220]. Starting from the 1970s, there has been a shift in the architectural field, where feminist critiques have tried to deconstruct the traditional narrative, raising awareness of the necessity of women's empowerment through design. MATRIX Feminist Design Cooperative and the Women's Design Service are among the precursors that tried to recast the discipline of architecture, intervening directly in the design of spaces. Both established in London in the 1980s, they promoted women's empowerment through active participation in the design process, focusing their attention on the relation of gender and space from and through a feminist perspective, co-exploring new building types.

Building on this framework and recognizing the relevance of design experimentation that modifies the 'hardware and the software,' namely the deep formal urban structure and the ethical-cultural dimension, this contribution embraces the concepts of "care" [19], "complexity and multiplicity" [20] and the "right to the city" [21]. It questions the traditional narrative of the male standard, understanding whose perspectives are valued, how they are represented in the built environment, and which are the uses. This approach re-weaves the relationship between gender, body, and architecture, spatializing them in the materiality of space and unveiling the networks of power and how people carve out space, which is negotiated and appropriated. Body as perception, movement, appropriation, and measure, where through its multiple interactions with space, it becomes an active medium between design and space [22].

## 3 Methodology

This contribution is part of a doctoral research developed through a research-by-design approach [23, 24] and grounded within a feminist and gendered critique of architecture. Specifically, the article contributes to this discussion on the one hand through the presentation of already existing design experimentations. On the other, it stimulates a reflection upon the necessity of translating the question of gender equality in space, fostering a more democratic and humanistic approach to design.

The two projects, Plaça d'en Baró by Equal Saree and Ruskin Square by muf, grounded within this framework, show how the intersection between space, architecture, body, and uses displays a stratification of both tangible and intangible values in consideration of gender as a crucial aspect to integrate. In this sense, the encounter of feminist methodologies with a design-driven approach could inform the design process and outcome reframing the relations between theory and practice encompassing creativity and co-design. Conceiving space as a relational reality [1, 15, 25] in which multiple identities and bodies come together and where "collectivity, interiority, alterity, materiality and performativity" [26: p. 24] take place, suggests a shift from being-in-space to becoming-in-space [27], introducing a transformative dynamism, correlating the "actual" with the "possible" [20]. This idea of openness, diversity, and open-ended future encompassed in feminist critique, where the two presented case studies are exemplary, assumes indeterminacy, change, adaptability, progress, and creativity, allowing different temporalities and bodies in motion to create the conditions for an 'architecture as process.' By considering temporality as a design material, it holistically recasts the intersections between sociality, spatiality, and temporality. Directing the attention towards the process means enabling an architecture of possibilities, open to modification and different temporalities where "the future is that openness of becoming that enables divergences from what exists" [28: p. 142].

The two projects interpret a feminist critique to design, centering the attention on everyday activities and creative practices of resistance and advocacy of the often marginalized groups: women, children, and the elderly. Hence, they recall the notion of "Critical Spatial Practice: projects located between art and architecture, that both critiqued the sites into which they intervened as well as the disciplinary procedures through which they operated" [29]. The contribution presents two case studies by drawing from the theoretical framework developed in the 1970s. It encompasses the conception of urban experience as simultaneously space and event emerged—showing the entanglement between space, body, and event [30]—the Lefebvrian concept of "social space," which is reciprocal—created by, but also generating social interactions [1]—and de Certeau's fluid appropriation of space which he defines as "tactical" [31]. They become a terrain of discussion grounded within this theoretical and critical framework, disclosing the intersections and contaminations between feminist methodologies and architectural design and how they can impact how we observe, describe, represent, inhabit, and design spaces.

## 4 Plaça d'en Baró by Equal Saree: A Collaborative Design

The project Plaça d'en Baró (2016–2019) can be considered a design manifesto. It is located in Santa Coloma de Gramenet, a densely populated city in the metropolitan area of Barcelona, which was undergoing a process of pedestrianization to foster greater accessibility and understanding of the needs and uses of the community (Fig. 1).

**Fig. 1** "We believe it is necessary to give people a voice to detect the uses, needs and specific wishes of the communities that inhabit the different territories. We are committed to the active participation of groups that are generally excluded from decisions about their environment." Equal Saree. © Conchi Berenguer Urrutia

The aim was to design an open square for the neighborhood where different uses, activities, and temporalities could be hosted synchronically. What emerges is an inclusive and safe place for children, women, and communities, reconstructing the social fabric through the materiality of space. Before the intervention, the square was considered neglected, unsafe, and inhabitable, a place not to identify.

Their working method was centered around developing a collective reflection on how uses and bodies could shape space and how they could define design criteria fostering the coexistence of people of different ages and gender. Accordingly, they set up a process in three phases: open participation, co-design with the students of the Torre Balldovina School, and elaboration of the project. In the first one, the architects and the community set up a dialogue to detect the needs and aspirations of different groups (Fig. 2).

In the second one, the Torre Balldovina School students have been acting as young urban planners, developing different proposals for the Plaça. By becoming both users and makers of space, people interact with architecture in a more thoughtful perspective, increasing their willingness to take care of a place.

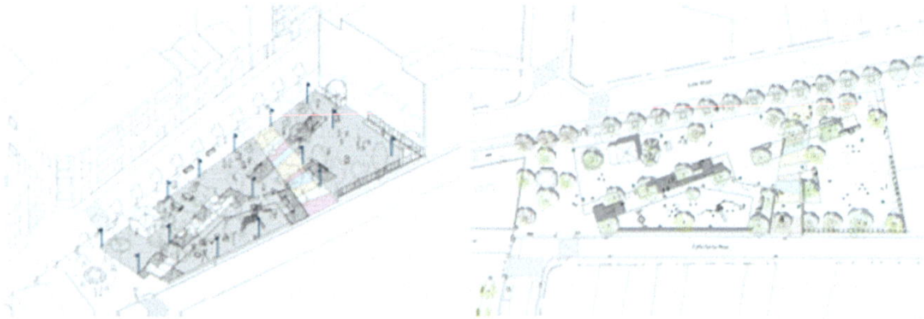

**Fig. 2** A proposal for new uses and spaces, which have been located together with the community: rest and quiet activities, low-intensity motor activities and symbolic play, high-intensity motor activities and space for artistic expression. © Equal Saree

The last phase was the actual design of the Plaça. The layout of the square is conceived to answer the main objectives that emerged within the community. The Di-verse Square can host different uses synchronically, encouraging non-stereotyped modes of play, allowing the coexistence of adults and children; each activity occupies a similar surface, not to create spatial and social hierarchies. The Caring Square em-bodies the reproductive and care tasks. The architects have designed a central space, shaded and with natural materials, encouraging social interaction and well-being.

The Plaça Vital aims at increasing social cohesion and improving the accessibility and perception of safety. One of the decisions has been to demolish a large section of the perimeter wall to increase visibility and permeability. Moreover, the great variety of connections allows one to freely cross the square, avoiding corners or residual spaces, which could become dangerous alleys (Fig. 3).

**Fig. 3** First column: open dialogue with the community (13 women, 7 men, 12 girls and 20 boys). Second column: drawings by the students of Torre Balldovina School. Third column: the design of the final proposal encompassing the initial needs and expectations of the community. © Equal Saree © Conchi Berenguer Urrutia (bottom left picture)

## 5 Ruskin Square by Muf: A Project Open to Modification

"The temporary inscribes the ambitions for the permanent." These words perfectly describe muf's methodological approach.

Their working method displays an interdisciplinary approach intersecting art, architecture, community engagement, and participation of people as active makers to generate mutual knowledge. They stress the importance of the architectural design process, not as an activity leading to a direct outcome but rather as a project in its own right. In this vision, Ruskin Square in Croydon (2012–2018) exemplifies several aspects of their attitude. The developers conceived it as an open space accessory to the area's future development by Foster + Partners. However, muf wanted to provide the community already living there with a gathering place encompassing future modifications, a meanwhile-use concept. They explored the spatial implications of how temporary features could become permanent and how a snapshot of the possible could be envisioned, pursuing the values of complexity and multiplicity of experiences in space (Fig. 4).

**Fig. 4** Ruskin Square is a meanwhile-use concept which occupies an area in Croydon where Foster-Partner will locate their housing and tertiary building intervention. Muf provided the local community with a gathering space. They have installed a series of temporary and permanent structures, exemplary is the Refugee Cricket Project for Afghan refugees. © Muf

This project frames the ambition of creating an 'open infrastructure' based on human activities in space, growing out of what is present and welcoming several uses and temporalities, and designing a place that could be appropriated. Their design method was based on a progressive discovery of the site's features and characters, drawing inspiration from the objects they found and the nature growing there. It is not a project designed looking at the site from above, but it is based on continuous learning through a sensorial experience (Fig. 5).

They have installed a series of temporary and permanent structures, exemplary are the Refugee Cricket Project, for Afghan refugees who did not have any, along with child-sized benches, fictional ruins, and sculptures, which contribute to creating a new reality, and a place in which to find oneself (Fig. 6).

**Fig. 5** Prior to the construction, the site was first occupied and animated by a rehearsal performed by the cast of "The Girls" exploring the "tactical" role of events in conveying design actions, for a more inclusive, fair and democratic modification of the place. Moreover, they explored the character of the place and how it could inform the design by letting the inspiration come from the objects they found and the nature growing there. © Muf

**Fig. 6** The question of temporality and the regenerative power of a place has been explored through the design of a resilient landscape, which could reduce flooding risk and climate change effects. To do so, they have composed a landscape where vegetation is at the center, reconstructing local biodiversity, and framing the coexistence between human-nature-architecture. © Muf

## 6 Gendered Bodies: A Discussion on Feminist Methodologies in Architecture

What emerges from the two design experiences is an architectural sensibility that tries to recast the discipline by questioning the outcomes and especially the process. By going beyond the traditional idea of architecture as a building, they shift towards an idea of 'architecture as process,' which also occurs through using, inhabiting, and experiencing space through time. By assuming the ephemeral, the performative, the occupation and appropriation, the experience of space in all its multiple declinations—for they reflect the dynamic transformations our societies and cultures are undergoing—architecture is called upon to rethink the intersections between sociality, spatiality, and temporality, holistically. Both the concepts of "situatedness" [32] and "throwntogetherness" [33] become crucial in describing the two projects, to design starting from somewhere in space and time (here and now), where gendered bodies embody a material dimension in space, instead of starting from nowhere and everywhere. Therefore, considering the movement of bodies through time as a proactive moment in the construction of the urban space—suggesting a shift from being-in-space to becoming-in-space—allows considering architecture not only as a "scena fissa delle vicende dell'uomo" [34: p. 12], but introduces a transformative dynamism, correlating the "actual" with the "possible" [20]. These ways of producing spaces and architecture, due to their extended temporality (Plaça d'en Baró 2016–2019, Ruskin Square 2012–2018), stress the importance of the process, where through the active involvement of the community, especially women and children, through co-creative processes, can encapsulate needs, aspirations, and knowledge without the risk of imposing a project from above.

The two case studies suggest an interdisciplinary attitude that sees the entanglement of architecture, art, performance, and theory and practice, where the attention is directed toward understanding which voices and needs are considered and how bodies move and perceive space. The methodological approaches help ground the intervention within a complex palimpsest of tangible and intangible values, translating them into spaces. Expressly, the discussion about space and power relations, and how they help shape gendered cities, encompassed in feminist critique is understood and translated into the design by breaking down spatial hierarchies between parts, activities, and uses. Both Placa d'en Baro and Ruskin Square decline the notion of diversity and equality by creating a sequence of spaces that flow one into the other, generating flexible spaces designed in a continuous tension between free appropriation and answered needs. Designing through a feminist critique helps the designers to give an aesthetic and a formal character to gendered bodies and voices, associating each kind of user (women, children, and elderly) with an ambiance.

Taking as an example Placa d'en Baro, it becomes the occasion to design the co-existence between genders and ages, creating a protected (through vegetation and light structures) but still visible, accessible, and comfortable space for women; a calm, enclosed, circle-shaped space for elderly and caregivers; an open, dynamic, colorful space for children to play. These have been designed looking at how those different bodies moved in space, how they appropriated it through time, which movement could be favored through spatial configurations, and how this impacted their sense of belonging. In this sense, social and Gender Equality are explored through a multifaceted perspective, encompassing the notions of care—for people and the environment—of the right to appropriate and inhabit space—where people can participate as equals in its conception and construction, and of multiple experiences in space through time—where bodies through their movement, can generate spatial configurations. It suggests a shift from being-in-space to becoming-in-space, introducing a transformative dynamism given by bodies in motion, recalling Herzberger's words about the objective of architecture that should allow users to become inhabitants [35]. "Bodies not only move in space but generate space produced by and through their movements" [36: p. 111]. This continuous interaction makes the social character of space explicit, where "everyday perception, spatial theory and lived space merge" [37: p. 120].

## 7 Conclusion

One of the main contributions of this paper is to open up the discussion on design experiences that can foster and translate gender equality into space. The framework promoted by the New European Bauhaus, with the three principal axes "Sustainability, Aesthetics, and Inclusion," and the SDGs certainly can give a valid impulse in promoting a more fair and democratic design process and approach, tracing a path that architecture should explore in its material and immaterial dimension. However, it is necessary to reflect on how the several targets can be translated into spaces and architectures and how they can inform the design process. In this sense, this contribution highlights the importance of an interdisciplinary perspective encompassed in a feminist critique and displayed by the two design experiences to observe, represent and transform the built environment, substantiating the spatial solutions with gender-aware theoretical and conceptual reflection.

The contribution considers 'architecture as process' as a critical issue for designing a feminist spatial critique. This conceptualization allows us to bridge the intersection between a critical and theoretical reflection and a pragmatical one related to how bodies move, inhabit, modify, and experience space. Working on the idea of the process allows us to reframe the social, spatial, and temporal dimensions altogether, where the different moments of the project coexist and inform one another. In conclusion, it opens up new horizons to creatively rethink the discipline of architecture. By weaving together tenets of care, democracy, and inclusion, as well as co-creativity, engagement, embodiment, and everyday experience, they can act directly on both the 'hardware and the software'—physical spaces and cultural dimension—promoting a lasting change, which would subvert the patriarchal system, any longer functioning under it.

**Acknowledgements.** The author is grateful to Equal Saree and muf architecture/art to have granted the use of graphic materials of which they remain copyright holders.

## References

1. Lefebvre H (1974) La production de l'espace, 1st edn. Éditions Anthropos, Paris
2. Rendell J, Penner B, Borden I (2000) Gender space architecture. An interdisciplinary introduction, 1st edn. Routledge, London
3. Schalk M, Kristiansson T, Ramia M (2017) Feminist futures of spatial practice: materialisms activisms dialogues pedagogies projections, 1st edn. AADR—Art Architecture Design Research, Baunach
4. Hooks B ([1984] 2000) Preface to the second edition seeing the light: visionary feminism. In: Feminist theory: from margin to center. South End Press, Cambridge, MA, pp X–XV
5. Bassanini G (2008) Per amore della città: donne, partecipazione, progetto, 1st edn. Franco Angeli, Milano
6. Petrescu D (2007) Altering practices: feminist politics and poetics of space, 1st edn. Routledge, London
7. Brown L (2016) Feminist practices. Interdisciplinary approaches to women in architecture, 1st edn. Routledge, New York

8. Grosz E (1995) Space, time and perversion. Essays on the politics of bodies, 1st edn. Routledge, New York
9. Pansera A (2021) 494: Bauhaus al femminile: 475 studentesse, 11 docenti, 6 donne intorno a Gropius, 1 manager, 1 fotografa, 1st edn. Nomos Edizioni, Busto Arsizio
10. Roessler P (2019) Bauhausmädels, 1st edn. Taschen, Koeln
11. European Commission (2021) New European Bauhaus—shaping more beautiful, sustainable and inclusive forms of living together. https://europa.eu/new-european-bauhaus/index_en. Accessed 25 Mar 2022
12. Darke J (1996) The man-shaped city. In: Booth C, Darke J, Yeandle S (eds) Changing places: women's lives in the city. Paul Chapman Publishing Ltd., London, pp 88–99
13. Stratigakos D (2008) A woman's Berlin, 1st edn. The University of Minnesota Press, Minneapolis
14. Spain D (1992) Gendered spaces, 1st edn. University of North Carolina Press, Chapel Hill
15. Massey D (1994) Space, place, and gender, 1st edn. University of Minnesota Press, Minneapolis
16. McDowell L (1999) Gender, identity, and place, 1st edn. University of Minnesota Press, Minneapolis
17. Scott JW (1986) Gender: a useful category of historical analysis. Am Hist Rev 91(5):1053–1075. https://doi.org/10.2307/1864376
18. Santus K, Scaioli A (2021) Designing the urban commons through gender and nature-based approach. A renewed project for public space in times of crisis. Ri-vista Res Landsc Archit 19(2):208–221. https://doi.org/10.36253/rv-11426
19. Tronto JC (2005) An ethic of care. In: Cudd AE, Andreasen RO (eds) Feminist theory: a philosophical anthology. Blackwell Publishing, Malden, MA, Oxford, pp 251–263
20. Deleuze G, Guattari FL (1987) A thousand plateaus: capitalism and schizophrenia, 1st edn. University of Minnesota Press, Minneapolis
21. Lefebvre H (1968) Le droit à la ville, 1st edn. Editions Anthropos, Paris
22. Bianchetti C (2020) Bodies. Between space and design. Jovis, Berlin
23. Roggema R (2017) Research by design: proposition for a methodological approach. Urban Sci 1(1):2. https://doi.org/10.3390/urbansci1010002
24. Schoonderbeek M (2017) A theory of "design by research"; mapping experimentation in architecture and architectural design. Ardeth (1):62–79. https://doi.org/10.17454/ARDETH01.05
25. Rose G (1993) Feminism & geography: the limits of geographical knowledge, 1st edn. Polity Press, Cambridge, UK
26. Rendell J (2011) Critical spatial practices: setting out a feminist approach to some modes and what matters in architecture. In: Brown LA (ed) Feminist practices: interdisciplinary approaches to women in architecture. Ashgate, Farnham, pp 17–56
27. Merleau-Ponty M (2002) Phenomenology of perception (trans: Smith C). Routledge. https://doi.org/10.4324/9780203994610
28. Grosz E (2001) Architecture from the outside: essays on virtual and real space, 1st edn. MIT Press, Cambridge, MA
29. Rendell J (2006) Art and architecture: a place between, 1st edn. IB Tauris, London
30. Tschumi B (1983) Spaces and events. In: Tschumi B, Coates N (eds) The discourse of events. Architectural Association, pp 6–11
31. de Certeau M (1984) The practice of everyday life, 1st edn. University of California Press, Berkeley
32. Haraway D (1988) Situated knowledges: the science question in feminism and the privilege of partial perspective. Fem Stud 14(3):575–599
33. Massey D (2005) For space. Sage, London

34. Rossi A (1978) L'architettura della città, 1st edn. Clup
35. Hertzberger H (1991) Lessons for students in architecture. 010 Publishers, Rotterdam
36. Tschumi B (1994) Architecture and disjunction. MIT Press, Cambridge, MA, London
37. Shields R (1999) Lefebvre, love and struggle: spatial dialectics. Routledge

# Critical Inroads: Toward a "European-African Bauhaus"

Anja Isabel Schneider(✉)

University of Castilla-La Mancha/ARTEA, Cuenca/Madrid, Spain
anjaisabel.schneider@uclm.es

**Abstract.** This paper takes as its departure point a 2021 recorded Digital Life Design conference with Swiss curator and artistic director of London's Serpentine Galleries, Hans Ulrich Obrist, Malian writer, film director and cultural theorist Manthia Diawara and German author, producer and filmmaker Alexander Kluge. One of the leading intellectual voices in Germany, Kluge turned to the discussion of the need for a New European Bauhaus for this century. He specified that a new Bauhaus should not be so rational and geometric as the historic Bauhaus. Rather, Kluge made an appeal for a "European-African Bauhaus". Rebuilding our lives, for Kluge, does not only imply the architectural structures of houses or landscapes, but also our inner faculties: our thinking, our emotions, our souls. Cooperation, for Kluge is key in the coming together of all the arts from distinct socio-cultural paradigms. Furthermore, what is needed, according to Manthia Diawara, are new forms of aesthetics, specifically an aesthetics of solidarity to address the ecological crisis of the Anthropocene. Addressing Kluge's and Diawara's critical contributions, this paper seeks to open up further inroads and new imaginaries. How to sustain the critical stance advocated here in the conception of a new Bauhaus, all the while attending to its inclusiveness?

**Keywords:** European-African Bauhaus · Aesthetics of solidarity · Orientation · Decolonising · Lighthouses

## 1 Introduction

At a 2021 Digital Life Design (DLD) conference, moderated by DLD founder Stephanie (Steffi) Czerny and bringing together Hans Ulrich Obrist, Manthia Diawara and Alexander Kluge, the discussion of the need for a New European Bauhaus for this century was addressed [1].[1] Notably, Kluge made an appeal for a European-African Bauhaus, all the while challenging the very idea of the historic Bauhaus as a temple of rationality.[2] Speaking about his film *A Letter from Yene* (2022), Manthia Diawara called for new forms of aesthetics, specifically an aesthetics of solidarity, as imagined by the Martinique-born poet and philosopher Édouard Glissant. This paper focuses on Kluge's and Diawara's critical work to open up further inroads and new imaginaries. It asks: How can the New European Bauhaus be brought into an African context? What solidarities can be forged? How can the New European Bauhaus actively contribute to decolonizing projects? And


M. Blanco Lage et al. (eds.), *Towards a New European Bauhaus—Challenges in Design Education*,
https://doi.org/10.1007/978-3-031-40188-6_38

if so, how can a relationship between Europe and Africa be imagined, based on reciprocity rather than on dominance? How to create a critical meeting place, a platform for intercultural exchange in which designers, architects, and artists from different backgrounds work together and exchange views with one another? At the same time, this paper seeks to delineate a dialogic space for (speculative) reflection. It does so, more precisely, through a reading of two films by Alexander Kluge and Manthia Diawara, related to the discussions of the old and new Bauhaus.

## 2 Orientation

Alexander Kluge's *Hommage für das Bauhaus* [Homage to the Bauhaus] was screened in the framework of projekt Bauhaus, as part of the project's culminating event *Ciao Bauhaus!*, held at Berlin's Volksbühne on May 29, 2019 [2]. The aim of projekt Bauhaus was to critically take stock of the ideas of the historic Bauhaus, contributing to a reassessment of the historic Bauhaus in past years.[3] In what follows, I will focus on the recurring motif in Kluge's work: the beacon or lighthouse. While the film's first half foregrounds the figure of László Moholy-Nagy, offering a kaleidoscopic view of some aspects of the Bauhaus as assessed by Kluge, such as utopia/heterotopia, the second half of the film revolves around the theme of the beacon/lighthouse. The film's intertitles announce its theme as follows: "Leuchtfeuer Mathematik, Leuchtfeuer Vernunft, Jean-Jacques Lequeu Revolutionsarchitekt von 1789", framing the figure of French architect Lequeu (1757–1826) in the cultural and architectural landscape of the French Revolution.

As a core point of attention, Kluge forges links between different 'lighthouses' he holds dear. He brings them into a constellation.[4] It is Lequeu's drawing *Colonne cochlide, Phare qui sert à guider dans le desert* (1777–1814) that Kluge incorporates into his film, here seen through stained glass panels by the artist Kerstin Brätsch (Fig. 1).

Lequeu's drawing is part of *Architecture civile* [Civil Architecture], which lies at the heart of his graphic work, held in the collection of the National Library of France. What is more, Kluge has reserved a special place for Lequeu's "Phare" in selected writings and recent exhibition projects.[5]

"Le Phare", serves as a beacon. Not far past the film's midpoint, the intertitles note: Lequeu built a lighthouse for wanderers in the deserts of Africa [3].[6] Elsewhere, Kluge writes:

> Anyone today who has gotten lost in the Sahel, in search of his happiness to the North, would need such a beacon – such a model of thought. Because the solitary wanderer runs the risk of drowning in the Mediterranean. The signal from a lighthouse in the sea of sand could warn him, give him an orientation. Such an installation is rare. Lequeu's radiant object has never touched the soil of Africa, never faced the "praxis of the desert" [4].[7]

With Kluge, we might concede that the chances for anyone to come across a lighthouse in a desert are very small. This notwithstanding, Kluge is very fond of Lequeu's idea. In this sense, what is crucial here is the idea and potential Lequeu's drawing communicates, as a means of exploring these potentialities. For Kluge, beacons, and

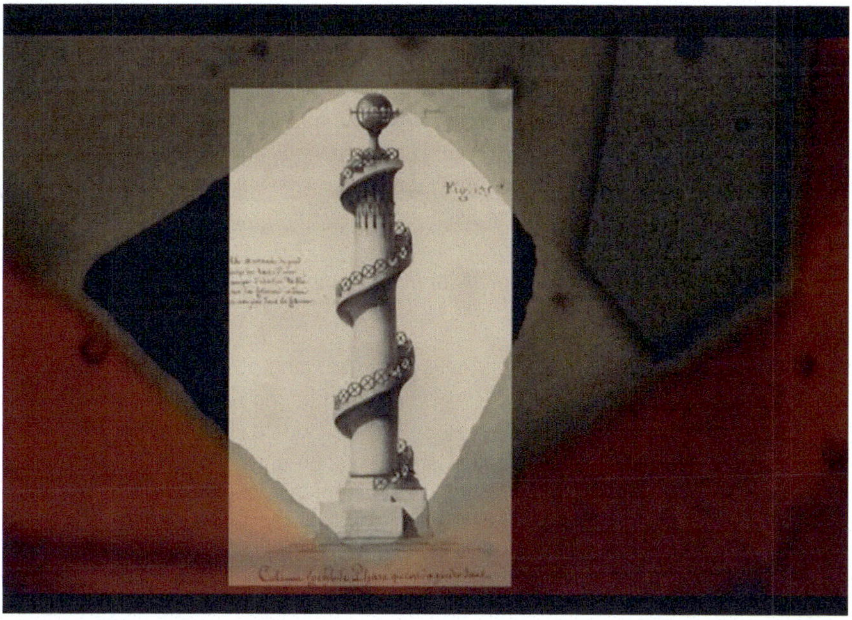

**Fig. 1** Alexander Kluge, *Hommage für das Bauhaus* (still), 2019. Film. Courtesy of Alexander Kluge

lighthouses as figures of thought are vital aids for orientation within our crises-driven present (Fig. 2).

**Fig. 2** Alexander Kluge, *Hommage für das Bauhaus* (still), 2019. Film. Courtesy of Alexander Kluge

Kluge's interest in Lequeu's lighthouse to orient in the desert carries with it the potential of rescue: to find our bearings, even in the most unlikely of places in the expectation that someone might be rescued still [5]. Kluge speaks of the faculty to orient oneself. This is achieved not only through reason, he insists, but also by way of letting our senses speak to us. In our world in which everything gets translated into algorithms, Kluge offers counter-algorithms. His counter-production, be it understood as an aesthetic strategy or practical critique, as literary scholar Langston underlines, "ultimately boils down to a matter of orientation" [6].

Throwing some light on Immanuel Kant's statement: "What does it mean to orient oneself in thinking?" Kluge explains:

> When I talk about orienting oneself, I mean it in the sense of lighthouses, of which there are two kinds. Using lanterns, wreckers based on the shores [...] would lure ships into running aground, so that they could be looted. [...] The second sort of orientation is the orientation toward objects we humans can neither reach nor touch, objects that cannot be displaced. One example of such objects is stars in the firmament. Stars are a means of orientation for seafarers at sea and wanderers in the desert alike [7].

And it is here within this ambiguous simultaneity produced between the difficulty "to discern a wrecker's lantern from a legitimate lighthouse" and the wanderer in the desert who "can never judge firsthand the source of light" that an additional faculty is needed "in ensuring [...] relevance for orienting oneself", as Langston notes, "one responsible for assigning trust to orientation that difference can provide" [6].

## 3 Consciousness

"(...) how do we begin a new planetary consciousness, where we see the landscape as character," Diawara asks in presenting his 50-min film *A Letter from Yene* [8]. Yene is a coastal village in Senegal, West Africa, where Diawara lives for part of the year. The film unfolds in a sequence of dialogic encounters between Diawara and local community members. It comes through as a compelling letter written to the viewer. In it, Diawara tells us that spending time in Yene has turned him into "an activist on behalf of the environment". There are many pressing issues Yene and its inhabitants are facing, such as the increasing heavy rains brought forth by climate change, an eroding coastline, urbanisation, the demise of long-standing traditions and rituals in a changed coastal economy and ecosystem. A sea that is scarce in fish yet loaded with waste. Diawara, who in 2010 bought one of Yene's seafront houses is critically aware that he, too, is implicated in the ways humans have intervened in the landscape and continue to intervene in it (Fig. 3).

*A Letter from Yene* is about the mutual changes brought forth by an interplay of relations, including our relation to other beings, the land and sea. Long takes accompanied by narration include the close-up of a fisherman's face, Aliou Diouf, who Diawara calls a "kind of philosopher". He cites Diouf saying: "[...] when you throw death at the ocean, the ocean will throw death back at you." Diawara grants us slow(er) viewing. And yet, these shots are also about listening. They are personal-collective accounts,

**Fig. 3** Manthia Diawara, *A Letter from Yene* (still), 2022. Film. Commissioned by Serpentine, MUBI and PCAI Polygreen Culture and Art Initiative, as part of Serpentine's Back to Earth project. Courtesy of the artist and Maumaus/Lumiar Cité, Lisbon

mediated by the filmmaker, on everyday labour and survival. Some of Diawara's critical reflections are triggered by his encounters with a group of women who used to work as fishmongers. To survive, they turned to pebble collecting instead. The heaps of stones, shells, pebbles, and sand are sold for use in coastal home construction; in part to protect against the winds and saltwater exposure. Diawara expresses concern about the women's new profession, as their daily labour further contributes to coastal erosion. "We are all implicated in what is happening," Diawara resumes. Accordingly, the urgencies that the film addresses translate into shared urgencies that resonate beyond the local in its links to other parts of the world.

In the DLD conference, introduced in the beginning of this text, Kluge called for a novel start for cooperation, one that is grounded in praxis. Cooperation, for Kluge, is key in the coming together of all the arts from distinct socio-cultural paradigms. Circling back to Manthia Diawara, and the notion of renewing our sense of aesthetics toward the world, the old aesthetics, as Diawara asserts, having already defined our positions through power relations and capitalism. How to fathom gestures that offer alternatives to a world driven by inequality, ecological and social crises? Taking Diawara's proposition seriously for the need of a new aesthetics to intervene in the world, the concern is toward taking collective responsibility: "If my work can have an echo with your work, or the work of somebody somewhere else," Diawara tells us toward the end of his film, "then we will begin to have this new solidarity. Édouard Glissant calls it 'planetary consciousness'" (Fig. 4).

**Fig. 4** Manthia Diawara, *A Letter from Yene* (still), 2022. Film. Commissioned by Serpentine, MUBI and PCAI Polygreen Culture and Art Initiative, as part of Serpentine's Back to Earth project. Courtesy of the artist and Maumaus/Lumiar Cité, Lisbon

## 4 Conclusion

If Diawara sheds new light on themes in Édouard Glissant's work, he upholds Glissant's "trembling thinking", to explore the relations between different people and places. *La pensée du tremblement*, according to Glissant, "is first of all the instinctive feeling that we must reject all categories of fixed thought and all categories of imperial thought" [9]. To conclude, the New European Bauhaus must be decolonizing in itself. It must be decolonizing through an aesthetics of solidarity. With this choice of path, the image, or motif of the lighthouse, we may associate it with heightened modes of perceiving, acting, and being in the world. How to apply the concepts of a critical orientation and consciousness to this new lighthouse of a European-African Bauhaus that I am invoking here? What different light and what image will it be projecting? How to generate a multiplicity of lighthouses that are related and that weave networks of solidarity and new forms of collaboration?[8]

The concepts developed in this text serve as fertile grounds for proposition. Bringing together both speculative thought and action, a situated and speculative proposal for a renewed European-African Bauhaus would be to contribute to reactivating a network of decommissioned lighthouses on the Mediterranean coasts as sites of solidarity. A series of linkage points between Africa and Europe, built around the axis of dialogue, care and agency. Each beacon and its surrounding environment would be reactivated by local communities through co-creative cultural and social engagements to establish new forms of collaboration, placing the need for decolonizing practices at its core. As Édouard Glissant would say, speaking to Hans Ulrich Obrist about his poetics of trembling:

> [I]t allows us to be in real contact with the world and with the peoples of the world. For me, that's what trembling thinking is. An instinct, an intuition of the world that

we can't achieve with imperial thoughts, with thoughts of domination, thoughts of a systematic path toward a truth that we've posited in advance. It's metaphorical, but it's also real, concrete [10].

The speculative proposal presented here operates to undermine the oppressive force associated with colonial history. The image of the lighthouse as a projection device can then function as a site for other collective imaginaries.

**Notes**

1. Kluge founded in 1953, together with Edgar Reitz and Detten Schleiermacher the Institut für Filmgestaltung at the Hochschule für Gestaltung Ulm, a successor of the Bauhaus.
2. Interestingly, Rem Koolhaas made a statement in one of the panel discussions to The Festival of the New European Bauhaus that echoes Kluge's appeal for a European-African Bauhaus. For Koolhaas and his collaborators, it is crucial "that Europe [...] establishes very creative," that is "deliberate" and "eager connections with Africa." This, he continued, "is a very important part what Europe could be and what Africa could be, maybe even a hybrid between the two." Rem Koolhaas in "Toward a diverse Europe: How can art, architecture, and technology contribute to a new vision of Europe?", Niklas Maak (moderator), with Rem Koolhaas, Francesca Bria, Hans Ulrich Obrist, June 10, 2022, https://www.youtube.com/watch?v=sIMSsUKalGc, last accessed 2023/01/10.
3. See for instance the exhibition *50 Years after 50 Years of the Bauhaus 1986*, held from May 5 to September 23, 2018 at the Württembergischer Kunstverein, Stuttgart. It focused on a contextual rereading of its own exhibition history, calling into question the narratives that understand modernism and the Bauhaus "as synonyms for progress, freedom and democracy," seeking to address the ambivalences inscribed in the Bauhaus with regard to colonial modernism. Alexander Kluge contributed to the exhibition with a special intervention, among other invited artists. See https://www.wkv-stuttgart.de/en/program/2018/exhibitions/bauhaus/, last accessed 2023/01/12.
4. Contrary to the film projector, the light of the lighthouse moves circularly. Its Fresnel lenses reflect the light, bundle it, and carry it many kilometers out to the sea.
5. Lequeu's "Le Phare" not only finds a central place in Kluge's "Hommage für das Bauhaus," but also, among others, in the exhibition *Pluriversum*, Literaturhaus München, 2019. Zirnstein, Michael (2019) Leuchtfeuer im Nichts, Süddeutsche Zeitung, September 10, 2019, https://www.sueddeutsche.de/kultur/podiumsdiskussion-leuchtfeuer-im-nichts-1.4595836, last accessed 2023/01/12.
6. My translation from the German: "Lequeu baute einen Leuchtturm für Wanderer in Afrikas Wüsten." Cf. https://www.dctp.tv/filme/leuchtfeuer-mathematik-leuchtfeuer-vernunft?thema=qr-codes-zum-buch-russland-kontainer, last accessed 2023/04/18.
7. My translation from the French: "Quiconque aujourd'hui s'est égaré dans le Sahel, à la recherche de son bonheur vers le Nord, aurait besoin d'un tel phare – d'un tel modèle de pensée. Car le marcheur solitaire court le risque de se noyer dans la Méditerranée. Le signal d'un phare dans la mer de sable pourrait le mettre en garde, lui donner

une orientation. Une telle installation est rare. L'objet radieux de Lequeu n'a jamais touché le sol de l'Afrique, jamais affronté la 'praxis du désert'."

8. I thank the artist Raúl Hidalgo for generously engaging with the topic of my essay and for spinning this thought further with me.

**Acknowledgements.** The author acknowledges a postdoctoral research grant for scientific excellence within the framework of the University of Castilla-La Mancha's own R&D and Innovation Plan, co-financed by the European Social Fund. This research inscribes itself within the research-creation group ARTEA (http://artea.uclm.es/en/). The author thanks Manthia Diawara, Jürgen Bock and Carlos Alberto Carrilho at Maumaus, Lumiar Cité, Alexander Kluge, Gülsen Döhr and Barbara Barnak at dctp, Giulio Bursi at Atelier Impopulaire and Anh-Linh Ngo at ARCH+.

## References

1. Kluge A, Diawara M, Obrist HU, Czerny S (2021) Framing the world: storytelling through music and film, DLD all stars. https://dldnews.com/the-art-of-planetary-storytelling/. Accessed 10 Jan 2023
2. Projekt Bauhaus homepage. https://scheringstiftung.de/en/programm/kultur/foerderprojekte/projekt-bauhaus/vorkurs-projekt-bauhaus-from-bauhaus-to-silicon-valley-2/. Accessed 15 Nov 2022
3. Kluge A. Hommage für das Bauhaus. https://www.youtube.com/watch?v=TbsueG9qAFg. Accessed 12 Jan 2023
4. Kluge A (2023) Phare qui sert à guider dans le désert. (À un ami). In: Critique 2023/1–2, no 908–909, p 162
5. Kluge A, Gawan F (2014) Die Fliege im Bernstein. Alexander Kluge über Rudolf Steiner und Andrei Tarkowski, Teil II. In: All-over, no 7, pp 15–26
6. Langston R (2019) Introduction. The guardian of difference. The essayist Alexander Kluge. In: Langston R (ed) Difference and orientation: an Alexander Kluge reader. Cornell University Press, Ithaca, NY, pp 3–22
7. Kluge A (2019) Theory of storytelling. Lecture one. In: Langston R (ed) Difference and orientation: an Alexander Kluge reader. Cornell University Press, Ithaca, NY, pp 100–110
8. Diawara M. https://www.tiktok.com/@mubi/video/7125395457286655238?lang=enlast. Accessed 10 Jan 2023
9. Glissant É, Obrist HU (2012) Le 21ème siècle est Glissant. In: Glissant É, Obrist HU (eds) 100 notes, 100 thoughts: documenta series 038. Hatje Cantz, Ostfildern, pp 1–6
10. Glissant É, Obrist HU (2021) The Archipelago conversations, isolarii. Excerpt, New York. https://032c.com/magazine/edouard-glissant-and-hans-ulrich-obrist. Accessed 12 Jan 2023

# Positioning Research in Architecture

Jörg Schröder[1(✉)], Martin Luce[2], Mieke Pfarr-Harfst[3], Judith Reeh[4], and Oliver Tessmann[3]

[1] Faculty of Architecture and Landscape, Leibniz Universität Hannover, Hannover, Germany
schroeder@staedtebau.uni-hannover.de
[2] School of Engineering and Design, Technische Universität München, Munich, Germany
[3] Department of Architecture, Technische Universität Darmstadt, Darmstadt, Germany
[4] Faculty of Architecture, Karlsruhe Institute of Technology, Karlsruhe, Germany

**Abstract.** The architectural disciplines are called for a strong and inspiring position towards current societal challenges in Europe. In public and political debates, their role for the ecological reconfiguring of the whole building sector, for the transition of cities to climate-neutrality and resilience, but also for social inclusion, accessibility, and affordability is increasingly acclaimed. And, in particular by the initiative New European Bauhaus, the holistic approach of architecture is seen as an asset, not least for a cultural and behavioural change to sustainability that stresses the value of everyday living spaces and new aesthetics. Recent discussion strongly points to the research dimension of architecture both in academia and in professional practice, and its interaction with other disciplines and society. The authors of this contribution are currently launching an initiative of the German TU9 architecture faculties to establish together discipline-specific standards for research evaluation. Starting from a research project on the German and European context in this regard, this article advocates for a joint effort to position research in architecture. It reviews debates and initiatives in order to identify decisive factors in funding and evaluation, research practice, interactions with practice, education, and society, and the qualification of methods and research outputs.

**Keywords:** New European Bauhaus · Research funding · Qualitative research assessment · Inquiry-based learning · Open Science · Problem-oriented research

## 1 Introduction

The architectural disciplines are called for a strong and inspiring position towards current societal challenges in Europe. In public and political debates, their role for the ecological reconfiguring of the whole building sector, for the transition of cities to climate-neutrality and resilience, but also for social inclusion, accessibility, and affordability is increasingly acclaimed. The aim of resilience regarding energy and resources as well as building and transformation processes currently becomes even more relevant due to the global consequences of the Russian invasion in Ukraine. In particular, by the initiative New European Bauhaus, the holistic approach of architecture is seen as an asset not least for a cultural and behavioural change to sustainability that stresses the quality and value


M. Blanco Lage et al. (eds.), *Towards a New European Bauhaus—Challenges in Design Education*,
https://doi.org/10.1007/978-3-031-40188-6_39

of everyday public and private living spaces and new aesthetics. For these challenges architecture needs not only to develop novel concepts and models to change cities, but also visions, theories, and paradigms which inspire and steer innovation—architecture is called as a research discipline. In fact, recent discussion strongly points to the research dimension of architecture [1–3], both in academia and in professional practice. The role of architecture as a research discipline is captured at European level already by the EAAE's Charter on Architectural Research [4]. At the same time, architecture is challenged to express excellence in increasingly competitive academic environments. From this background, this article aims to discuss how to foster research in architecture through identifying factors for a positioning and approaches that link between factors until now discussed in separate contexts.

The argumentation of this article refers a positioning of research in architecture to the initiative New European Bauhaus [5]. NEB aims at merging creativity, arts, and technology for climate-neutrality, sustainability, and resilience—highlighting inclusiveness, affordability, new aesthetics and new culture as transversal and interlinked fields of action. In particular, NEB calls for the cooperation of creative and design disciplines with other sciences, economic and social actors. Still, in many discussions one gets the impression that the creative and design disciplines are not understood as researching but more as practical supporters to deliver innovation. This is astonishing since the need for an ongoing progress in research for the built environment in the framework of the New European Bauhaus is already discussed in regard to Horizon Europe [3], and architecture should promote its research qualities. A further starting point is that not only in the NEB context but also in regard to discussions on how to deliver the UN Sustainability Goals [6], more holistic approaches are called for transversal approaches and, as NEB puts it, for a cultural change for the Green Deal. In both qualities, architecture can be said to have extraordinary strengths that need to be expressed—based on an understanding as a research and innovation discipline, connecting research-driven creativity [7] with design-driven research [8]. This background implies that efforts for a positioning of research in architecture, besides targeting the academic context and the practice sector, need to set innovative methods and new models in dialogue with the different sectors and disciplines involved in transforming the built environment and in urban transition to sustainability, and, furthermore, to address society, culture, and politics where architecture often is not yet seen as a protagonist of research and innovation. In order to contribute to a positioning of research in architecture, this article reviews debates and initiatives to identify fields for joint action towards academia and in society. Specific objectives are (1) to identify decisive contextual factors in funding and evaluation for research in architecture, (2) to detect the range of current research practice, (3) to include considerations about the interaction of research with practice, education, and society, and (4) to discuss consequences for sharpening research methods and outputs.

## 2 Materials and Methods

The article is based on the initiative of a working group of the German TU9 technical universities' architecture schools on standards for research evaluation in the context of the national excellence competition for their universities. This initiative includes research into performance evaluation in Germany and several European countries and into pathways to foster research [9]. Materials come from literature review and the selective evaluation of current research projects and initiatives in a European perspective, analysed and synthesised in a hermeneutic approach. The review is organised in four steps, starting with a comparative evaluation of different contexts of funding programmes and research assessment schemes and proceeds with an evaluation of the range of current research projects, exemplarily chosen from the TU9 schools. Then, based on literature and selected initiatives, implications from the interaction between research in architecture with practice, education, and society are discussed and pathways to qualify research methods and outputs are proposed.

## 3 Contexts for Funding and Evaluation

Positioning research in architecture needs to consider the extended range of approaches in architecture—a multidisciplinary discipline, so to say—including methods and paradigms from technology, arts, social sciences and humanities, working on different scales from buildings to cities and regions, and bound together by a design perspective.

Funding for research differs in European countries in regard of excellence and relevance funding (Fig. 1) and is influenced through the context of classification: in some countries (e.g. Italy or Germany) architecture is seen as part of engineering, in others (e.g. France or UK) as part of arts and humanities, with remarkable consequences for the access to funding from different sub-fields. At European level, the scheme for ERC funding reflects this apparently ambiguous character of architecture as a discipline pragmatically, attributing different sub-field to no less than seven subpanels in three different panels (Fig. 2). The consequence of this classification is that applications from architecture have always to succeed at the margin of decisions focused mainly on approaches and methods of other subjects and that more comprehensive approaches in architecture struggle to find an allocation in the funding system. In particular, approaches with an orientation in design research are difficult to place on European as well as on national levels. This gap in funding is a problem for the holistic approaches called for in NEB. With its projective and performative orientation, design research [10–13] is seen as an important field of innovation to combine the different approaches in the architectural disciplines and to offer intersections for new collaboration with other disciplines and sectors. Still, design research is not addressed in the Horizon Europe—NEB Nexus report [3].

Connected to the funding gap, the statement of an evaluation gap [14] can be seen as still valid, due to the increasing influence of metric assessment based on journal percentiles that is, at least until now, not corresponding to the research strengths in architecture. Even if not for architecture itself, metrics are applied for large interdisciplinary research applications, nationally and on European level, and in the context of performance budgeting in the universities. A further challenge is the subject-specific culture of research outputs and methods, e.g. exhibitions or prototypes, that are not in the portfolio of evaluations dominated by a perspective from natural sciences. In the EU-funded project CA$^2$RE+, quality criteria for design-driven doctoral training are currently being developed, especially on subject-specific forms of peer review [8]. Still, in

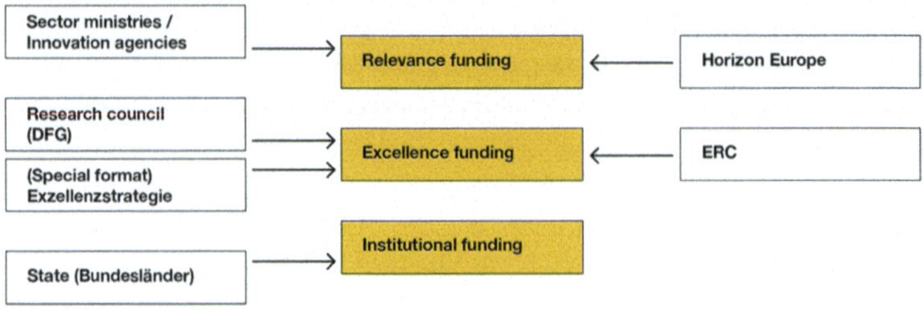


**Fig. 1** Overview of national (here as example: Germany) and European funding streams, graphic by Jörg Schröder, based on Debackere et al. 2018, p. 37/Technopolis Group

| | |
|---|---|
| **Arts & Humanities** | **SH5_4** Visual and performing arts, film, design and architecture<br>**SH5_6** History of art and architecture, arts-based research |
| **Engineering** | **PE8_3** Civil engineering, architecture, offshore construction, light weight construction |
| **Social Sciences** | **SH7_7** Cities; urban, regional and rural studies<br>**SH7_8** Land use and planning<br>**SH7_9** Energy, transportation and mobility<br>**SH7_6** Environmental and climate change, societal impact and policy |

**Fig. 2** ERC subpanels relevant for research in architecture, graphic by the authors

many cases joint definitions are lacking to include discipline-specific research outputs into qualitative assessment. By comparing different international assessment practices and discussions [9], a standard of qualitative, peer-review-based evaluation of research performance in architecture could be claimed based on subject-specific definitions, a discussion already started in 2010 in the Netherlands [15, 16]. Corresponding to international initiatives such as DORA [17] or the Leiden Manifesto [18], metrics such as h-index or journal percentiles can support qualitative assessment, with a reflective and critical use of indicators, taking the cultures of sub-areas into consideration, and aiming "to avoid creating any undesirable behavioural incentives" [19, p. 11], in particular regarding performance-based budgeting.

## 4 A Range of Existing Research Practices

A selection of current research projects at the TU9 schools is able to show the range of methods and topics as well as different funding avenues, each with their own logic of evaluation and of conditioning methodological access. In 2022, the first ERC Grant has been awarded to a member of one of the TU9 schools [20]. One of the first MSCA-ITN in architecture overall is TACK Communities of Tacit Knowledge [21], aiming at an understanding specific knowledge production in architectural and urban design processes. In Horizon, a prominent project is the NEB Lighthouse in Munich Neuperlach [22], addressing the sustainable transformation and renovation of a mass housing quarter in a multidisciplinary approach. In the Creative Europe Programme, Creative Food Cycles [23] addresses the interaction between design processes, circular economy, and sustainable urban transformation starting from the topic of food and the city. The most prominent national funding is currently the, so far only, DFG Excellence Cluster at one of the TU9 schools, in cooperation with other disciplines: IntCDC [24] works on integrative computational design and construction, rethinking the design and building process and related building systems, based on interdisciplinary research. In terms of strategic development of schools, the SABRE Erasmus+ Strategic Partnership [25] aims at fostering research in architecture through innovation in Ph.D. education, entrepreneurial thinking, collaboration between academia and practice, and linkages with research by SME in the field.

Each of these projects can be analysed for their impact on the development of research in architecture in general, which leads to the observation that while technically oriented projects are well established for (national) excellence funding, design research approaches still have to rely on European relevance and excellence funding. Nevertheless, this extension into different funding programmes can be also seen as an opportunity for positioning, and it can support the development of design-driven research through linking to research in other creative and cultural disciplines. Inter- and transdisciplinary projects, at the same time, also offer strong opportunities for positioning research in architecture in the broader academic field and in society, in particular when the coordination comes from architecture.

## 5 Interaction with Practice, Education, and Society

The discussion for the European Charter on Architectural Education [1] highlights that reciprocity between academia and practice needs to be strengthened as a prerequisite for innovation in architecture and design. Actually, architecture can be characterised as a field where innovation is already in a large extent developed in multi-stakeholder constellations between academia, architectural and other enterprises, cities and public bodies, civil society and cultural initiatives and organisations [9]. This is due to the specific relation of architecture to context and relies on developed capacities in transdisciplinary research, but poses on the other hand challenges to achieve transferability and to innovate theory, to articulate and qualify multi-stakeholder approaches and practice-based research. At the same time, the role of research in a changing professional practice is seen as increasing [26, 27].

Consequently, practice in architecture would need to be redefined, extending definitions of architecture as a protected profession bound to qualities of the built environment [28, 29] towards a research dimension. The increasing awareness of professional organisations for research is already manifest in frontrunning initiatives such as the RIBA Medal for Research [30]. On the other hand, demand for research-oriented approaches in education linked to creative professional practice in a design research paradigm is discussed [2, 9, 27, 31]. In exchange with other disciplines the role of scientific components and paradigms in design processes can be strengthened [32], methods and concepts [33] reflected, and design components and paradigms be induced in research processes [34], supporting a definition of academic design education as inquiry-based learning by EAAE [1]. To position research in architecture, communication and dissemination are highly important, not only university-wide and to targeted non-academic stakeholders, but also towards the wider public. As part of Open Science, transfer activities can at the same time support to establish interactive research tools and formats, as shown e.g. with the initiatives of Pavillon 333 (Fig. 3) in the Museum Quarter in Munich [35], the fair stand Ressource.Architektur (Fig. 4) at the Real Estate Fair in Hannover [36], or the urban event space Architekturschaufenster Karlsruhe [37].

**Fig. 3** Pavillon 333 has been realised as a DesignBuild project at the TUM Department of Architecture, next to the Pinakothek der Moderne in Munich, as a temporary and collaboratively used mediation space for architecture, art and design. The Pavillon promotes cooperation between the local cultural institutions and offers low-threshold services, exchange and encounters for visitors to the Munich art area. Photo by Matthias Kestel for TUM, 2022

**Fig. 4** The fair stand Ressource.Architektur at the Real Estate Fair in Hannover has been realised as a DesignBuild project of Leibniz University Hannover in cooperation with regional architecture organisations. The installation shows the amount of 3 t steel used in conventional concrete-based housing construction per person (corresponding to 6 t $CO_2$). Photo by Max Passgang for LUH, 2022

## 6 Qualification of Methods and Outputs

While in Germany technical or historic research in architecture is well established in schemes for excellence funding (as fundamental research), more holistic and advanced approaches of design-driven research [8, 38], even in conjunction with traditional research methods of the discipline, are still encountering scepticism. Research in architecture needs to argue for not being classified as research services (applied research; even if funding from industry is very rare in architecture) and thus excluded from excellence funding. Hence, the challenge is to position a design cognition, "the 'designerly' ways of knowing and thinking" [39] and to articulate the specific relation between research, context and practice in architecture as design-based and practice-based research [11, 40]. These characteristics do not correspond to the scheme valid for other disciplines which separates (knowledge-oriented) fundamental research and (practical) application research (research services). They also do not comply with deductive research methods: in architecture, the identification of a problem—since it is embedded in further, larger problem entanglements ('wicked problems' [41])—must itself be part of the research process and necessarily requires open questions. Thus, problem-oriented research in architecture [9], in reference to the Frascati Manual [42], gains new knowledge and contributes to innovation—at the same time—precisely in the methodological connection between knowledge-oriented with transferable and practice-oriented research, observing the principles of novelty, creativity, uncertainty, systemic approach, transferability/reproducibility and leading to new forms of open and evolving research processes. A further contested field are research outputs: many formats of research outputs in architecture are in need for qualification to be recognised outside of the discipline, such as exhibitions, performances, material/digital artefacts, and designs [9]; additionally, they can even assume the role of research tools in inter- and transdisciplinary research processes. Still, in many cases architectural research outputs and activities are considered as transfer or third mission, not as research. Here, efforts in qualification are necessary on national as well as on European level. This regards, also, the field of publication, with the establishment of peer-reviewed journals for architecture in Europe, is a major field for improvement, even if remarkable initiatives already started, such as the ARENA Journal of Architectural Research [43], the journal Spool at TU Delft [44], or the journal Dimensions at Technical University of Munich [45].

## 7 Conclusions

Based on these findings, pathways for a positioning of research in architecture regarding different factors and their interdependencies can be discussed: in funding and evaluation, in the use of a range of programmes and methods, the interaction with practice, education, and society, and, not least, in qualification of methods and research outputs. It becomes clear that a joint effort to position research needs to extend from architecture schools to professional organisations, civil society, and politics—even more, architecture can become a frontrunner for Open Science. Cooperation between the schools should target national as well as European levels since they are interlinked in regard to all identified factors. Four fields for joint approaches can be summarised: (1) advocating and influencing funding schemes and framework as well as qualitative evaluation based on subject-specific standards; (2) extending and linking existing research and funding practices; (3) fostering interaction of research with education, practice, and society; (4) promoting problem-oriented and design driven research and the value of subject-specific research outputs. In particular, the introduction of dedicated funding lines for architecture, of its positioning in interdisciplinary projects, and of definitions for subject-specific evaluation criteria can be seen as promising fields for cooperation to position research in architecture.

## References


1. EAAE European Association for Architectural Education (2018) Towards a charter on architectural education: principles and practices of architectural education. Position paper of the EAAE Education Academy. http://www.eaae.be/wp-content/uploads/2018/10/EAAE-EA-Porto-position-paper-180901.pdf. Accessed 22 Apr 2023
2. Hay R, Samuel F, Farrelly L (2020) Demonstrating the value of design through research in architecture practice. University of Reading, Architects' Council of Europe, co-funded by the Creative Europe Programme. http://www.ace-cae.eu/fileadmin/user_upload/EURO_report_-_Photography_Cover_20_5_neu_20210406.pdf. Accessed 22 Apr 2023
3. European Commission, Directorate-General for Research and Innovation (2022) Horizon Europe—New European Bauhaus Nexus report. https://doi.org/10.2777/49925
4. EAAE European Association for Architectural Education (2022) EAAE charter on architectural research. http://www.eaae.be/about/statutes-and-policypapers/eaae-charter-architectural-research. Accessed 22 Apr 2023
5. http://europa.eu/new-european-bauhaus/index_en. Accessed 22 Apr 2023
6. EAAE European Association for Architectural Education (2021) EAAE Oslo pledge for climate crisis and sustainable future. http://www.eaae.be/wp-content/uploads/2019/09/Pledge_07052021.pdf. Accessed 22 Apr 2023
7. NEBC New European Bauhaus Collective (2021) Common ground: making the renovation wave a cultural project. Conference report. http://www.ace-cae.eu/fileadmin/user_upload/NEB_report_v3.pdf. Accessed 22 Apr 2023
8. Pedersen CP, Zupančič T, Schwai M, Van den Berghe J, Lagrange T (eds) (2021) $CA^2RE+$ strategies of design-driven research. Collective evaluation of design-driven doctoral training. Aarhus School of Architecture, ARENA Architectural Research European Network Association, EAAE European Association for Architectural Education, ELIA European League of Institutes of Arts. http://ca2re.eu/wp-content/uploads/2021/10/2021_10_03_CA2RE_STRATEGIES_screen.pdf. Accessed 22 Apr 2023
9. Luce M, Pfarr-Harfst M, Reeh J, Schröder J, Tessmann O (2022) Forschungsexzellenz in der Architektur: Standards zur Bewertung und Förderung der Forschung in den Architekturfachbereichen der Technischen Universitäten in Deutschland (TU9). Repositorium Leibniz Universität Hannover, Hannover. https://doi.org/10.15488/11827
10. Fraser M (ed) (2013) Design research in architecture: an overview. Ashgate, Chichester
11. Fraser M (2016) A new deal for architectural research. ARENA J Archit Res 1(1):1. https://doi.org/10.5334/ajar.12
12. Buchert M (ed) (2014) Reflexive design. Design and research in architecture [Reflexives Entwerfen. Entwerfen und Forschen in der Architektur]. Jovis, Berlin
13. Buchert M (ed) (2018) Processes of reflexive design. Design and research in architecture and landscape [Prozesse Reflexiven Entwerfens. Entwerfen und Forschen in Architektur und Landschaft]. Jovis, Berlin
14. van der Hoeven F (2011) Mind the evaluation gap: reviewing the assessment of architectural research in the Netherlands. Arq Archit Res Q 15(2):177–187. https://doi.org/10.1017/S1359135511000595
15. KNAW (2010) Quality assessment in the design and engineering disciplines. A systematic framework. Royal Netherlands Academy of Arts and Sciences. Available online at: http://www.knaw.nl/en/news/publications/quality-assessment-in-the-design-and-engineering-disciplines. Accessed 22 Apr 2023

16. van der Meulen B, Daemen F, Van Drooge L, De Jong S, Spaapen J, Wamelink F, Van den Besslaar P (2010) Pilot study at Faculty of Architecture TU Delft. ERiC Evaluating Research in Context, publication 1002. Rathenau Institute. http://resolver.tudelft.nl/uuid:f0a713f1-1564-4b79-be66-4f5299ebba2c. Accessed 22 Apr 2023
17. DORA (2012) Declaration on research assessment. http://sfdora.org/read/. Accessed 22 Apr 2023
18. Hicks D, Wouters P, Waltman L, de Rijcke S, Rafols I (2015) Bibliometrics: the Leiden Manifesto for research metrics. Nature 520:429–431. https://doi.org/10.1038/520429a
19. Jonkers K, Zacharewicz T (2016) Research performance based funding systems: a comparative assessment. European Commission, JRC Science for Policy Report. http://op.europa.eu/en/publication-detail/-/publication/b0e081dc-4cbe-11e7-a5ca-01aa75ed71a1/language-en. Accessed 22 Apr 2023
20. ERC Starting Grant, CONSTEMO. Recurring elements of modern facades (1960–1990), at Technical University of Munich (TUM)
21. TACK communities of tacit knowledge: architecture and its ways of knowing (funded MSCA-ITN in Horizon 2020, 4 years, 2.7 mio). ETH Zurich, LUH Hannover, TU Delft, KTH Stockholm, AHO Oslo, Uni Wuppertal, Polimi Milan, ADBK Vienna, Uni Antwerpen, UCL Bartlett, in cooperation with nine distinguished architecture offices. http://tacit-knowledge-architecture.com. Accessed 22 Apr 2023
22. Creating NEBourhoods together (funded in the Horizon Mission NEB Lighthouse Projects, 2 years, 5 mio). City of Munich, Technical University of Munich (TUM), Unternehmer-TUM with 11 more partners. http://www.arc.ed.tum.de/arc/ueber-uns/aktuelles/news-single-view/article/eu-kommission-foerdert-leuchtturmprojekt-creating-nebourhoods-together-muenchen-neuperlach. Accessed 22 Apr 2023
23. CFC Creative Food Cycles (funded in the Creative Europe Programme, 2 years, 330,000). LUH Hannover, IAAC Barcelona, Uni Genova. https://doi.org/10.15488/10074
24. IntCDC Cluster of Excellence Integrative Computational Design and Construction for Architecture (funded by DFG, 7 years, up to 5 mio/year). University of Stuttgart. http://www.intcdc.uni-stuttgart.de. Accessed 22 Apr 2023
25. SABRE Strengthening Architecture and Built Environment Research (Erasmus+ Strategic Partnership, 3 years, 275.000). Technical University of Munich (TUM), TU Delft, UCL Bartlett, Chalmers Göteborg, ETH Zürich. http://www.bauhow5.eu. Accessed 22 Apr 2023
26. Dye A (2014) How architects use research—case studies from practice. RIBA, London
27. Hensel MU, Nilsson F (eds) (2016) The changing shape of practice: integrating research and design in architecture. Routledge, London
28. ACE Architects' Council of Europe (2019) The value of design and the role of architects. http://www.ace-cae.eu/uploads/tx_jidocumentsview/Value_of_Design.pdf. Accessed 22 Apr 2023
29. ACE Architects' Council of Europe (2020) The architectural profession in Europe 2020. A sector study. http://www.ace-cae.eu/fileadmin/user_upload/2020ACESECTORSTUDY.pdf. Accessed 22 Apr 2023
30. https://www.architecture.com/awards-and-competitions-landing-page/awards/riba-presidents-awards-for-research. Accessed 22 Apr 2023
31. Zupancic T, Pedersen CP (eds) (2017) Relational knowledge and creative practice. ADAPT-r Horizon Project. Brussels, KU Leuven. https://issuu.com/adapt-r/docs/deliverable-14. Accessed 22 Apr 2023
32. Atalay Franck O (2021) Reflexions on the plurality of methods in architecture. Dimensions 1(1):55–62
33. Silberberger J (2021) Against and for method: revisiting architectural design as research. Zürich, gta Verlag

34. Schröder J (2021) Cosmopolitan design. In: Schröder J, Carta M, Scaffidi F, Contato A (eds) Cosmopolitan habitat. Berlin, Jovis, pp 12–26. https://doi.org/10.1515/9783868599626-002
35. Technical University of Munich (TUM). http://pavillon333.de. Accessed 22 Apr 2023
36. Schröder J, Wandt R (eds) (2023) Ressource Architektur magazin. Leibniz Universität Hannover, Hannover. https://doi.org/10.15488/13188
37. http://www.architekturschaufenster.de. Accessed 22 Apr 2023
38. Schröder J, Cappeller R, Diesch A, Scaffidi F (2023) Circular design. Jovis, Berlin
39. Cross N (2012) From a design science to a design discipline: understanding designerly ways of knowing and thinking. In: Michel R (ed) Design research now: essays and selected projects. Birkhäuser, Berlin/Boston, pp 41–54. https://doi.org/10.1007/978-3-7643-8472-2_3
40. Verbeke J (ed) (2017) Impact by designing. In: Proceedings of the 3rd ARENA annual conference, KU Leuven, Faculty of Architecture, Campus Sint-Lucas, Brussels, 6–7 Apr 2017. http://arch.kuleuven.be/onderzoek/publicaties/copy_of_2018-proceedings-impact-by-designing-sec.pdf. Accessed 22 Apr 2023
41. Rittel H, Webber M (1973) Dilemmas in a general theory of planning. Policy Sci 4:155–169
42. OECD (2015) Frascati manual 2015: guidelines for collecting and reporting data on research and experimental development. In: The measurement of scientific, technological, and innovation activities. OECD Publishing, Paris. https://doi.org/10.1787/9789264239012-en
43. http://ajar.arena-architecture.eu. Accessed 22 Apr 2023
44. http://spool.ac. Accessed 22 Apr 2023
45. http://www.dimensions-journal.eu. Accessed 22 Apr 2023

# Correction to: Towards a Future Past: The New and the European in the Bauhaus

Sophie Mak-Schram(✉)

Zeppelin University, Friedrichshafen, Germany
sophie.mak-schram@zu.de

## Correction to: Chapter 34 in: M. Blanco Lage et al. (eds.), *Towards a New European Bauhaus—Challenges in Design Education*, https://doi.org/10.1007/978-3-031-40188-6_34

The chapter "Towards a Future Past: The New and the European in the Bauhaus" written by Sophie Mak-Schram, was originally published without open access. Following the authors' decision to opt for open access, the copyright of the chapter changed on 23 April 2025 to © The Author(s) 2025 and the chapter is now distributed under the terms of the Creative Commons Attribution 4.0 International License (https://creativecommons.org/licenses/by/4.0/), which permits use, sharing, adaptation, distribution and reproduction in any medium or format, as long as you give appropriate credit to the original author(s) and the source, provide a link to the Creative Commons license and indicate if changes were made.

The images or other third party material in this chapter are included in the chapter's Creative Commons license, unless indicated otherwise in a credit line to the material.

If material is not included in the chapter's Creative Commons license and your intended use is not permitted by statutory regulation or exceeds the permitted use, you will need to obtain permission directly from the copyright holder. Grant Number: 2936234898

The updated version of this chapter can be found at https://doi.org/10.1007/978-3-031-40188-6_34


M. Blanco Lage et al. (eds.), *Towards a New European Bauhaus—Challenges in Design Education*,
https://doi.org/10.1007/978-3-031-40188-6_40

# Author Index




M. Blanco Lage et al. (eds.), *Towards a New European Bauhaus—Challenges in Design Education*,
https://doi.org/10.1007/978-3-031-40188-6

The manufacturer's authorised representative in the EU is Springer Nature Customer Service Centre GmbH, Europaplatz 3, 69115 Heidelberg, Germany. If you have any concerns regarding our products, please contact ProductSafety@springernature.com

Printed and bound by CPI Group (UK) Ltd, Croydon, CR0 4YY
05/10/2026
02228132-0007